Marine Bioactive Compounds against Oxidative Stress and Inflammation

Marine Bioactive Compounds against Oxidative Stress and Inflammation

Editors

Elena Talero
Javier Ávila-Román

MDPI

Basel • Beijing • Wuhan • Barcelona • Belgrade • Novi Sad • Cluj • Manchester

Editors
Elena Talero
Department of Pharmacology
University of Seville
Seville
Spain

Javier Ávila-Román
Department of Pharmacology
University of Seville
Seville
Spain

Editorial Office
MDPI
St. Alban-Anlage 66
4052 Basel, Switzerland

This is a reprint of articles from the Special Issue published online in the open access journal *Marine Drugs* (ISSN 1660-3397) (available at: https://www.mdpi.com/journal/marinedrugs/special_issues/MNP_OxidativeStress_Inflammation).

For citation purposes, cite each article independently as indicated on the article page online and as indicated below:

Lastname, A.A.; Lastname, B.B. Article Title. *Journal Name* **Year**, *Volume Number*, Page Range.

ISBN 978-3-7258-0333-0 (Hbk)
ISBN 978-3-7258-0334-7 (PDF)
doi.org/10.3390/books978-3-7258-0334-7

Contents

About the Editors

Elena Talero

Elena Talero, a full professor in pharmacology, received her Ph.D. degree with a European mention from the University of Seville (Spain) in 2009 under the supervision of Prof. V. Motilva and Dr. S. Sánchez-Fidalgo. During her predoctoral period, Elena joined the laboratory of Prof. Alfredo Martinez (Cellular, Molecular, and Developmental Neurobiology Department, Instituto Cajal, Madrid) for 5 months as well as the Division of Pre-Clinical Oncology (School of Medical and Surgical Sciences, University of Nottingham, U.K.) for 3 months under the supervision of Prof. S. Watson. In 2011, she worked as a postdoc on several projects with Prof. Chris Paraskeva at Bristol University (U.K.) for 6 months.

Her current research interests focus on the understanding of inflammation as an essential component in the multifactorial origin of different diseases, with a special emphasis on intestinal bowel disease and skin inflammation, as well as associated tumor processes. Her goal is to find new bioactive compounds of marine or terrestrial origin for the treatment and/or prevention of these pathologies.

Javier Ávila-Román

Javier Ávila Román, a researcher, graduated with a degree in biology in 2008 and received his Ph.D. degree with an international mention from the University of Seville (Spain) in 2014. He worked as an assistant professor in the Faculty of Biology and as a researcher in the Faculty of Pharmacy from 2010 to 2017. During this period, he studied the role of natural products isolated from microalgae, mainly oxylipins (OXLs), in inflammatory bowel diseases and associated colon cancer. Furthermore, he joined the laboratory of Prof. Philip Calder (Human Development and Health Unit, Faculty of Medicine, Southampton, United Kingdom) for 3 months in 2012. He moved to the Universitat Rovira i Virgili in Tarragona, Spain, from 2018 to 2021, and worked as a postdoc in the Nutrigenomic Group, under the supervision of Prof. Begoña Muguerza in the Department of Biochemistry and Biotechnology of the Faculty of Chemistry, and he focused on the low-grade inflammation and oxidative stress in obesity and metabolic syndrome, as well as the deciphering of their role in circadian and circannual rhythms and their molecular mechanisms. Since 2021, he has worked as an assistant professor in the Faculty of Pharmacy, University of Seville, and his current research interests focus on the comprehension of the inflammatory process in atopic dermatitis as a comorbidity of inflammatory bowel disease. His goal is to investigate natural and semi-synthesized bioactives from marine and terrestrial environments for the prevention of this growing pathology.

marine drugs

MDPI

Article

In Vitro Anti-Inflammatory Activity and Structural Characteristics of Polysaccharides Extracted from *Lobonema smithii* Jellyfish

Thitikan Summat [1], Sutee Wangtueai [1], SangGuan You [2,3], Weerawan Rod-in [2,4], Woo Jung Park [2,3], Supatra Karnjanapratum [5], Phisit Seesuriyachan [6] and Utoomporn Surayot [1,*]

1 College of Maritime Studies and Management, Chiang Mai University, Samut Sakhon 74000, Thailand; thitikan.ts@gmail.com (T.S.); sutee.w@cmu.ac.th (S.W.)
2 Department of Marine Bio Food Science, Gangneung-Wonju National University, Gangneung 25457, Gangwon, Republic of Korea; umyousg@gwnu.ac.kr (S.Y.); weerawan.ve@gmail.com (W.R.-i.); pwj0505@gwnu.ac.kr (W.J.P.)
3 East Coast Life Sciences Institute, Gangneung-Wonju National University, Gangneung 25457, Gangwon, Republic of Korea
4 Department of Agricultural Science, Faculty of Agriculture Natural Resources and Environment, Naresuan University, Phitsanulok 65000, Thailand
5 Division of Marine Product Technology, Faculty of Agro-Industry, Chiang Mai University, Chiang Mai 50100, Thailand; supatra.ka@cmu.ac.th
6 Faculty of Agro-Industry, Chiang Mai University, Chiang Mai 50100, Thailand; phisit.s@cmu.ac.th
* Correspondence: utoomporn.su@cmu.ac.th; Tel.: +66-34-870709

Citation: Summat, T.; Wangtueai, S.; You, S.; Rod-in, W.; Park, W.J.; Karnjanapratum, S.; Seesuriyachan, P.; Surayot, U. In Vitro Anti-Inflammatory Activity and Structural Characteristics of Polysaccharides Extracted from *Lobonema smithii* Jellyfish. *Mar. Drugs* **2023**, *21*, 559. https://doi.org/10.3390/md21110559

Academic Editors: Elena Talero, Javier Ávila-Román and Kazuo Umezawa

Received: 5 July 2023
Revised: 22 October 2023
Accepted: 23 October 2023
Published: 26 October 2023

Abstract: Crude polysaccharides were extracted from the white jellyfish (*Lobonema smithii*) using water extraction and fractionated using ion-exchange chromatography to obtain three different fractions (JF1, JF2, and JF3). The chemical characteristics of four polysaccharides were investigated, along with their anti-inflammatory effect in LPS-stimulated RAW264.7 cells. All samples mainly consisted of neutral sugars with minor contents of proteins and sulphates in various proportions. Glucose, galactose, and mannose were the main constituents of the monosaccharides. The molecular weights of the crude polysaccharides and the JF1, JF2, and JF3 fractions were 865.0, 477.6, 524.1, and 293.0 kDa, respectively. All polysaccharides were able to decrease NO production, especially JF3, which showed inhibitory activity. JF3 effectively suppressed iNOS, COX-2, IL-1β, IL-6, and TNF-α expression, while IL-10 expression was induced. JF3 could inhibit phosphorylated ERK, JNK, p38, and NF-κB p65. Furthermore, flow cytometry showed the impact of JF3 on inhibiting CD11b and CD40 expression. These results suggest that JF3 could inhibit NF-κB and MAPK-related inflammatory pathways. The structural characterisation revealed that (1→3)-linked glucopyranosyl, (1→3,6)-linked galactopyranosyl, and (1→3,6)-linked glucopyranosyl residues comprised the main backbone of JF3. Therefore, *L. smithii* polysaccharides exhibit good anti-inflammatory activity and could thus be applied as an alternative therapeutic agent against inflammation.

Keywords: *Lobonema smithii*; jellyfish; polysaccharides; macrophages; anti-inflammatory

1. Introduction

The polysaccharides of marine organisms, which are bioactive natural products with medicinal properties, have been shown to possess a large range of biological properties, such as antiviral, antioxidant, antitumoral, anti-cancer, anti-obesity, and cardioprotective activities [1–5], especially in immune responses [4–7]. Marine-derived polysaccharides can also be extensively used in the field of biomedical engineering [8]. Recent studies have reported on the anti-inflammatory activities of natural polysaccharides against stimulated inflammation in RAW264.7 macrophages [9–12]. Lipopolysaccharides (LPS) are a major component of the cell wall of Gram-negative bacteria and are one of the most potent inflammatory agents [10]. Indeed, LPS-stimulated macrophages are a useful model for studying inflammation, possible anti-inflammatory agents, and their action mechanisms [13,14]. Numerous factors can be used for monitoring the anti-inflammatory effects of natural products in in vitro cell culture models, such as inflammatory cytokines (tumour necrosis factor-α (TNF-α), interleukin-1β (IL-1β), IL-4, IL-6, and IL-10), inflammatory enzymes such as nitric oxide (NO), nitric oxide synthases (iNOS), prostaglandin E2 (PGE_2), cyclooxygenase-2 (COX-2), and cell surface receptor proteins (CD11b and CD40), which contribute to modulating the inflammatory response and process [10,14–16].

Jellyfish are marine invertebrate animals consisting of around 200 described species of the class Scyphozoa of the phylum Cnidaria [17]. There are several species of edible jellyfish, including *Lobonema smithii*, *Rhopilema hispidum*, *Rhopilema esculentum*, *Nemopilema nomurai*, and *Lobonemoides gracilis* [18]. Jellyfish contain a variety of nutrients, including protein, amino acids, carbohydrates, vitamins, and inorganic elements, making them valuable economic and nutritional resources [7,19]. Previous studies have shown that jellyfish polysaccharides isolated from *R. esculentum* have anti-inflammatory, antioxidant, and immunomodulatory activities in macrophages and C57BL/6 mice [7,19,20]. An aqueous extract of the jellyfish *N. nomurai* exhibited an anti-inflammatory effect on RAW264.7 macrophages activated by LPS and a zebrafish model [21].

The white jellyfish (*L. smithii*) belongs to the family Lobonematidae, which is an important fishery commodity in Southeast Asia [22]. This jellyfish mainly consists of collagen protein [23,24]. The protein hydrolysate of *L. smithii* has been revealed to possess antioxidant, antibacterial, and tyrosinase inhibitory activity [25,26]. However, information on the anti-inflammatory properties of polysaccharides from *L. smithii* has not yet been reported. In this study, we isolated and fractionated polysaccharides from *L. smithii* before investigating the protective action of *L. smithii* on inflammation triggered by LPS and probable signalling pathways implicated through in vitro studies using RAW264.7 cells. Furthermore, the polysaccharides were also evaluated physiochemically and structurally.

2. Results

2.1. Chemical Composition of Polysaccharides Isolated from L. smithii

The yields of polysaccharides from *L. smithii* as crude polysaccharides and as fractions are presented in Table 1. The yield of crude polysaccharides was 1.3%, which mainly consisted of carbohydrates (62.5%) and proteins (24.1%) with minor amounts of sulphates (10.2%) and uronic acids (3.17%). Analysis of monosaccharide composition showed that L-arabinose (15.0%), D-mannose (15.2%), D-glucose (33.1%), and D-galactose (26.2%) were the predominant sugars in the crude polysaccharides, with significant levels of L-rhamnose (8.30%) and L-fucose (2.23%), indicating a heterogeneous monosaccharide composition.

The crude polysaccharides of *L. smithii* were further separated on a DEAE Sepharose column, producing three different fractions, JF1, JF2, and JF3. The elution yields of JF1 (distilled water), JF2 (1.0 M NaCl), and JF3 (1.5 M NaCl) were 77.2%, 13.2%, and 9.7%, respectively. All fractions contained carbohydrates as the major component, while the protein, sulphate, and uronic acid contents varied. The crude polysaccharides and all fractions contained carbohydrates as the major component (62.5%), while protein (24.1%) was present in considerable amounts in the crude polysaccharides. Fraction JF1 possessed a chemical composition comparable to the crude polysaccharides (carbohydrate 72.2% and protein 18.1%), while the JF2 fraction contained carbohydrates (66.0%) alongside proteins (13.6%), sulphates (17.3%), and lesser amounts of uronic acid (2.97%). On the other hand, carbohydrates (67.6%), as well as considerable amounts of sulphates (22.7%) and proteins (7.12%), were included in the JF3 fraction. The uronic acid contents of all fractions were minor. The monosaccharide composition of the JF1 fraction consisted of D-galactose (41.0%) and D-glucose (40.2%), which were the main sugars in the fraction, alongside considerable amounts of L-arabinose (11.7%). The majority of the unit sugars in the JF2 fraction also comprised D-glucose (52.4%) and minor amounts of L-arabinose (15.4%), D-galactose (12.5%), and D-mannose (10.6%), as well as minor levels of L-rhamnose (6.36%) and L-fucose (2.53%). The JF3 polysaccharides contained D-glucose, D-galactose, and D-mannose (56.7%, 28.4%, and 13.7%, respectively) as the main sugars, while L-arabinose, L-rhamnose, and L-fucose represented minor sugar units (0.60%, 0.54%, and 0.10%, respectively). The separation of crude polysaccharides using a DEAE Sepharose fast flow column indicated that the different fractions of polysaccharides were characterised by different ionic strengths and chemical properties.

Table 1. Chemical compositions of polysaccharides isolated from *Lobonema smithii*.

Components	*L. smithii* Polysaccharides			
	Crude	**JF1**	**JF2**	**JF3**
Yield (%)	1.30 ± 0.10 [x]	77.2 ± 0.61 [y]	13.2 ± 0.22 [y]	9.70 ± 0.05 [y]
Total carbohydrate (%)	62.5 ± 0.12 [a]	72.2 ± 0.06 [a]	66.0 ± 0.06 [a]	67.6 ± 0.06 [a]
Protein (%)	24.1 ± 0.06 [b]	18.1 ± 0.00 [b]	13.6 ± 0.04 [c]	7.12 ± 0.13 [c]
Sulphate (%)	10.2 ± 0.88 [c]	8.27 ± 0.21 [c]	17.3 ± 0.21 [b]	22.7 ± 0.21 [b]
Uronic acid (%)	3.17 ± 0.06 [d]	1.47 ± 0.03 [d]	2.97 ± 0.02 [d]	3.40 ± 0.10 [d]
Monosaccharide content (%)				
D-galactose	26.2 ± 0.20 [b]	41.0 ± 1.00 [a]	12.5 ± 0.02 [c]	28.4 ± 0.00 [b]
D-glucose	33.1 ± 0.10 [a]	40.2 ± 0.21 [b]	52.4 ± 0.36 [a]	56.7 ± 0.05 [a]
L-arabinose	15.0 ± 0.00 [c]	11.7 ± 0.07 [c]	15.4 ± 0.15 [b]	0.60 ± 0.15 [d]
D-mannose	15.2 ± 0.10 [c]	3.83 ± 0.06 [d]	10.6 ± 0.11 [d]	13.7 ± 0.06 [c]
L-rhamnose	8.30 ± 0.20 [d]	1.80 ± 0.03 [e]	6.36 ± 0.15 [e]	0.54 ± 0.03 [d]
L-fucose	2.23± 0.25 [e]	1.30 ± 0.06 [e]	2.53 ± 0.06 [f]	0.10 ± 0.10 [e]

[x] Yield, (the amount weight of crude/powdered sample) × 100. [y] Yield, (the amount weight of fractions/crude injected into column) × 100. % dry basis. Different letters [a,b,c,d,e,f] indicate significant differences ($p < 0.05$) between the groups in each column.

2.2. Molecular Weight (M_w) Analysis

Figure 1 displays the combined UV and RI chromatograms of the crude and fractions. A significant amount of crude polysaccharide was eluted from the high-performance size-exclusion chromatography (HPSEC) column between elution times, with broad primary peaks revealing their heterogeneous polymer distributions (Figure 1A). According to the UV chromatogram, proteins were found in peak quantity at an elution duration of 25 to 40 min. The M_w of the peaks obtained with the MALLS technique was 865.0 kDa (Table 2). The estimated radius of gyration (R_g) from the peaks also served as a method of approximating the size of the crude polysaccharides (Table 2). The R_g values of the peaks were 124.0 nm. Crude polysaccharides showed a significant UV chromatogram, in agreement with the chemical constitution of their 24.1% protein content (Table 1).

The chromatograms of the fraction profiles showed a single dominant peak for JF1 at the elution duration ranging from 27.3 to 45.2 min, as well as a major peak at 25.6 to 45.1 min for JF2 (Figure 1B,C, respectively). However, the elution time for JF3 ranged between 28.5 and 40.0 min, exhibiting a narrower peak than the other fractions, thus indicating that the polymer was more homogeneous (Figure 1D).

Table 2. Molecular weight of polysaccharides isolated from *Lobonema smithii*.

L. smithii Polysaccharides	M_w (kDa)	R_g (nm)	SV_g (cm^3/g)
Crude	865.0 ± 8.71 [a]	124.0 ± 3.90 [a]	5.57 ± 0.57 [a]
JF1	477.6 ± 7.94 [c]	77.7 ± 3.61 [c]	2.49 ± 0.39 [c]
JF2	524.1 ± 7.55 [b]	94.4 ± 2.53 [b]	4.05 ± 0.26 [b]
JF3	293.0 ± 6.63 [d]	56.3 ± 7.20 [d]	1.59 ± 0.66 [d]

Different letters [a,b,c,d] refer to significant differences ($p < 0.05$) between the groups in each column.

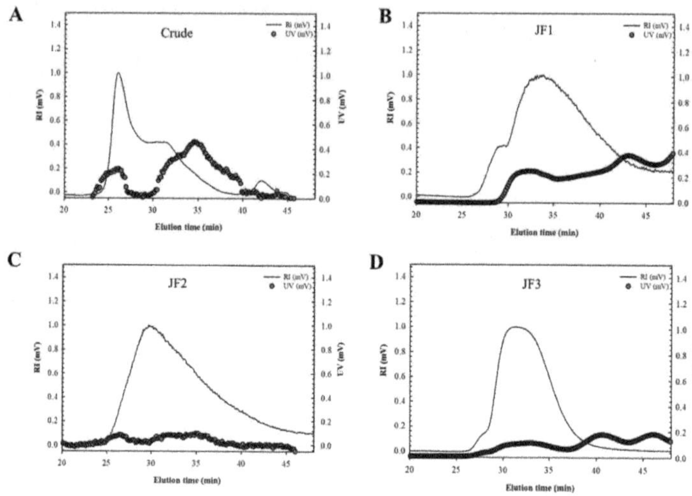

Figure 1. HPSEC chromatograms of crude polysaccharides (**A**) and fractions JF1 (**B**), JF2 (**C**), and JF3 (**D**) isolated from *L. smithii* jellyfish.

The relatively low value of the UV chromatogram compared with the marked RI signal in fractions JF2 and JF3 suggested the presence of small quantities of protein in these fractions. Conversely, the significant UV reaction shown in the JF1 fraction was consistent with its 18.1% protein level. The M_w values of JF1, JF2, and JF3 were significantly less than those of the crude polysaccharides. Table 2 shows that the M_w values were 477.6, 524.1, and 293.0 kDa for JF1, JF2, and JF3, respectively. It was previously reported that the M_w values of jellyfish skin polysaccharides (RP-JSP1 and RP-JSP2) were 121 and 590 kDa, respectively [27]. On the other hand, the M_w values differed from those reported by Li et al., who found that the M_w of jellyfish polysaccharides (JSP1) was 1250 kDa [7]. R_g is explained as the distribution of units of a polysaccharide around its axis and can define the size of the polymer, which can be calculated from light scattering up to various angles of the MALLS system. Table 2 shows that the R_g values were 124.0, 77.7, and 94.4 nm for the crude, JF1, and JF2 fractions and 56.3 nm for fraction JF3, indicating that in this result, the size of the polymer was in accordance with the M_w.

The individual volume for gyration (SV_g) of the SPs may be derived from the values of M_w and R_g using the following equation, as described by You and Lim [28].

$$SV_g = \frac{\frac{4}{3}\pi \left(R_g \times 10^8\right)^3}{M_w / N} = \frac{2.522\ R_g^3}{M_w}$$

where the values for SV_g, M_w, and R_g are cm^3/g, g/mol, and nm, respectively, and N is Avogadro's number (6.02×10^{23}/mol). SV_g is inversely related to the degree of molecular compactness, giving the theoretical gyration volume per unit of molar mass and providing mass-based information on polysaccharide density [28]. The overall SV_g values of the crude and fractionated polysaccharides are shown in Table 2. The SV_g values for the crude polysaccharides were 5.57 cm^3/g, and the values for their fractions were 5.57, 2.49, 4.05, and 1.59, respectively. The crude SV_g values were significantly higher than those of the fractions. This finding indicated that the crude possessed less compact and more extended conformational structures than JF2, following JF1 and JF3, which were the most compact.

These discrepancies in M_w, R_g, and SV_g were most likely caused by a combination of factors, including the extraction and purification processes as well as the analytical methods used in each investigation.

2.3. Effects of L. smithii Polysaccharides on the Cell Viability and NO Production

This study evaluated the anti-inflammatory effects of crude and fractionated (JF1, JF2, and JF3) polysaccharides isolated from *L. smithii*. Initially, the potential cytotoxicity of *L. smithii* polysaccharides on RAW264.7 cells, varying from 125 to 1000 µg/mL, was examined using a WST assay. As shown in Figure 2A, *L. smithii* polysaccharides significantly increased cell proliferation up to 1000 µg/mL relative to the negative control (RPMI medium). This finding suggests that the polysaccharides were not cytotoxic to RAW264.7 cells at the concentrations tested.

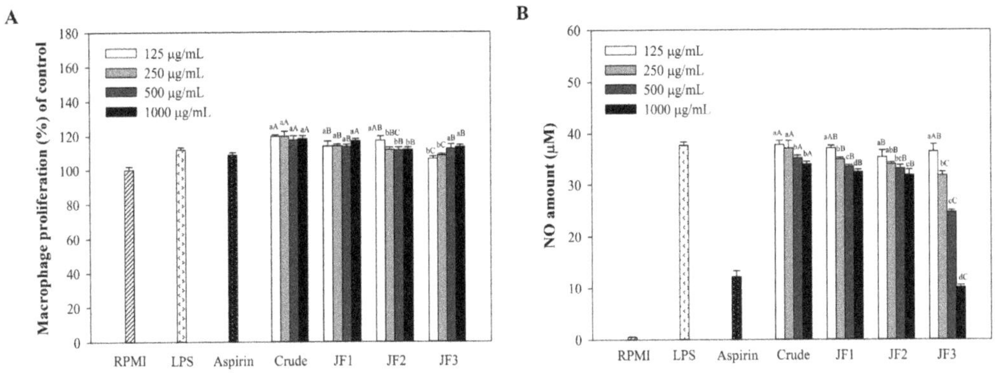

Figure 2. Effect of crude and fractions of *L. smithii* polysaccharides on the viability of RAW264.7 macrophages. The cells (1×10^6 cells/mL) received treatments with varying polysaccharide concentrations (125–1000 µg/mL) and were triggered with LPS (1 µg/mL). The cell viability was determined with a WST assay (**A**). The production of NO was determined with a Griess reagent (**B**). The values are displayed as the mean ± SD (n = 3). Different letters ^{a,b,c,d} refer to a significant difference between samples at each concentration, while $^{A,\,B,\,C}$ refer to a significant difference within concentrations at $p < 0.05$.

In addition, the immunoinhibiting effect was preliminarily evaluated by measuring NO accumulation (Figure 2B). All *L. smithii* polysaccharides decreased NO production in a dose-dependent manner. Among the polysaccharides, fraction JF3 showed a higher production of NO than the other polysaccharides. Increasing the concentrations of JF3 (125 to 1000 µg/mL) enhanced NO generation by 10.05 ± 0.45 µM, while the same concentrations of the crude polysaccharides, JF1, and JF2 reduced the NO production by 33.94 ± 0.54, 32.45 ± 0.45, and 31.82 ± 1.08 µM, respectively. According to these results, the most immunoinhibitory fraction, JF3, was chosen from the four different *L. smithii* polysaccharides to be assessed for their anti-inflammatory activity on macrophages at concentrations ranging from 125 to 1000 µg/mL.

2.4. JF3 Inhibits LPS-Induced Cell Viability and NO Production

To further investigate the effect of the most immunoinhibiting JF3 on the inhibition of macrophage activation, RAW264.7 cells were pretreated with or without JF3 (125–1000 µg/mL) and aspirin (200 µg/mL, as a positive drug) before being induced with LPS (1 µg/mL). As shown in Figure 3A, fraction JF3 (125–1000 µg/mL) significantly enhanced macrophage viability in RAW264.7 cells, and the enhancement increased with the JF3 concentration. Importantly, JF3 did not cause any significant toxicity to these cells. Therefore, these doses were used in the following experiments to treat JF3. As shown in Figure 3B, the NO secretion levels in the RAW264.7 cells were measured to be 0.55 ± 0.05 µM in the negative control and 37.70 ± 0.67 µM in the LPS-induced cells alone. In the presence of JF3 at concentrations of 125 to 1000 µg/mL, the levels of NO production were significantly inhibited, reducing the NO secretion levels to 35.31 ± 1.34, 31.94 ± 0.27, 23.28 ± 0.83, and

9.66 ± 0.72 μM. In addition, the NO contents at 1000 μg/mL JF3 concentrations were lower than those of the positive drug (aspirin).

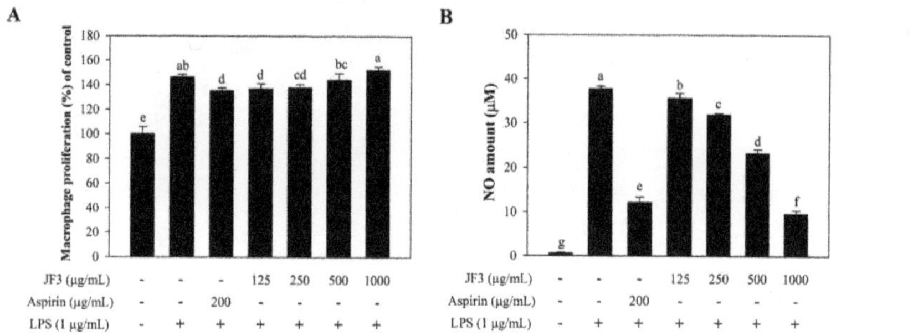

Figure 3. Effect of polysaccharide fraction JF3 on the viability of RAW264.7 macrophages. The cells (1×10^6 cells/mL) received treatments with varying polysaccharide concentrations (125–1000 μg/mL) and were triggered with LPS (1 μg/mL). The cell viability was determined by WST assay (**A**). The production of NO was determined by a Griess reagent (**B**). The values are displayed as the mean ± SD (n = 3). The letters $^{a, b, c, d, e, f, g}$ indicate significant differences ($p < 0.05$) between treatment groups.

2.5. *JF3 Inhibits LPS-Induced Expression of iNOS, COX-2, and Cytokines*

To determine whether the JF3 fraction inhibited the LPS-induced inflammatory response, the levels of inflammatory mediators (iNOS and COX-2) and inflammatory cytokines (IL-1β, IL-6, IL-10, and TNF-α) were measured. As shown in Figure 4A,B, the mRNA expression levels of iNOS and COX-2 were reduced in a dose-dependent manner by JF3 in the LPS-induced RAW264.7 cells. In addition to inhibiting the iNOS and COX-2 expressions, a dose-dependent suppression of JF3 on the IL-1β, IL-6, and TNF-α expression was observed (Figure 4C–E). Meanwhile, the effects of JF3 on IL-10 expression in the LPS-induced RAW264.7 cells were evaluated. As shown in Figure 4F, the mRNA level of IL-10 expression was dramatically raised following JF3 treatment at dosages ranging from 125 to 1000 μg/mL. These results show that the JF3 fraction activated the expression of inflammatory mediators and cytokines at the transcriptional level.

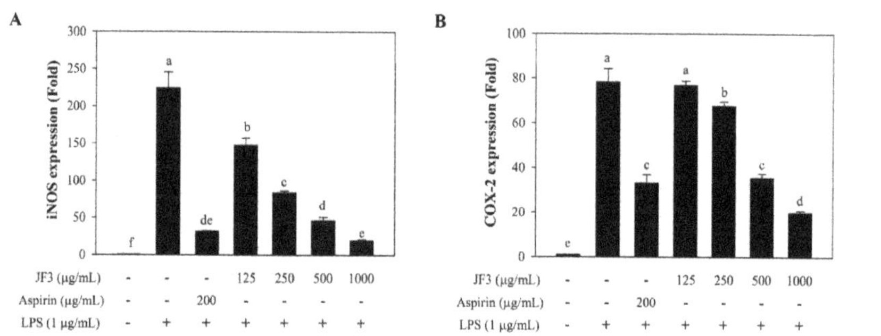

Figure 4. *Cont.*

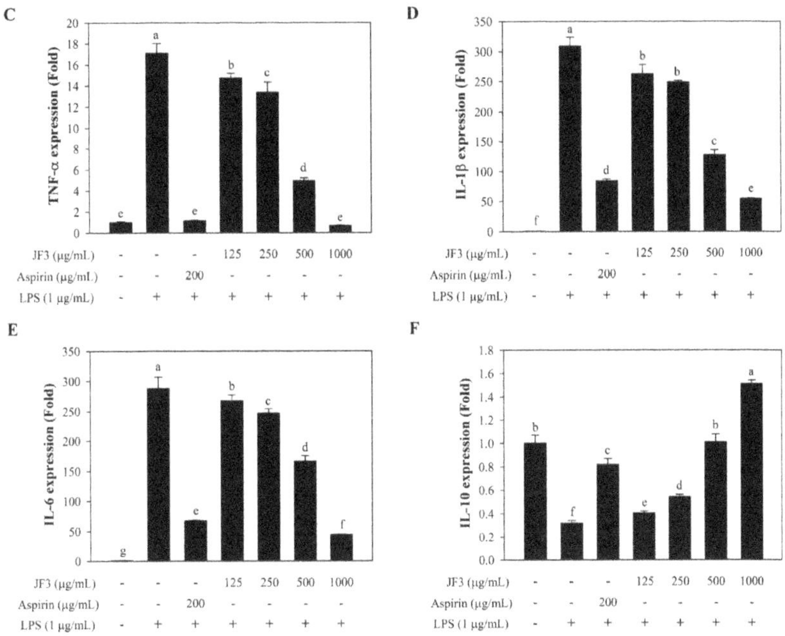

Figure 4. Effects of the JF3 polysaccharide fraction on the secretion of inflammatory mediators and cytokines in RAW264.7 macrophages. The cells (1×10^6 cells/mL) received treatments with varying polysaccharide concentrations (125–1000 μg/mL) and were triggered with LPS (1 μg/mL). The mRNA expressions of iNOS (**A**), COX-2 (**B**), TNF-α (**C**), IL-1β (**D**), IL-6 (**E**), and IL-10 (**F**) were determined by qPCR. The values are displayed as the mean ± SD (n = 3). The letters $^{a, b, c, d, e, f, g}$ indicate significant differences ($p < 0.05$) between treatment groups.

2.6. JF3 Suppresses LPS-Induced Nuclear Factor-κB (NF-κB) Activation

To investigate the possible inhibitory effects of JF3 through the suppression of LPS-induced activation of NF-κB signalling, the effect of JF3 on the nuclear translocation of the NF-κB p65 subunit was assessed using Western blot analyses. Figure 5A shows that LPS stimulation induced the phosphorylation of the NF-κB-p65 subunit. The JF3 fraction dramatically reduced phosphorylated NF-κB-p65, depending on the concentration. These findings imply that JF3 may prevent LPS-induced NF-κB activation.

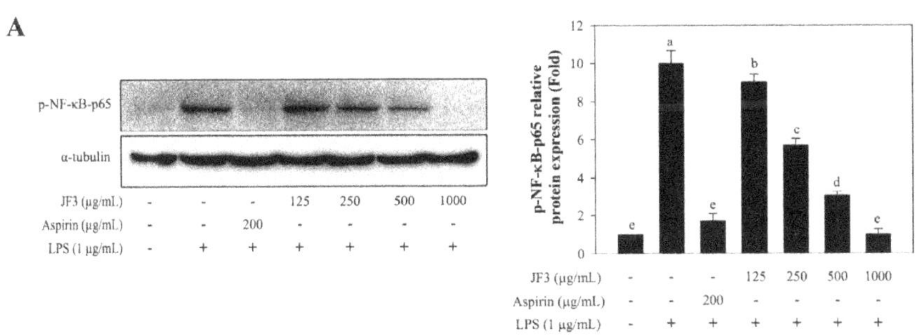

Figure 5. *Cont.*

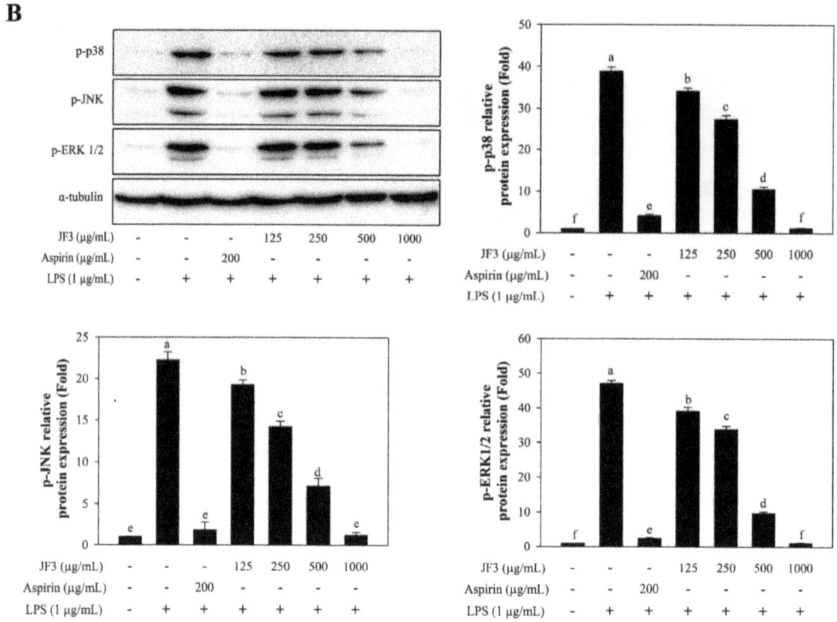

Figure 5. Effects of the JF3 fractionated polysaccharides on the phosphorylated NF-κB subunit 65, ERK1/2, JNK, and p38 MAPK. The cells (2×10^6 cells/mL) received treatments with varying polysaccharide concentrations (125–1000 μg/mL) and were triggered with LPS (1 μg/mL). The levels of protein expression were determined by Western blotting with phospho-NF-κB p65 antibodies (**A**) and specific antibodies to MAPKs (**B**) across three independent experiments. The values are displayed as the mean ± SD ($n = 3$). The letters [a, b, c, d, e, f] indicate significant differences ($p < 0.05$) between treatment groups.

2.7. JF3 Suppresses LPS-Induced Mitogen-Activated Protein Kinase (MAPK) Activation

It is well known that MAPK signalling pathways play an essential role in LPS-activated RAW264.7 cells [14,29,30]. The effects of JF3 on the phosphorylation levels of extracellular signal-regulated kinase (ERK), c-Jun N-terminal kinase (JNK), and p38 MAPK were determined in LPS-stimulated cells. As shown in Figure 5B, all MAPKs were stimulated by LPS treatment, whereas JF3 suppressed the LPS-induced phosphorylation of ERK, JNK, and p38 MAPK. Moreover, JF3 (125–1000 μg/mL) was shown to suppress the phosphorylation of three MAPKs in a dose-dependent manner, suggesting that MAPK pathways might contribute to the anti-inflammatory effects of JF3.

2.8. JF3 Inhibits LPS-Induced Cell Surface Expression

Fluorescence-activated cell sorting (FACS) analysis was utilised to investigate the effects of JF3 on LPS binding to the surface of RAW264.7 cells. As shown in Figure 6A,B, JF3 exerted an inhibitory effect on cell surface molecules such as CD11b and CD40. LPS treatment increased both cell surface molecules compared with RPMI. Compared with LPS, the LPS-induced CD11b expression was significantly reduced by JF3 at 125–1000 μg/mL in a dose-dependent manner, whereas the LPS-induced CD40 expression showed no differences in significance ($p < 0.05$) at 125 μg/mL concentrations and was substantially decreased at doses ranging from 250 to 1000 μg/mL.

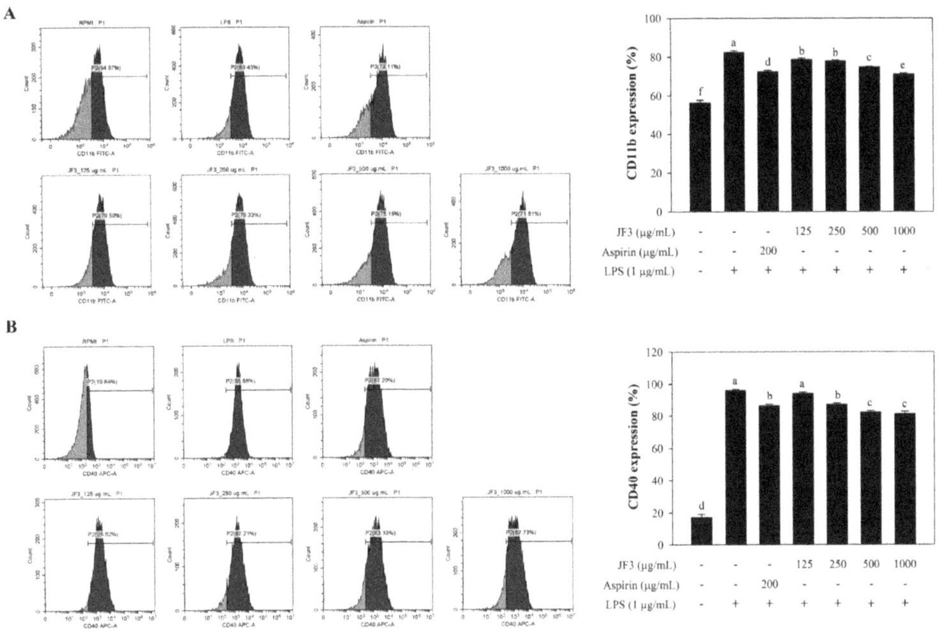

Figure 6. Evaluation of CD11b and CD40 expressions in JF3-treated RAW264.7 macrophages. The cells (2×10^6 cells/mL) received treatments with varying polysaccharide concentrations (125–1000 μg/mL) and were triggered with LPS (1 μg/mL). The expression of CD11b (**A**) and CD40 (**B**) was determined by flow cytometry. The values are displayed as the mean ± SD (n = 3). The letters [a, b, c, d, e, f] indicate significant differences ($p < 0.05$) between treatment groups.

2.9. Methylation Analysis of JF3

The glycosidic linkage of the most anti-inflammatory JF3 polysaccharides was analysed by GC–MS. Table 3 reveals that the alditol acetate products of JF3 mainly consisted of 2,4,6-Me3-Glc*p*, 2,4-Me2-Glcl*p*, and 2,4-Me2-Gal*p*, indicating the presence of (1→3)-linked glucopyranosyl, (1→3,6)-linked glucopyranosyl, and (1→3,6)-linked galactopyranosyl residues. A small amount of 2,3,4-Me3-Gal*p* was also detected, suggesting a (1→6)-linked galactopyranosyl residue. In addition, JF3 also included 2,3,4,6-Me4-Man*p*, implying a terminal residue. After desulphating fraction JF3 (D-JF3) by the solvolytic desulphation method, the methylation analysis of D-JF3 exhibited a marked decrease in 2,4-Me2-glucitol acetate with a concomitant increase in 2,4,6-Me3-glucitol acetate and no changes in the proportions of other methylated alditol acetates (Table 3). This suggests that the sulphate groups were mostly linked at the O-6 position. Overall, the above data indicate that the polysaccharide chain JF3 could be a (1→3)-glucose partially sulphated at O-6. Furthermore, (1→6)-linked galactose could be a branched polysaccharide, and other units and terminals could be in the side chains. Future studies need to elucidate the connection between the linkage through 1D and 2D NMR analyses. Li et al. (2017) reported the linear linkage of the fractionated polysaccharides of jellyfish with a M_w of 1250 kDa (JSP-11), which was previously reported as (1→3,6)-Man*p*, (1→6)-Gal*p*, and (1→)-Glc*p*A as terminal [7].

Table 3. Glycosidic linkage analysis of JF3.

Characteristic Fragment Ions (m/z)	Methylation Product	Glycosidic Linkage	JF3 (%)	D-JF3 (%)
84, 102, 118, 129, 162, 207	1,5-di-O-acetyl-2,3,4,6-tetra-O-methyl-Man	Man*p*-(1→	9.1	10.5
87, 101, 118, 129, 161, 206, 234	1,3,5-tri-O-acetyl-2,4,6-tri-O-methyl-Glc	→3)-Glc*p*-(1→	16.4	48.4
87, 101, 118, 129, 162, 189, 234	1,5,6-tri-O-acetyl-2,3,4-tri-O-methyl-Gal	→6)-Gal*p*-(1→	1.00	1.30
87, 101, 118, 129, 189, 234	1,3,5,6-tetra-O-acetyl-2,4-di-O-methyl-Glc	→3,6)-Glc*p*-(1→	45.2	13.5
87, 101, 118, 129, 189, 234	1,3,5,6-tri-O-acetyl-2,4-di-O-methyl-Gal	→3,6)-Gal*p*-(1→	28.3	26.3

3. Discussion

This study focused on the extraction, isolation, structure characterisation, and anti-inflammation properties of polysaccharides from the white jellyfish (*L. smithii*). After fractionation using ion-exchange chromatography to obtain three fractions (JF1, JF2, and JF3), the crude and three fractions were found to contain carbohydrates (62.5–72.2%) with some proteins (7.12–24.1%), sulphate (8.27–22.7%), and uronic acids (1.47–3.40%). The M_w and R_g of these polysaccharides showed ranges of 293.0–865.0 kDa and 56.3–124.0 nm, respectively, indicating variations in their molecular size. Moreover, the monosaccharide composition revealed differences in levels, including galactose (12.5–41.0%), glucose (33.1–56.7%), mannose (3.83–15.2%), and arabinose (0.60–15.4%), with a small amount of fucose (0.10–2.53%) and rhamnose (0.54–8.30%). The backbones of the most immunoinhibiting polysaccharide, JF3, were mainly linked through (1→3)-glucose, with some sulphate groups attached at position O-6. In addition, there is a possibility that (1→6)-linked galactose could form a branching structure within a polysaccharide, and various units and terminals could be present in the side chains.

The research of hot water extraction of polysaccharides from *Gracilaria rubra* (GRPS) and isolation with a DEAE-52 cellulose column and a Sephadex G-50 column yielded three fractions, GRPS-1-1, GRPS-2-1, and GRPS-3-2, which were heteropolysaccharides consisting of galactose and fucose at ratios of 1.79:1, 2.16:1, and 2.76:1, respectively. In addition, the M_w of GRPS-1-1, GRPS-2-1, and GRPS-3-2 were 1310, 691, and 923 kDa, respectively [31]. Another study performed enzyme-assisted extraction of sulphated polysaccharide (SCVP-1) from the sea cucumber, which consisted of mannose, glucosamine, glucuronic acid, *N*-acetyl-galactosamine, glucose, galactose, and fucose; had a relative molecular weight of 180.8 kDa; and was composed of total carbohydrates, uronic acid, proteins, and sulphate groups. The structure showed glycosaminoglycan with sulphation linked to the fucose residue [32]. In general, the structure and molecular properties of polysaccharides differ greatly depending on the type, species, growing conditions, extraction process, and analytical techniques. Among these parameters, the variation in their proximate compositions appears to be significantly impacted by the type, species variations, and extraction processes [33,34].

Numerous polysaccharides have been demonstrated to possess anti-inflammatory effects in macrophages, suppressing the production of inflammation mediators, such as NO, PGE_2, iNOS, COX-2, IL-1β, IL-6, and TNF-α [9,11,31,35]. The inflammation process involves multiple molecular mechanisms. For example, iNOS and COX-2 are two of the most important mediators of NO production and prostaglandin synthesis [11,13]. First, we observed that *L. smithii* polysaccharides, both in crude form and when fractionated (JF1, JF2, and JF3), could effectively inhibit the LPS-induced production of NO; the highest inhibitory effect on NO production was shown by fraction JF3. Our results demonstrate that JF3 inhibited LPS-induced NO production in macrophages, which was associated with their ability to downregulate iNOS mRNA expression. TNF-α, IL-1β, and IL-6 are pro-inflammatory cytokines that play essential roles in immunological responses to a range of inflammatory stimuli [36]. JF3 also significantly reduced the expression of COX-2, IL-1β, IL-6, and TNF-α induced by LPS. Conversely, IL-10 is one of the most effective anti-inflammatory cytokines, with numerous immunomodulatory properties [10,12,37]. Our findings revealed that JF3 dramatically boosted the levels of IL-10, an anti-inflammatory cytokine. Furthermore, these results are consistent with those of previous reports, which showed that polysaccharides regulated LPS-induced inflammatory mediators and cytokine production [38–40]. Overall, this study suggests that JF3 polysaccharides have anti-inflammatory activity by inhibiting the expression of pro-inflammatory cytokines and enhancing anti-inflammatory cytokines.

A variety of signalling pathways related to cellular immunological responses have been studied to elucidate the molecular mechanisms of *L. smithii* polysaccharides. Polysaccharides may be involved in immunity and inflammation through MAPK and NF-κB signalling processes, which are critical intracellular signalling pathways with complicated connections [15]. NF-κB is an important transcription factor that regulates the generation of various inflammation-related mediators and cytokines [36,39]. Here, pretreatment with JF3 significantly reduced LPS-induced p65 phosphorylation. The MAPK pathway is additionally recognised to serve as a critical mediator in the promotion of inflammatory responses, which plays a role in regulating several cellular processes [41]. The pathway consists of three-tiered cascades, including JNK, ERK, and p38 [10,42]. The phosphorylation activates pro-inflammatory transcription factors, such as activator proteins (AP)-1 (cFos/cJun), Runt-related transcription factor (RUNX)-2, hypoxia inducible factor (HIF)-2α, and CCAAT-enhancer-binding protein (C/EBP)-β [42]. In this work, JF3 markedly decreased the phosphorylation of p38, JNK, and ERK1/2. A previous study found that sulphated polysaccharides from *Sargassum cristaefolium* had anti-inflammatory effects by suppressing NF-κB and downregulating the phosphorylation of p-38, ERK, and JNK kinases in the MAPK signalling cascade [6].

Additionally, markers such as CD68, CD86, CD80, F4/80, CD11b, and CD40 are expressed by activated macrophages [43,44]. The CD40 molecule interacts with antigen-presenting cells and plays a costimulatory role in immune regulation [16]. CD11b/CD18 is a receptor on innate immune cells that recognises pathogens [45]. Here, the flow cytometric analysis revealed that JF3 inhibited the expression of CD11b and CD40 in LPS-treated RAW264.7 cells. Thus, our findings suggest that JF3 could inhibit the initial phase of LPS-activated cellular signalling pathways by decreasing LPS binding to cell surface receptors and CD11b and CD40 expression on RAW264.7 cells.

Recently, research demonstrated that the structural characteristics of a polysaccharide have a close relationship with anti-inflammatory activities, including chemical composition, monosaccharide, molecular weight, structure, functional groups, glycosidic linkage, and conformations. Many reports have shown that the sulphate groups and M_W of polysaccharides markedly affect their anti-inflammatory properties [46]. Brown algal polysaccharides (*S. cristaefolium*) with an M_W of 386.1 kDa and a sulphate concentration of 9.42% reduced NO production in LPS-stimulated RAW264.7 cells [6]. It seems that the dependence of suitable M_W and sulphate groups might enhance the binding affinity with the cell receptor [6,47].

According to our results, the JF3 chain contained (1→3)-linked glucopyranosyl, (1→3,6)-linked glucopyranosyl, and (1→3,6)-linked galactopyranosyl residues. The sulphate groups were mostly presented at the position of O-6. In addition, the 1→6-linked galactose units could be a branched polysaccharide, and other units and terminals could be in the side chains. JF3 had a molecular weight of 293.0 kDa and a high content of sulphates. This is similar to the previously mentioned research, which found that glycosidic linkages could affect anti-inflammatory activities. α-D-(1→3)-linked glucose shows anti-inflammatory, antiangiogenic activities, and anti-CAG (chronic atrophic gastritis) activity [48–51]. Similarly, the existence of monosaccharides could be a factor that affects the anti-inflammatory activities of polysaccharides, glucose, and fucose, and has been found to have a good effect on inflammatory activity [5,52]. Numerous studies have been conducted on a fucose-rich fucoidan that strongly affects both in vitro and in vivo anti-inflammatory activities. In brown seaweed, the fucoidan's effects could potentially inhibit anti-inflammatory activities via the inhibition of protein denaturation, primarily due to its fucose content and the moderation of sulphate content [53,54]. An analysis of monosaccharides of JF3 revealed that they mostly contained glucose, galactose, and mannose, while arabinose, rhamnose, and fucose represented a minor sugar unit. However, the limited quantities of fucose may be responsible for the anti-inflammatory activities. As such, investigation into the monosaccharide, molecular weight, and fine structures is essential to understand the relationships between structural elements and biological processes and realise their maximum effects.

4. Materials and Methods

4.1. Materials and Chemicals

Fresh white jellyfish (*L. smithii*) were obtained from local fisheries in La-ngu District, Satun Province, Thailand. After collection, fresh jellyfish were individually packed into a polyethylene bag, frozen in the freezer, packed into a carton box, and transported by a temperature-controlled container car (−18 to 20 °C) to the laboratory of the College of Maritime Studies and Management, Chiang Mai University, Samut Sakhon Province, Thailand, within 12 h. The collected jellyfish was rinsed with distilled water (DW) and dried in a hot air oven at 55 °C until the final moisture content of the sample was 8%. The dried sample was then powdered with a grinder. The jellyfish was placed into a polyethylene bag, frozen at −20 °C, and used for polysaccharide extraction after two months. The Roswell Park Memorial Institute (RPMI) 1640 Medium was purchased from Gibco (Thermo Fisher Scientific Inc., Waltham, MA, USA). Mouse macrophage RAW264.7 cells were purchased from the Korean Cell Line Bank (KCLB; Seoul, Republic of Korea). Fetal bovine serum (FBS) and 1% penicillin/streptomycin were procured by Welgene (Daegu, Republic of Korea). All other cell culture chemicals and reagents were obtained from Sigma Chemical Co. (St. Louis, MO, USA). All solvents used were of high-performance liquid chromatography (HPLC) grade.

4.2. Polysaccharide Extraction and Fractionation

The extraction of *L. smithii* polysaccharides was performed using water extraction [7]. The dried sample powder of *L. smithii* was extracted two times in DW at 90 °C for 2 h. All water extracts were combined and precipitated from the supernatant using four volumes of cold 98% (*v*/*v*) ethanol and stored overnight at 4 °C, which was then subjected to filtration. The resulting pellet was collected and dried under a fume hood overnight to obtain the crude polysaccharide powder. The crude polysaccharide was then fractionated using ion-exchange chromatography on a DEAE Sepharose fast flow column (17-0709-01; GE Healthcare Bio-Science AB, Uppsala, Sweden). The polysaccharides were eluted with DW and different concentrations of NaCl (0.5–2 M). The elution profile of carbohydrates was evaluated using the phenol-H_2SO_4 method by determining 490 nm absorbance [55]. The chromatography yielded three fractions, namely, JF1 (eluted with DW), JF2 (eluted with 1.0 M NaCl), and JF3 (eluted with 1.5 M NaCl), as shown in Figure 7.

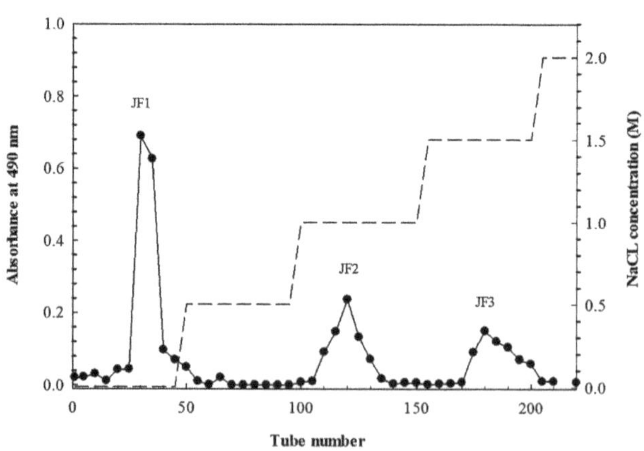

Figure 7. Elution profile of *L. smithii* polysaccharides (JF1, JF2, and JF3) fractionated in a DEAE Sepharose fast flow column.

4.3. Chemical Composition

The chemical compositions were examined, including the total carbohydrate content, which was determined with the phenol-sulfuric acid method using D-glucose as a standard [55]. The protein content was evaluated with the Lowry method using a DC protein assay kit (Bio-Rad, Hercules, CA, USA) [56]. The determination of the sulphate content was carried out with the $BaCl_2$ gelatin method using K_2SO_4 as a standard [57]. The uronic acid contents were analysed with a sulphamate/m-hydroxydiphenyl assay using glucuronic acid as a standard [58].

4.4. Monosaccharide Characterisation

Gas chromatography–mass spectrometry (GC–MS) was used to examine the monosaccharide composition of *L. smithii* polysaccharides. The sample was hydrolysed with trifluoroacetic acid (TFA, 4 M) at 100 °C for 6 h, then reduced in water with sodium borodeuteride ($NaBD_4$) and acetylated with acetic anhydride. Finally, the sample was analysed by GC–MS (6890 N/MSD 5973, Agilent Technologies, Santa Clara, CA, USA) coupled with an HP-5MS capillary column (30 m × 0.25 mm × 0.25 μm). As a carrier gas, nitrogen was applied. To certify the monosaccharide content, monosaccharide standards were employed.

4.5. Measurement of M_w

The M_w of the four polysaccharides was estimated using the HPSEC-UVMALLS-RI system), which included a pump (Waters 510, Milford, MA, USA), an injector valve with a 200 μL sample loop (model 7072, Rheodyne, Rohnert Park, CA, USA), SEC columns (TSK G5000 PW, 7.5 mm × 600 mm; TosoBiosep, Mongomeryville, PA, USA), a UV detector at 280 nm (Waters 2487), a multi-angle laser light scattering detector (HELEOS, Wyatt Technology Corp, Santa Barbara, CA, USA), and a refractive index detector (Waters 2414). At a flow rate of 0.4 mL/min, the mobile phase of this system contained 0.15 M $NaNO_3$ and 0.02% NaN_3. Each sample was immersed in DW and then heated for 15 min at 75 °C before being injected into MALLS. The M_w and R_g were estimated using the ASTRA version 6.0 software (Wyatt Technology Corp., Beijing, China).

4.6. Measurement of the Anti-Inflammatory Activity of L. smithii Polysaccharides

4.6.1. Cell Culture and Treatment

The RAW264.7 cells were grown in RPMI-1640 medium, which was supplied with 10% FBS and 1% penicillin/streptomycin and kept at 37 °C in 5% CO_2 humidified incubators. The cells were incubated with 100 μL of various doses of polysaccharides (125, 250, 500, and 1000 μg/mL) or aspirin (200 μg/mL) as a positive drug for 1 h. Following sample treatment, the cells were stimulated with 100 μL of LPS (1 μg/mL) for a further 24 h.

4.6.2. Cell Viability Analysis

The cell proliferative ability was assessed through the EZ-Cytox Cell Viability Assay Kit (DaeilLab Service, Seoul, Republic of Korea). Briefly, the RAW264.7 cells (1×10^6 cells/mL) were incubated in a 96-well cell culture plate with various concentrations of *L. smithii* polysaccharides for 24 h, and the amount of tetrazolium salt was quantified using a microplate reader (EL-800; BioTek Instruments, Winooski, VT, USA) at 450 nm absorbance.

4.6.3. Determination of NO Release

Griess reagent (Sigma-Aldrich, St. Louis, MO, USA) was applied to quantify the concentration of NO in the supernatant. RAW264.7 cells (1×10^6 cells/mL) in a 96-well cell culture plate were subjected to treatment with *L. smithii* polysaccharides prior to LPS stimulation. After incubation, 100 μL of cell culture supernatant was combined with Griess reagent (0.1% *N*-1-napthyl ethylenediamine dihydrochloride in DW and 1% sulphanilamide in 5% phosphoric acid) and incubated for 10 min. The nitrite accumulation was measured using 540 nm absorbance and a standard curve for sodium nitrite.

4.6.4. Real-Time Polymerase Chain Reaction (PCR) Analysis

The RAW264.7 cells (1×10^6 cells/mL) were placed in a 24-well cell culture plate and subjected to treatment with polysaccharides induced by LPS. Total RNA extraction was carried out using the TRIzol reagent (Invitrogen, Carlsbad, CA, USA), and cDNA was created using a High-Capacity cDNA Reverse Transcription Kit (Applied Biosystems, Foster City, CA, USA). Then, the fragments were amplified using a real-time PCR system on a QuantStudio™ 3 FlexReal-Time PCR System (Thermo Fisher Scientific Inc., Waltham, MA, USA) with TB Green® Premix Ex Taq™ II (Takara Bio, Inc., Shiga, Japan) and the specified primers (Table 4).

Table 4. The sequences of primers used in real-time PCR.

Target Genes	Sequences of the Primers (5′ to 3′)	
	Forward	Reverse
IL-1β	GGGCCTCAAAGGAAAGAATC	TACCAGTTGGGGAACTCTGC
IL-6	AGTTGCCTTCTTGGGACTGA	CAGAATTGCCATTGCACAAC
IL-10	TACCTGGTAGAAGTGATGCC	CATCATGTATGCTTCTATGC
TNF-α	ATGAGCACAGAAAGCATGATC	TACAGGCTTGTCACTCGAATT
iNOS	TTCCAGAATCCCTGGACAAG	TGGTCAAACTCTTGGGGTTC
COX-2	AGAAGGAAATGGCTGCAGAA	GCTCGGCTTCCAGTATTGAG
β-actin	CCACAGCTGAGAGGGAAATC	AAGGAAGGCTGGAAAAGAGC

4.6.5. Western Blotting Analysis

The RAW264.7 cells (2×10^6 cells/mL) in a 6-well cell culture plate were subjected to treatment with polysaccharides and LPS for 24 h at 37 °C. Then, the cells were harvested and lysed in radioimmunoprecipitation assay (RIPA) buffer for 30 min before being collected by centrifugation at 12,000× *g* for 10 min at 4 °C. The protein contents were determined using the Micro BCA™ Protein Assay Kit (Thermo Fisher Scientific Inc., Waltham, MA, USA). A total of 30 μg of protein per sample was separated by electrophoresis on 10% SDS-polyacrylamide gel electrophoresis and transferred onto a polyvinylidene fluoride (PVDF) membrane. Primary antibodies against phospho-NF-κB p65, phospho-p38 MAPK, phospho-SAPK/JNK, phospho-p44/42 MAPK (Erk1/2), and alpha-tubulin, along with an anti-rabbit IgG HRP-linked antibody (Abcam, Cambridge, UK), were used to incubate the membrane. The proteins were identified using the Pierce ECL Plus Western Blotting Substrate (Thermo Fisher Scientific Inc., Waltham, MA, USA), and the expression was visualised using the Bio-Rad image analysis program (Bio-Rad, Hercules, CA, USA).

4.6.6. Flow Cytometry Analysis

RAW264.7 cells at a density of 1×10^6 cells/mL in a 6-well cell culture plate treated with the polysaccharides (125–1000 μg/mL) were rinsed with a flow cytometry buffer. Then, 10 μL of anti-CD40-APC (Anti-CD40 (1C10) Allophycocyanin (APC), Thermo Fisher Scientific Inc., Waltham, MA, USA) and 10 μL of anti-CD11b (anti-CD11b (M1/70) fluorescein isothiocyanate (FITC), Thermo Fisher Scientific Inc., Waltham, MA, USA) monoclonal antibodies were stained to each sample and incubated for 30 min at 4 °C in the dark. The unconjugated antibodies were removed by washing the cells with FACS buffer and then resuspended with 1% paraformaldehyde. In order to analyse the flow cytometry, a CytoFLEX Flow Cytometer (Beckman Coulter, Inc., Brea, CA, USA) was used.

4.7. Desulphation of Polysaccharide

The desulphation of polysaccharide was based on the method reported by Tarbasa et al. [59]; the JF3 fraction (100 mg) was dissolved in 10 mL of distilled water and eluted with pyridine from a Dowex 50 W resin column, yielding lyophilised polysaccharide–pyridinium salts. Then, the polysaccharide–pyridinium salts were processed at 120 °C and for 40 min to obtain the desulphation sample. The solution was dialysed against distilled water and finally lyophilised to obtain the desulphated JF3. The glycosidic linkage of native JF3 and desulphated JF3 was analysed using GC–MS.

4.8. Methylation Analysis of JF3

The samples were methylated according to Ciucanu's method [60], which was slightly modified. The sample was dissolved in 0.5 mL of dimethyl sulfoxide (DMSO) and methylated with methyl iodide (CH_3I) and sodium hydroxide powder (NaOH). Partially methylated alditol acetates were created by methylating polysaccharides using acid hydrolysis with 4 M TFA at 100 °C for 6 h. After that, the hydrolysates were decreased with $NaBD_4$ and acetylated with acetic anhydride. GC–MS (6890 N/MSD 5973, Agilent Technologies, Santa Clara, CA, USA) was used for the analysis of partly methylated alditol acetates on an HP-5MS capillary column (30 m × 0.25 mm × 0.25 µm) (Agilent Technologies, Santa Clara, CA, USA). The carrier gas was helium, with a constant flow rate of 1.2 mL/min. The oven settings incorporated a temperature program that proceeded from 160 to 210 °C in 10 min and then to 240 °C in 10 min. Thus, a temperature gradient was used at 5 °C/min, the inlet temperature was maintained at 250 °C, and the mass range was set to 35 to 450 m/z. Peaks were assigned by considering retention times and mass spectra.

4.9. Statistical Analyses

The experimental design was completed with a completely randomised design (CRD). The analysed data were reported as the mean ± standard deviation (SD) of three independent experiments ($n = 3$) using SPSS version 23.0 software (SPSS, Inc, Chicago, IL, USA) for data analysis. The significance was determined using one-way ANOVA, and Duncan's new multiple range tests (DMRTs) were used to test the differences between treatment groups ($p < 0.05$).

5. Conclusions

The water-soluble polysaccharides isolated from jellyfish (*Lobonema smithii*) were fractionated into three fractions with different sugar, protein, and sulphate contents, as well as different sugar units, exhibiting wide ranges of M_w values. The crude and fractionated (JF1, JF2, and JF3) polysaccharides could activate RAW264.7 cells by promoting cell viability and reducing the secretion of NO. In particular, JF3 inhibited the LPS-induced release of factors and cell surface molecules that promote inflammation by decreasing NF-κB and MAPK signalling. Furthermore, JF3 also contributed to a reduction in pro-inflammatory cytokines and an increase in anti-inflammatory cytokines in response to LPS. The increased activity of JF3 could have resulted from its lower molecular weight relative to the crude polysaccharides and the other two fractions. In addition, (1→3)-linked glucopyranosyl, (1→3,6)-linked glucopyranosyl, and (1→3,6)-linked galactopyranosyl residues comprised the main chain of JF3, partially sulfated at O-6 of glucose; furthermore, the other units and terminals could be in the side chains. Therefore, this study suggests that *L. smithii* polysaccharides potentially possess anti-inflammatory activity and could be useful as effective therapeutic agents against inflammation according to the chemical composition and primary molecular structure results, especially for JF3 polysaccharides. To maximise the utilisation of this marine species as a bioactive resource in the pharmaceutical industry, additional research is required on the purification and identification of the active polysaccharides in JF3, and the impact of the functional group and molecular structure on its anti-inflammatory properties should be investigated in future studies.

Author Contributions: Conceptualisation, U.S.; methodology, T.S. and W.R.-i.; validation, S.W., P.S., S.K. and U.S.; formal analysis, T.S.; investigation, T.S. and U.S.; data curation, T.S. and U.S.; writing—original draft preparation, T.S. and W.R.-i.; writing—review and editing, S.W., S.Y., W.J.P., P.S., S.K. and U.S.; supervision, S.W., P.S. and U.S.; project administration, U.S.; funding acquisition, U.S. All authors have read and agreed to the published version of the manuscript.

Funding: This research project was supported by the Fundamental Fund 2022, Chiang Mai University.

Institutional Review Board Statement: Not applicable.

Data Availability Statement: The datasets used and/or analysed during the current study are available from the corresponding author upon reasonable request.

Conflicts of Interest: The authors declare no conflict of interest.

References

1. Fedorov, S.N.; Ermakova, S.P.; Zvyagintseva, T.N.; Stonik, V.A. Anticancer and cancer preventive properties of marine polysaccharides: Some results and prospects. *Mar. Drugs* **2013**, *11*, 4876–4901. [CrossRef] [PubMed]
2. Wang, W.; Wang, S.-X.; Guan, H.-S. The antiviral activities and mechanisms of marine polysaccharides: An overview. *Mar. Drugs* **2012**, *10*, 2795–2816. [CrossRef] [PubMed]
3. Cheong, K.-L.; Yu, B.; Chen, J.; Zhong, S. A comprehensive review of the cardioprotective effect of marine algae polysaccharide on the gut microbiota. *Foods* **2022**, *11*, 3550. [CrossRef] [PubMed]
4. Hu, X.; Tao, N.; Wang, X.; Xiao, J.; Wang, M. Marine-derived bioactive compounds with anti-obesity effect: A review. *J. Funct. Foods* **2016**, *21*, 372–387. [CrossRef]
5. Castro, L.S.E.P.W.; Pinheiro, T.S.; Castro, A.J.G.; Dore, C.M.P.G.; da Silva, N.B.; Faustino Alves, M.G.d.C.; Santos, M.S.N.; Leite, E.L. Fucose-containing sulfated polysaccharides from brown macroalgae *Lobophora variegata* with antioxidant, anti-inflammatory, and antitumoral effects. *J. Appl. Phycol.* **2014**, *26*, 1783–1790. [CrossRef]
6. Wu, G.J.; Shiu, S.M.; Hsieh, M.C.; Tsai, G.J. Anti-inflammatory activity of a sulfated polysaccharide from the brown alga *Sargassum cristaefolium*. *Food Hydrocoll.* **2016**, *53*, 16–23. [CrossRef]
7. Li, Q.-M.; Wang, J.-F.; Zha, X.-Q.; Pan, L.-H.; Zhang, H.-L.; Luo, J.-P. Structural characterization and immunomodulatory activity of a new polysaccharide from jellyfish. *Carbohydr. Polym.* **2017**, *159*, 188–194. [CrossRef]
8. D'Ayala, G.G.; Malinconico, M.; Laurienzo, P. Marine derived polysaccharides for biomedical applications: Chemical modification approaches. *Molecules* **2008**, *13*, 2069–2106. [CrossRef]
9. Chen, H.; Sun, J.; Liu, J.; Gou, Y.; Zhang, X.; Wu, X.; Sun, R.; Tang, S.; Kan, J.; Qian, C.; et al. Structural characterization and anti-inflammatory activity of alkali-soluble polysaccharides from purple sweet potato. *Int. J. Biol. Macromol.* **2019**, *131*, 484–494. [CrossRef]
10. Chu, H.; Tang, Q.; Huang, H.; Hao, W.; Wei, X. Grape-seed proanthocyanidins inhibit the lipopolysaccharide-induced inflammatory mediator expression in RAW264.7 macrophages by suppressing MAPK and NF-κB signal pathways. *Environ. Toxicol. Pharmacol.* **2016**, *41*, 159–166. [CrossRef]
11. Hwang, P.A.; Chien, S.Y.; Chan, Y.L.; Lu, M.K.; Wu, C.H.; Kong, Z.L.; Wu, C.J. Inhibition of lipopolysaccharide (LPS)-induced inflammatory responses by *Sargassum hemiphyllum* sulfated polysaccharide extract in RAW 264.7 macrophage cells. *J. Agric. Food Chem.* **2011**, *59*, 2062–2068. [CrossRef] [PubMed]
12. Li, S.-T.; Dai, Q.; Zhang, S.-X.; Liu, Y.-J.; Yu, Q.-Q.; Tan, F.; Lu, S.-H.; Wang, Q.; Chen, J.-W.; Huang, H.-Q.; et al. Ulinastatin attenuates LPS-induced inflammation in mouse macrophage RAW264.7 cells by inhibiting the JNK/NF-κB signaling pathway and activating the PI3K/Akt/Nrf2 pathway. *Acta Pharmacol. Sin.* **2018**, *39*, 1294–1304. [CrossRef] [PubMed]
13. Lee, H.H.; Han, M.H.; Hwang, H.J.; Kim, G.Y.; Moon, S.K.; Hyun, J.W.; Kim, W.J.; Choi, Y.H. Diallyl trisulfide exerts anti-inflammatory effects in lipopolysaccharide-stimulated RAW264.7 macrophages by suppressing the Toll-like receptor 4/nuclear factor-kB pathway. *Int. J. Mol. Med.* **2015**, *35*, 487–495. [CrossRef] [PubMed]
14. Li, M.Y.; Sun, L.; Niu, X.T.; Chen, X.M.; Tian, J.X.; Kong, Y.D.; Wang, G.Q. Astaxanthin protects lipopolysaccharide-induced inflammatory response in *Channa argus* through inhibiting NF-kB and MAPKs signaling pathways. *Fish Shellfish Immunol.* **2019**, *86*, 280–286. [CrossRef] [PubMed]
15. Chen, Q.; Qi, C.; Peng, G.; Liu, Y.; Zhang, X.; Meng, Z. Immune-enhancing effects of a polysaccharide PRG1-1 from *Russula griseocarnosa* on RAW264.7 macrophage cells via the MAPK and NF-kB signalling pathways. *Food Agric. Immunol.* **2018**, *29*, 833–844. [CrossRef]
16. Li, Y.; Meng, T.; Hao, N.; Tao, H.; Zou, S.; Li, M.; Ming, P.; Ding, H.; Dong, J.; Feng, S.; et al. Immune regulation mechanism of Astragaloside IV on RAW264.7 cells through activating the NF-κB/MAPK signaling pathway. *Int. Immunopharmacol.* **2017**, *49*, 38–49. [CrossRef] [PubMed]
17. Nishimoto, S.; Goto, Y.; Morishige, H.; Shiraishi, R.; Doi, M.; Akiyama, K.; Yamauchi, S.; Sugahara, T. Mode of action of the immunostimulatory effect of collagen from jellyfish. *Biosci. Biotechnol. Biochem.* **2008**, *72*, 2806–2814. [CrossRef] [PubMed]
18. Sugahara, T.; Ueno, M.; Goto, Y.; Shiraishi, R.; Doi, M.; Akiyama, K.; Yamauchi, S. Immunostimulation effect of jellyfish collagen. *Biosci. Biotechnol. Biochem.* **2006**, *70*, 2131–2137. [CrossRef]
19. Cao, Y.; Gao, J.; Zhang, L.; Qin, N.; Zhu, B.; Xia, X. Jellyfish skin polysaccharides enhance intestinal barrier function and modulate the gut microbiota in mice with DSS-induced colitis. *Food Funct.* **2021**, *12*, 10121–10135. [CrossRef]
20. Zhang, H.-L.; Cui, S.-H.; Zha, X.-Q.; Bansal, V.; Xue, L.; Li, X.-L.; Hao, R.; Pan, L.-H.; Luo, J.-P. Jellyfish skin polysaccharides: Extraction and inhibitory activity on macrophage-derived foam cell formation. *Carbohydr. Polym.* **2014**, *106*, 393–402. [CrossRef]
21. Hwang, S.J.; Ahn, E.-Y.; Park, Y.; Lee, H.-J. An aqueous extract of Nomura's jellyfish ameliorates inflammatory responses in lipopolysaccharide-stimulated RAW264.7 cells and a zebrafish model of inflammation. *Biomed. Pharmacother.* **2018**, *100*, 583–589. [CrossRef] [PubMed]

22. Omori, M.; Nakano, E. Jellyfish fisheries in southeast Asia. *Hydrobiologia* **2001**, *451*, 19–26. [CrossRef]
23. Assaw, S.; Ahmad, A.S.; Abd Wahid, M.E. Potential of Malaysian white type edible jellyfish, *Lobonema smithii* as antioxidant and collagen promoter in dermal wound of Sprague Dawley rats. *Middle East J. Sci. Res.* **2016**, *24*, 2137–2144.
24. Rodsuwan, U.; Thumthanaruk, B.; Kerdchoechuen, O.; Laohakunjit, N. Functional properties of type A gelatin from jellyfish (*Lobonema smithii*). *Int. Food Res. J.* **2016**, *23*, 507–514.
25. Upata, M.; Siriwoharn, T.; Makkhun, S.; Yarnpakdee, S.; Regenstein, J.M.; Wangtueai, S. Tyrosinase inhibitory and antioxidant activity of enzymatic protein hydrolysate from jellyfish (*Lobonema smithii*). *Foods* **2022**, *11*, 615. [CrossRef] [PubMed]
26. Muangrod, P.; Charoenchokpanich, W.; Roytrakul, S.; Rungsardthong, V.; Vatanyoopaisarn, S.; Charoenlappanit, S.; Wonganu, B.; Thumthanaruk, B. Effect of pepsin on antioxidant and antibacterial activity of protein hydrolysate from salted jellyfish (*Lobonema smithii* and *Rhopilema hispidum*) by-products. *E3S Web Conf.* **2022**, *355*, 02013. [CrossRef]
27. Migone, C.; Scacciati, N.; Grassiri, B.; De Leo, M.; Braca, A.; Puppi, D.; Zambito, Y.; Piras, A.M. Jellyfish polysaccharides for wound healing applications. *Int. J. Mol. Sci.* **2022**, *23*, 11491. [CrossRef] [PubMed]
28. You, S.; Lim, S.-T. Molecular characterization of corn starch using an aqueous HPSEC-MALLS-RI system under various dissolution and analytical conditions. *Cereal Chem.* **2000**, *77*, 303–308. [CrossRef]
29. Lu, C.; Zhu, W.; Wang, M.; Hu, M.; Chen, W.; Xu, X.; Lu, C. Polysaccharides from *Smilax glabra* inhibit the pro-inflammatory mediators via ERK1/2 and JNK pathways in LPS-induced RAW264.7 cells. *Carbohydr. Polym.* **2015**, *122*, 428–436.
30. Himaya, S.W.A.; Ryu, B.; Qian, Z.-J.; Kim, S.-K. Sea cucumber, *Stichopus japonicus* ethyl acetate fraction modulates the lipopolysaccharide induced iNOS and COX-2 via MAPK signaling pathway in murine macrophages. *Environ. Toxicol. Pharmacol.* **2010**, *30*, 68–75. [CrossRef]
31. Bezerra, I.L.; Caillot, A.R.C.; Palhares, L.; Santana-Filho, A.P.; Chavante, S.F.; Sassaki, G.L. Structural characterization of polysaccharides from Cabernet Franc, Cabernet Sauvignon and Sauvignon Blanc wines: Anti-inflammatory activity in LPS stimulated RAW264.7 cells. *Carbohydr. Polym.* **2018**, *186*, 91–99. [CrossRef] [PubMed]
32. Liang, Y.; Wan, X.; Niu, F.; Xie, S.; Guo, H.; Yang, Y.; Guo, L.; Zhou, C. *Salvia plebeia* R. Br.: An overview about its traditional uses, chemical constituents, pharmacology and modern applications. *Biomed. Pharmacother.* **2020**, *121*, 109589. [CrossRef] [PubMed]
33. Baker, E.J.; Valenzuela, C.A.; De Souza, C.O.; Yaqoob, P.; Miles, E.A.; Calder, P.C. Comparative anti-inflammatory effects of plant- and marine-derived omega-3 fatty acids explored in an endothelial cell line. *BBA-Mol. Cell Biol.* **2020**, *1865*, 158662. [CrossRef] [PubMed]
34. Zhu, Y.; Feng, X.; Guo, J.; Wang, L.; Guo, X.; Zhu, X. A review of extraction, purification, structural properties and biological activities of legumes polysaccharides. *Front. Nutr.* **2022**, *9*, 1021448. [CrossRef] [PubMed]
35. Wang, L.; Yu, X.; Yang, X.; Li, Y.; Yao, Y.; Lui, E.M.K.; Ren, G. Structural and anti-inflammatory characterization of a novel neutral polysaccharide from North American ginseng (*Panax quinquefolius*). *Int. J. Biol. Macromol.* **2015**, *74*, 12–17. [CrossRef] [PubMed]
36. Kim, K.N.; Heo, S.J.; Yoon, W.J.; Kang, S.M.; Ahn, G.; Yi, T.H.; Jeon, Y.J. Fucoxanthin inhibits the inflammatory response by suppressing the activation of NF-kB and MAPKs in lipopolysaccharide-induced RAW264.7 macrophages. *Eur. J. Pharmacol.* **2010**, *649*, 369–375. [CrossRef] [PubMed]
37. Zhang, W.; Yan, J.; Wu, L.; Yu, Y.; Ye, R.D.; Zhang, Y.; Liang, X. In vitro immunomodulatory effects of human milk oligosaccharides on murine macrophage RAW264.7 cells. *Carbohydr. Polym.* **2019**, *207*, 230–238. [CrossRef] [PubMed]
38. Sun, H.; Zhang, J.; Chen, F.; Chen, X.; Zhou, Z.; Wang, H. Activation of RAW264.7 macrophages by the polysaccharide from the roots of *Actinidia eriantha* and its molecular mechanisms. *Carbohydr. Polym.* **2015**, *121*, 388–402. [CrossRef]
39. Xie, Z.; Wang, Y.; Huang, J.; Qian, N.; Shen, G.; Chen, L. Anti-inflammatory activity of polysaccharides from *Phellinus linteus* by regulating the NF-κB translocation in LPS-stimulated RAW264.7 macrophages. *J. Biol. Macromol.* **2019**, *129*, 61–67. [CrossRef]
40. Wen, Z.S.; Xiang, X.W.; Jin, H.X.; Guo, X.Y.; Liu, L.J.; Huang, Y.N.; OuYang, X.K.; Qu, Y.L. Composition and anti-inflammatory effect of polysaccharides from *Sargassum horneri* in RAW264.7 macrophages. *Int. J. Biol. Macromol.* **2016**, *88*, 403–413. [CrossRef]
41. Thalhamer, T.; McGrath, M.A.; Harnett, M.M. MAPKs and their relevance to arthritis and inflammation. *Rheumatology* **2008**, *47*, 409–414. [CrossRef]
42. Lee, S.A.; Moon, S.-M.; Choi, Y.H.; Han, S.H.; Park, B.-R.; Choi, M.S.; Kim, J.-S.; Kim, Y.H.; Kim, D.K.; Kim, C.S. Aqueous extract of *Codium fragile* suppressed inflammatory responses in lipopolysaccharide-stimulated RAW264.7 cells and carrageenan-induced rats. *Biomed. Pharmacother.* **2017**, *93*, 1055–1064. [CrossRef] [PubMed]
43. Zailan, N.F.Z.; Jaganathan, N.; Sandramuti, T.; Sarchio, S.N.E.; Hassan, M. Inhibition of GSK-3 by Tideglusib suppresses activated macrophages and inflammatory responses in lipopolysaccharide-stimulated RAW264.7 cell line. *Malays. J. Med. Health Sci.* **2020**, *16*, 2–8.
44. Rhule, A.; Navarro, S.; Smith, J.R.; Shepherd, D.M. *Panax notoginseng* attenuates LPS-induced pro-inflammatory mediators in RAW264.7 cells. *J. Ethnopharmacol.* **2006**, *106*, 121–128. [CrossRef] [PubMed]
45. Hu, Y.A.-O.; Lu, S.A.-O.; Xi, L. Murine macrophage requires CD11b to recognize *Talaromyces marneffei*. *Infect. Drug Resist.* **2020**, *13*, 911–920. [CrossRef] [PubMed]
46. Hou, C.; Chen, L.; Yang, L.; Ji, X. An insight into anti-inflammatory effects of natural polysaccharides. *Int. J. Biol. Macromol.* **2020**, *153*, 248–255. [CrossRef] [PubMed]
47. Lake, A.C.; Vassy, R.; Di Benedetto, M.; Lavigne, D.; Le Visage, C.; Perret, G.Y.; Letourneur, D. Low molecular weight fucoidan increases VEGF165-induced endothelial cell migration by enhancing VEGF165 binding to VEGFR-2 and NRP1. *J. Biol. Chem.* **2006**, *281*, 37844–37852. [CrossRef] [PubMed]

48. Shi, Y.; Xiong, Q.; Wang, X.; Li, X.; Yu, C.; Wu, J.; Yi, J.; Zhao, X.; Xu, Y.; Cui, H. Characterization of a novel purified polysaccharide from the flesh of *Cipangopaludina chinensis*. *Carbohydr. Polym.* **2016**, *136*, 875–883. [CrossRef] [PubMed]
49. Xiong, Q.; Hao, H.; He, L.; Jing, Y.; Xu, T.; Chen, J.; Zhang, H.; Hu, T.; Zhang, Q.; Yang, X.; et al. Anti-inflammatory and anti-angiogenic activities of a purified polysaccharide from flesh of *Cipangopaludina chinensis*. *Carbohydr. Polym.* **2017**, *176*, 152–159. [CrossRef]
50. Wang, M.; Gao, Y.; Xu, D.; Gao, Q. A polysaccharide from cultured mycelium of *Hericium erinaceus* and its anti-chronic atrophic gastritis activity. *Int. J. Biol. Macromol.* **2015**, *81*, 656–661. [CrossRef]
51. Wang, D.; Zhang, Y.; Yang, S.; Zhao, D.; Wang, M. A polysaccharide from cultured mycelium of Hericium erinaceus relieves ulcerative colitis by counteracting oxidative stress and improving mitochondrial function. *Int. J. Biol. Macromol.* **2019**, *125*, 572–579. [CrossRef]
52. Lee, S.-H.; Ko, C.-I.; Jee, Y.; Jeong, Y.; Kim, M.; Kim, J.-S.; Jeon, Y.-J. Anti-inflammatory effect of fucoidan extracted from *Ecklonia cava* in zebrafish model. *Carbohydr. Polym.* **2013**, *92*, 84–89. [CrossRef] [PubMed]
53. Obluchinskaya, E.D.; Pozharitskaya, O.N.; Shikov, A.N. In vitro anti-inflammatory activities of fucoidans from five species of brown seaweeds. *Mar. Drugs* **2022**, *20*, 606. [CrossRef] [PubMed]
54. Ni, L.; Wang, L.; Fu, X.; Duan, D.; Jeon, Y.-J.; Xu, J.; Gao, X. In vitro and in vivo anti-inflammatory activities of a fucose-rich fucoidan isolated from *Saccharina japonica*. *Int. J. Biol. Macromol.* **2020**, *156*, 717–729. [CrossRef] [PubMed]
55. DuBois, M.; Gilles, K.A.; Hamilton, J.K.; Rebers, P.A.; Smith, F. Colorimetric method for determination of sugars and related substances. *Anal. Chem.* **1956**, *28*, 350–356. [CrossRef]
56. Lowry, O.; Rosebrough, N.; Farr, A.L.; Randall, R. Protein measurement with the folin phenol reagent. *J. Biol. Chem.* **1951**, *193*, 265–275. [CrossRef]
57. Dodgson, K.; Price, R.G. A note on the determination of the ester sulphate content of sulphated polysaccharides. *Biochem. J.* **1962**, *84*, 106–110. [CrossRef] [PubMed]
58. Filisetti-Cozzi, T.M.C.C.; Carpita, N.C. Measurement of uronic acids without interference from neutral sugars. *Anal. Chem.* **1991**, *197*, 157–162. [CrossRef] [PubMed]
59. Tabarsa, M.; Lee, S.-J.; You, S. Structural analysis of immunostimulating sulfated polysaccharides from *Ulva pertusa*. *Carbohydr. Res.* **2012**, *361*, 141–147. [CrossRef]
60. Ciucanu, I.; Kerek, F. A simple and rapid method for the permethylation of carbohydrates. *Carbohydr. Res.* **1984**, *131*, 209–217. [CrossRef]

marine drugs

MDPI

Article

Ultrasound-Based Recovery of Anti-Inflammatory and Antimicrobial Extracts of the Acidophilic Microalga *Coccomyxa onubensis*

Mari Carmen Ruiz-Domínguez [1,*], María Robles [2], Lidia Martín [2], Álvaro Beltrán [3], Riccardo Gava [3], María Cuaresma [2], Francisco Navarro [4] and Carlos Vílchez [2]

[1] Laboratorio de Microencapsulación de Compuestos Bioactivos (LAMICBA), Departamento de Ciencias de los Alimentos y Nutrición, Facultad de Ciencias de la Salud, Universidad de Antofagasta, Antofagasta 1240000, Chile

[2] Algal Biotechnology, CIDERTA-RENSMA, Faculty of Experimental Sciences, University of Huelva, 21007 Huelva, Spain; maria.robles@dqcm.uhu.es (M.R.); lidia.martin@dqcm.uhu.es (L.M.); maria.cuaresma@dqcm.uhu.es (M.C.); cvilchez@dqcm.uhu.es (C.V.)

[3] Bioplagen S.L., Av. Castilleja de la Cuesta, 20-22, Bollullos de la Mitación, 41110 Seville, Spain; alvarobeltran@bioplagen.com (Á.B.); riccardogava@bioplagen.com (R.G.)

[4] Cell Alterations by Exogenous Agents, RENSMA, Department of Integrated Sciences, Faculty of Experimental Sciences, University of Huelva, 21007 Huelva, Spain; fnavarro@uhu.es

* Correspondence: maria.ruiz@uantof.cl; Tel.: +56-552-633-660

Abstract: In the present study, the recovery of valuable molecules of proven anti-inflammatory and antimicrobial activity of the acidophilic microalga *Coccomyxa onubensis* (*C. onubensis*) were evaluated using green technologies based on ultrasound-assisted extraction (UAE). Using a factorial design (3 × 2) based on response surface methodology and Pareto charts, two types of ultrasonic equipment (bath and probe) were evaluated to recover valuable compounds, including the major terpenoid of *C. onubensis*, lutein, and the antimicrobial activity of the microalgal extracts obtained under optimal ultrasound conditions (desirability function) was evaluated versus conventional extraction. Significant differences in lutein recovery were observed between ultrasonic bath and ultrasonic probe and conventional extraction. Furthermore, the antimicrobial activity displayed by *C. onubensis* UAE-based extracts was greater than that obtained in solvent-based extracts, highlighting the effects of the extracts against pathogens such as *Enterococcus hirae* and *Bacillus subtilis*, followed by *Staphylococcus aureus* and *Escherichia coli*. In addition, gas chromatography–mass spectrometry was performed to detect valuable anti-inflammatory and antimicrobial biomolecules present in the optimal *C. onubensis* extracts, which revealed that phytol, sterol-like, terpenoid, and even fatty acid structures could also be responsible for the antibacterial activities of the extracts. Moreover, UAE displayed a positive effect on the recovery of valuable molecules, improving biocidal effects. Our study results facilitate the use of green technology as a good tool in algal bioprocess engineering, improving energy consumption and minimizing environmental impacts and process costs, as well as provide a valuable product for applications in the field of biotechnology.

Keywords: *Coccomyxa onubensis*; ultrasound-assisted extraction; lutein recovery; antimicrobial activity; sterols; food applications

Citation: Ruiz-Domínguez, M.C.; Robles, M.; Martín, L.; Beltrán, Á.; Gava, R.; Cuaresma, M.; Navarro, F.; Vílchez, C. Ultrasound-Based Recovery of Anti-Inflammatory and Antimicrobial Extracts of the Acidophilic Microalga *Coccomyxa onubensis*. *Mar. Drugs* **2023**, *21*, 471. https://doi.org/10.3390/md21090471

Academic Editors: Herminia Domínguez and Leonel Pereira

Received: 26 July 2023
Revised: 10 August 2023
Accepted: 25 August 2023
Published: 27 August 2023

1. Introduction

Microalgae are a diverse group of unicellular photosynthetic microorganisms found in various aquatic habitats, including marine, freshwater, and brackish water environments [1,2]. They are one of the most promising biomass sources for biotechnological applications owing to their high growth rates, ability to produce various valuable compounds, and low environmental impact. In particular, extremophile microalgae offer numerous biotechnological benefits owing to their unique ability to adapt to extreme environments. These microorganisms have garnered considerable attention because of their potential applications in various industries, including food, pharmaceutical, and cosmetics, owing to their ability to produce pigments, antioxidants, and other bioactive molecules [3,4]. *Coccomyxa onubensis* is an acidophilic microalga and a well-known producer of the proven anti-inflammatory and antimicrobial terpenoid lutein [5–7]. It exerts antibacterial activity [8] and is considered a safe food source for animals, improving the antihyperglycemic and antihyperlipidemic protective effects on rats [9,10]. This microalga was isolated from the acidic mine drainages of the Pyritic Belt located southwest of Andalusia (Huelva, Spain), an environment suitable for the growth of many extremophile microorganisms; this area has high concentrations of heavy metals, low pH, and is exposed to a high UV light irradiation [6].

Microalgae have been explored for their antimicrobial activities, which have been attributed to different chemical compounds, including indoles, terpenes, acetogenins, phenols, fatty acids, and volatile halogenated hydrocarbons [11]; most of these compounds have been also attributed anti-inflammatory properties through in vitro assays [12,13]. In general, to obtain high-value products, the extraction technique is an important step in algal bioprocess engineering. The selection of the extraction technique can substantially affect the energy consumption of the overall production process. A suitable extraction technique minimizes environmental effects and associated process costs and increases product value [14].

Ultrasound-assisted extraction (UAE) is a green extraction technology primarily based on the cavitation phenomenon [15]. The thermal and mechanical effects of ultrasound waves on the medium, trigger biomass rupture [16]. These characteristics of UAE significantly improve the mixing and high mass transfer of the solvent into the sample matrix and establish a greater surface contact area between the solid and liquid phases [17]. Two types of ultrasonic equipment are commonly used to perform UAE: probes (high ultrasound intensity) and baths (high ultrasound intensity) [18]. These aspects lead to considerable differences in the recovery of valuable molecules from the extraction processes [18,19]. Several studies have reported the use of ultrasound technology for carotenoid recovery from algal biomass using different ultrasound frequencies, from low frequencies of 18–200 kHz to high frequencies of 400–10 MHz [15,20]. However, only a few studies have described the effects of this technology on the antimicrobial activity of the resulting microalgae extracts [21,22]. The novelty of this study is not related to the use of UAE technologies but to their specific application to developing green extraction processes from acidophilic microalgae; extremophilic microalgae are gaining relevance in biotechnology [5,6,10], and extraction procedures must be developed which unveil novel active molecules and efficient recovery protocols.

Therefore, we studied the effects of UAE on lutein extraction yield and other valuable metabolites in *C. onubensis* extracts that can exert antimicrobial and anti-inflammatory activity; the former activity was determined in the extracts, and the latter activity was proposed based on the unveiled presence of specific compounds having been reported to exert anti-inflammatory activity. This study was performed by using two types of ultrasonic equipment: an ultrasonic bath and probe (50 Hz and 20 kHz of ultrasound frequency, respectively), and the results were compared with those obtained by conventional extraction using the non-green solvent methanol (maceration). All experiments were performed using factorial designs (3 × 2) based on response surface methodology (RSM) and Pareto charts, using ethanol as the green extractant.

2. Results and Discussion

2.1. Effects of the Ultrasonic Techniques on the Extraction Yield of *C. onubensis*

Ultrasonic techniques are known for improving the recovery of valuable molecules compared with other methods and have been tested for biomass of many origins such as fruits and vegetables, algae, or agri-food wastes among other sources [21,23]. In the present study, we elucidated the differences between an ultrasonic bath (Table 1) and an ultrasonic probe (Table 2) under two different factorial designs (3 × 2) to determine the effects of two factors with three levels in eleven runs (with two additional central points) on the extraction yield and recovery of the anti-inflammatory and antimicrobial terpenoid lutein, used for evaluating the recovery method efficiency of proven anti-inflammatory and antimicrobial lipophilic compounds of the acidophilic microalga *C. onubensis* extracted with ethanol (a type of green solvent).

Table 1. Experimental design to optimize the parameters for ultrasonic bath extraction of *C. onubensis*. The response variables were extraction yield and lutein recovery. Lutein was used as a marker to compare the extraction efficiency of different methods.

Run	T (°C)	Time (min)	Extraction Yield (% *w/w*)	Lutein Recovery (% *w/w*)
1	−1	−1	13.05	51.81
2	0	−1	21.50	85.17
3	1	−1	27.98	127.81
4	−1	0	21.15	128.51
5	0	0	31.27	108.68
6	1	0	35.15	129.93
7	−1	1	17.82	88.29
8	0	1	26.00	101.53
9	1	1	31.48	134.27
10	0	0	28.45	125.81
11	0	0	29.01	123.20

Standard deviations of the extraction yield result = 0.989558 and lutein recovery = 0.793382. High level (+1): 70 °C and 30 min; low level (−1): 30 °C and 10 min; and central point (0): 50 °C and 20 min.

Table 2. Experimental design to optimize the parameters for ultrasonic probe extraction of *C. onubensis*. The response variables were extraction yield and lutein recovery. Lutein was used as a marker to compare the extraction efficiency of different methods.

Run	Pulse * (s/s)	Time (min)	Extraction Yield (% *w/w*)	Lutein Recovery (% *w/w*)
1	−1	−1	15.43	37.85
2	0	−1	14.57	40.69
3	1	−1	14.98	69.88
4	−1	0	17.17	51.44
5	0	0	15.71	58.94
6	1	0	16.12	97.85
7	−1	1	16.97	65.36
8	0	1	16.78	54.19
9	1	1	17.34	54.93
10	0	0	14.97	60.64
11	0	0	15.24	57.90

* The pulse duration and pulse interval refer to "on" time (equal to 10 s) and "off" time (from 30 to 60 s) of the sonicator. Standard deviation of the extraction yield results = 0.849697 and lutein recovery = 0.817347. High level (+1): 10/60 s/s and 15 min; low level (−1): 10/30 s/s and 5 min; and central point (0): 10/45 s/s and 10 min.

The factors for the ultrasonic bath were extraction time (10, 20, and 30 min) and temperature (30 °C, 50 °C, and 70 °C). Table 1 indicates that the optimal condition for maximum extraction yield was run 6, i.e., extraction at 70 °C for 20 min, with an extraction yield of 35.15% *w/w*, followed by runs 9 and 5 (yields of 31.48% and 31.27% *w/w*, respectively). However, the extraction conditions of runs 1 and 7 (low temperature of 30 °C for low-middle extraction time of 10 and 20 min, respectively) obtained the poorest extraction yield of 13.05% and 17.82% *w/w*, respectively. Figure 1 illustrates the Pareto chart (Figure 1A) and RSM plot (Figure 1B) of *C. onubensis* after adjusting the effect of each factor (extraction time and temperature) on the extraction yield using the ultrasonic bath. The Pareto chart (Figure 1A) revealed that factors such as temperature, extraction time, and quadratic extraction time were statistically significant in the extraction ($p \leq 0.05$).

In particular, temperature and extraction time improved the process under high values (Table 1, runs 6 and 9), whereas quadratic extraction time provided opposite results. On the other hand, the statistical software complemented with the complete mathematical model (Equation (1)), with an R^2 value of 98.96%; this indicates that this model has a good fit (Table S1). In the equation given below, the extraction yield was represented using T as the extraction temperature (In °C) and t as the extraction time (In min) and their combinations.

$$\text{Yield} = 29.5221 + 7.09833 \times \text{T} + 2.12833 \times \text{t} - 1.29026 \times \text{T}^2 - 0.3175\text{T} \times \text{t} - 5.69026 \times \text{t}^2 \quad (1)$$

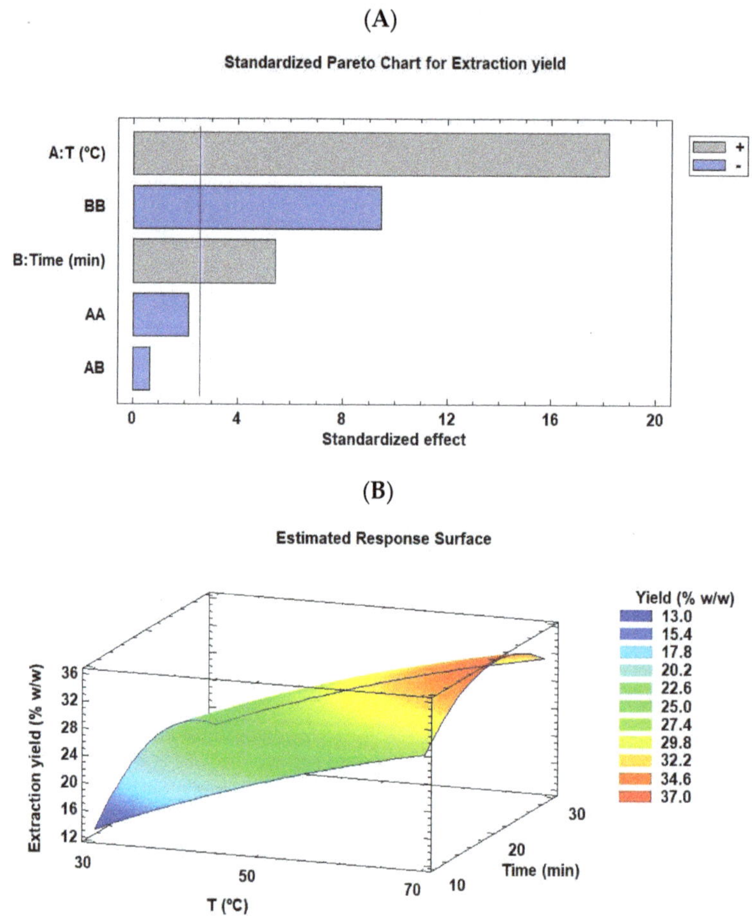

Figure 1. Pareto chart (**A**) and RSM (**B**) rationalizing the effect of each factor on the extraction yield of *C. onubensis*, using an ultrasonic bath. The vertical line in the Pareto chart indicates the 95% confidence level for the effects. Note: RSM, response surface methodology. The factors were extraction time (10, 20, and 30 min) and temperature (30 °C, 50 °C, and 70 °C).

The nonsignificant terms from the model were excluded and the mathematical model was refitted to obtain a new equation (Equation (2)) with a similar deviation (R^2 = 97.90%).

$$\text{Yield} = 29.006 + 7.09833 \times \text{T} + 2.12833 \times \text{t} - 6.03433 \times \text{t}^2 \quad (2)$$

Furthermore, Figure 1B (RSM) illustrates that the color intensity increased with extraction at high temperatures and moderately high extraction times, as explained above. The optimal conditions obtained using the statistical software were approximately 70 °C and

22 min (Supplementary Materials, Table S1), with an optimal extraction yield of 36.29% *w/w*, which is extremely close to the experimental data (run 6).

Table 2 presents the extraction yield data of *C. onubensis* after using an ultrasonic probe. In this experiment, the factors were pulse duration and pulse interval (including varying sonication times of ON equal to 10 s and OFF from 30 to 60 s) and extraction time (5, 10, and 15 min).

In general, the extraction yield data of the probe were lower than those obtained using an ultrasonic bath. The best yield was 17.34% *w/w* (run 9), closely followed by 17.17% *w/w* (run 4) (Table 2). In both cases, the extraction conditions were opposite, with run 9 being a high-level condition (10/60 s and 15 min) and run 4 being a low-middle level condition (10/30 s and 10 min). The worst extraction yield obtained was not very low (14.57% *w/w*) and was observed for run 2 (10/45 s and 5 min). Figure 2 summarizes the Pareto chart (Figure 2A) and RSM plot (Figure 2B) for the extraction yield of *C. onubensis* obtained using an ultrasonic probe. Extraction time and quadratic pulse interval were factors significantly and positively affecting the extraction process. In addition, a model equation (Equation (3)) was calculated for yield under the ultrasonic probe condition using statistical software (R^2 = 88.98%, Table S2).

$$\text{Yield} = 15.4821 - 0.188333 \times \text{Pi} + 1.01833 \times \text{t} + 0.899737 \times \text{Pi}^2 + 0.205\ \text{Pi} \times \text{t} - 0.0702632 \times \text{t}^2 \quad (3)$$

where Pi indicates the pulse intervals (s/s), and t indicates the extraction time (min). The nonsignificant terms of the initial model (pulse, quadratic extraction time, and their interactions) were excluded and the following equation was obtained (Equation (4)) with an R^2 of 84.97%.

$$\text{Yield} = 15.454 + 1.01833 \times \text{t} + 0.881 \times \text{Pi}^2 \quad (4)$$

As demonstrated in the RSM plot (Figure 2B), a high extraction time (15 min) and extreme pulse intervals (10/30 and 10/60 s/s) improved the extraction yield of *C. onubensis* extracts using an ultrasonic probe. For the optimal extraction conditions predicted using the statistical software for yield, the maximum yield was obtained at 10/60 s/s and 15 min, with a projected value of 17.35% *w/w*.

Studies have recommended using ultrasound as a pretreatment process for cell disruption in the microalgal biorefinery process because it confers advantages such as high efficiency, mild operating conditions, low toxicity, and time-saving methodology [24,25]. We observed that both ultrasound techniques increased the extraction yield of *C. onubensis* in lower extraction times, resulting in up to 2.9- and 1.4-fold higher yields for the ultrasonic bath and probe, respectively, compared with maceration (12.31% ± 0.02% *w/w*). Therefore, compared with conventional extraction, UAE allows greater permeability of the biomass with the ethanol solvent to recover the biomolecules present in *C. onubensis*.

Regarding the specific ultrasonic methods used in this work, ultrasonic bath improved the extraction yield compared with ultrasonic probe. This could be because of the formation and accumulation of radicals during the cavitation process under high ultrasound intensity (the frequency was 20 kHz or 226 W/cm^2 of acoustic power delivered into a liquid for the ultrasonic probe versus 50 Hz for the ultrasonic bath). Pingret et al. [26] comprehensively described the physicochemical effects of UAE in food processing. Acoustic cavitation is characterized by an increase in temperature and pressure conditions. It confers beneficial effects on the extraction of bioactive compounds; however, it can also alter the extraction conditions by producing radicals and molecules such as OH and H radicals, resulting in substantial quality defects in these products. Vintila et al. [27] also confirmed similar results using an ultrasonic bath. They increased the ultrasound intensity from 60% to 100% and demonstrated that the extraction yield of carotenoids and lipids considerably decreased. Therefore, the effects of ultrasound intensity or power input on the extractability of a target component are complex and warrant additional studies into the detailed extraction procedure.

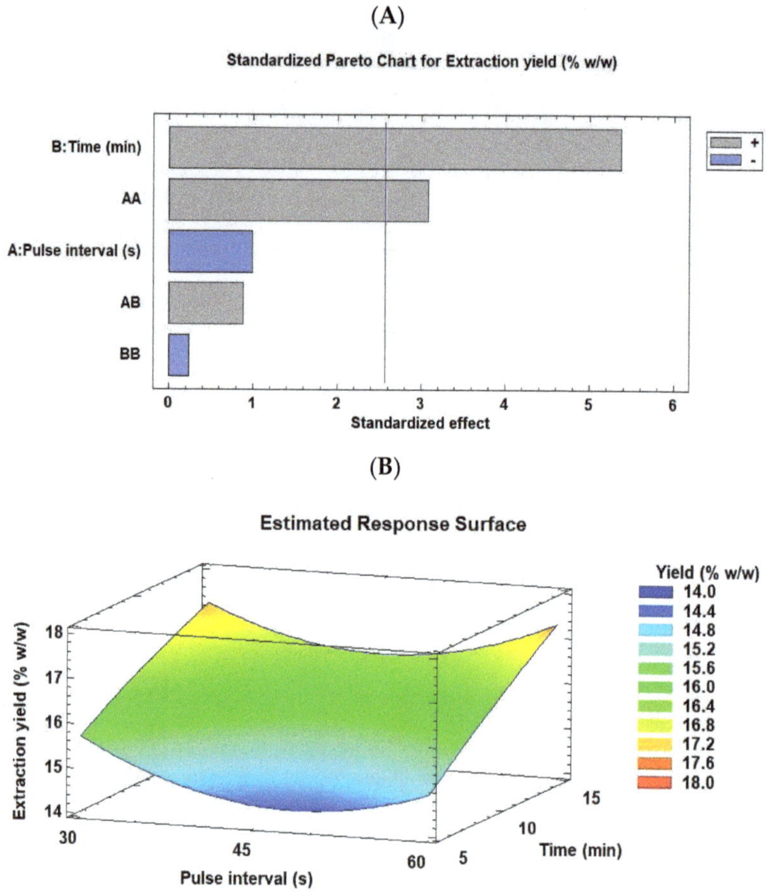

Figure 2. Pareto chart (**A**) and RSM (**B**) rationalizing the effect of each factor on the extraction yield of C. onubensis obtained using an ultrasonic probe. The vertical line in the Pareto chart indicates the 95% confidence level for the effects. Note: RSM, response surface methodology. Pulse intervals understanding as the "off" ultrasonic time ("on mode" was 10 s in all cases). The factors were pulse interval (10/30, 10/45, and 10/60 s/s) and extraction time (5, 10, and 15 min).

2.2. *Effects of the Ultrasonic Techniques on the Anti-Inflammatory Carotenoids Profile of C. onubensis*

Lutein is the main anti-inflammatory carotenoid present in C. *onubensis* (~70% of the total carotenoids quantified). Figure S1 illustrates the HPLC profile, which demonstrated that other major, anti-inflammatory carotenoids such as neoxanthin, violaxanthin, zeaxanthin, astaxanthin, and β-carotene are present in lower contents in this microalga. Studies have reported that C. *onubensis* accumulates high levels of lutein, which is improved by cultivating the microalgal cultures under different conditions, modifying the lutein synthesis route as an antioxidant protector [6,7,28]. We evaluated the effects of the two ultrasonic modes (bath and probe) on lutein recovery from C. *onubensis* using an experimental design (3 × 2). Lutein recovery represents the achieved percentage of lutein using UAE compared with conventional extraction (3.24 mg/g of lutein with respect to biomass grams, benchmark extraction). Table 1 suggests that the best conditions for lutein recovery using an ultrasonic bath were runs 9 and 6 (70 °C/30 min and 70 °C/20 min, respectively), with recoveries of 134.27% and 129.93% *w*/*w*, respectively. However, the condition of low temperature and extraction time (30 °C/10 min, run 1) was noted to be the worst for lutein recovery

(51.81% *w*/*w*). These data are complemented by the data illustrated in Figure 3, which presents the Pareto chart (Figure 3A) and RSM plot (Figure 3B) of lutein recovery from *C. onubensis* using an ultrasonic bath. The positively significant factor was temperature, followed by quadratic extraction time ($p \leq 0.05$). The optimal response obtained using the statistical software was at 70 °C for 20.43 min (148.96% *w*/*w*). In general, compared with conventional extraction (over 100% of lutein recovery), this ultrasonic technique improved lutein extraction. The completed regression equation (Equation (5)) was fitted to the data as follows:

$$\text{Lutein recovery} = 119.818 + 20.5667 \times \text{T} + 9.88333 \times \text{t} + 8.51895 \times \text{T}^2 - 7.505 \times \text{T} \times \text{t} - 27.3511 \times \text{t}^2 \quad (5)$$

where T is the extraction temperature (in °C), and t is the extraction time (in min). The R^2 value was 79.34% (Table S1). Equation (6) shows the mathematical model with the significant factors of the initial model (extraction temperature and quadratic extraction time). Other nonsignificant variables were excluded, and the following equation was obtained (Equation (6)) with an R^2 of 64.29%:

$$\text{Lutein recovery} = 123.226 + 20.5667 \times \text{T} - 25.0793 \times \text{t}^2 \quad (6)$$

(**A**) (**B**)

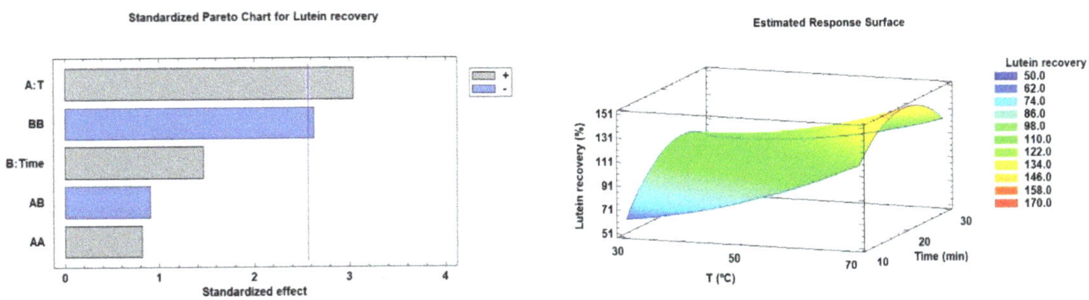

Figure 3. Pareto chart (**A**) and RSM (**B**) rationalizing the effect of each factor on lutein recovery from *C. onubensis*, using an ultrasonic bath. The vertical line in the Pareto chart indicates the 95% confidence level for the effects. Note: RSM, response surface methodology.

RSM (Figure 3B) was used to simultaneously optimize the levels of these variables to obtain the system with the best performance based on the fit of a polynomial equation to the experimental data [29]. In this case, the plot corresponded to the results described above, where the optimal condition was high extraction temperature (70 °C) and moderate extraction time (~20 min).

Next, we elucidated lutein recovery from *C. onubensis* using an ultrasonic probe. As shown in Table 2, the data ranged from 37.85% to 97.85% *w*/*w*. These values were lower than those obtained using the ultrasonic bath. Nevertheless, the best extraction condition was observed for run 6 (1/60 s/s of pulse and 10 min, 97.86% *w*/*w*), followed by run 3 (1/60 s/s of pulse and 5 min, 69.88% *w*/*w*). Figure 4A,B illustrate the Pareto chart and RSM plot, respectively. The Pareto chart revealed that pulse interval was the unique significant variable in the lutein extraction from *C. onubensis,* calculated as lutein recovery. This factor was defined as the "off" ultrasonic time (30, 45, and 60 s). The "on" mode was 10 s in all cases. By increasing the pulse interval (10/60 s), lutein recovery also improved. However, extraction time (in min) was a nonsignificant variable as well as the combination or quadratic of these factors. Figure 4B (RSM) demonstrates that intense color was observed under elevated pulse intervals (in s) and intermediate extraction time conditions. The optimal condition obtained using the software was a pulse of 10/60 s/s for 9 min, with an optimal recovery of 84.77% *w/w*. The regression equation (Equation (7)) was fitted to the data to obtain the mathematical model as follows:

$$\text{Lutein recovery} = 60.4063 + 11.335 \times \text{Pi} + 4.34333 \times \text{t} + 12.3692 \times \text{Pi}^2 - 10.615 \times \text{Pi} \times \text{t} - 14.8358 \times \text{t}^2 \quad (7)$$

where the values of the variables are specified in their original units (Pi, pulse intervals in s/s and t, extraction time in min) and R^2 was 81.73% (Table S2). To simplify the obtained mathematical model, the nonsignificant variables in the lutein recovery process were excluded. As a result, the following equation (Equation (8)) with a very low R^2 of 30.21% *w/w* was obtained.

$$\text{Lutein recovery} = 59.0609 + 11.335 \times \text{Pi} \quad (8)$$

(**A**) (**B**)

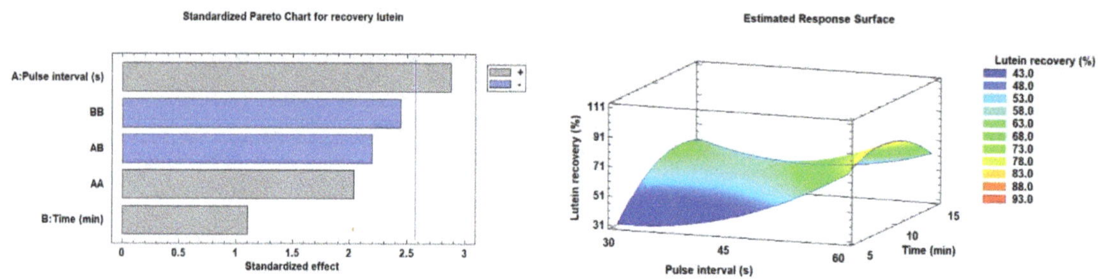

Figure 4. Pareto chart (**A**) and RSM (**B**) rationalizing the effect of each factor on the lutein recovery from *C. onubensis*, using an ultrasonic probe. The vertical line in the Pareto chart indicates the 95% confidence level for the effects. Note: RSM, response surface methodology. Pulse interval understanding as the "off" ultrasonic time (the "on" mode was 10 s in all cases).

In general, the effect of ultrasound intensity on the extractability of a target component from different natural sources is complex and warrants further investigation into the detailed extraction procedure. In the present study, the lutein recovery data of *C. onubensis* using an ultrasonic bath were higher than those using an ultrasonic probe. This could be because of the high ultrasound intensity used in the sonicator (50 kHz), resulting in the damaging or degrading effect of ultrasonic waves on the pigments. In fact, in the ultrasonic probe, when the pulsed interval was longer (10/30 s/s for 5 min vs. 10/60 s/s for 10 min), the lutein recovery increased from 37.85% to 97.85% *w/w*, respectively. On the other hand, for the ultrasonic bath, which used a less intense ultrasound frequency (50 Hz) but high temperatures (30 °C–70 °C), lutein recovery was 1.4-fold higher than the best result of the ultrasonic probe (run 6, Table 2). Extraction temperature can be another relevant factor in the extraction process. In the ultrasonic probe experiment, the temperature was controlled at 12 °C ± 4 °C to avoid the degradation of thermosensitive bioactive compounds and formation of vapor-filled bubbles (cushioning effect) [30]. Although the ultrasonic bath had a lower ultrasound intensity than the ultrasonic probe, the increase in temperature significantly affected the performance of the solvent by improving its diffusion rate and mass transfer capacity with the sample.

Several studies have described the importance of temperature in UAE and have highlighted that low temperatures (<30 °C) can exert a beneficial effect on the extraction process; in contrast, temperatures above 75 °C may increase the degradation of the obtained compounds [31,32]. As a result, the processing temperature should be optimized to obtain the highest extraction yield.

In a previous study, supercritical CO_2 extraction, another green extraction technique, was performed to optimize the extraction of valuable molecules in *C. onubensis* [33]. The lutein recovery value was less than that obtained using UAE in the present study (up to 50% of that of the optimal ultrasound bath condition). Similar findings were obtained for extraction yield, with 2-fold more yield under the bath condition. Deenu et al. [34] used RSM to optimize UAE experiments with or without enzymatic pretreatment to obtain the optimal conditions for lutein extraction from the microalga *Chlorella vulgaris*. They revealed that the optimal lutein recovery was 3.16 ± 0.03 mg/g wet weight of *Chlorella vulgaris* under the following UAE conditions: frequency, 35 kHz; ultrasound intensity, 56.58 W/cm^2; extraction temperature, 37.7 °C; extraction time, 5 h; and solvent–biomass ratio, 31 mL/g. This value was for the experiment without enzymatic pretreatment, which was similar to that with enzymatic pretreatment (3.36 mg/g wet weight of *Chlorella vulgaris*). Regarding the best lutein yield calculated using dry biomass, the optimal value was 12.38 mg/g dry weight of *Chlorella vulgaris* (approximately three times more than the optimal lutein content of *C. onubensis* using an ultrasonic bath, run 9). Although extraction was performed for 5 h in the previous study and for 15–30 min in the present study, studies can be performed to modify extraction times.

Another important concept is the stability of the bioactive compounds extracted using UAE. Sun et al. [35] investigated the effects of different UAE factors on the stability of all *trans*-β-carotene in a model system and the degradation kinetics and products. They varied the intensities from 5% to 85% (corresponding to 60.5 and 1028.9 W/cm^2) for 10 min on pulsed mode (2 s on and 2 s off) with temperature control. Their findings were consistent with those of our study; the concentration of β-carotene in dichloromethane decreased to approximately 70% when ultrasound intensity was varied from 60.5 to 302.5 W/cm^2.

Notably, culture conditions also play an important role in increasing lutein production in microalgae [36]. In summary, UAE improves lutein recovery from *C. onubensis* and can be a sustainable green process that can be applied in extreme microalgae biorefineries without overlooking that ultrasound power intensity is a critical parameter that requires optimization.

2.3. Desirability Function

The desirability function approach is one of the most widely used methods for optimizing multiple response processes. In the present study, it was based on maximizing variable responses such as extraction yield and lutein recovery. Using this approach, we identified the specific operating conditions that provide the "most desirable" response values from the UAE of *C. onubensis*. For the ultrasonic bath, the optimal conditions were an extraction temperature of ~70 °C and time of ~22 min (optimum value = 1.0, extraction yield of 35.33% *w*/*w* and lutein recovery of 147.34% *w*/*w*). For the ultrasonic probe, the factors were an extraction time of ~12 min and pulsed duration/interval of ~10/60 s of the sonicator (optimum value = 0.736, extraction yield of 16.8% *w*/*w* and lutein recovery of 78.7% *w*/*w*). The biomass–solvent ratio was the same (1:100) in all experiments. Figure 5A,B illustrate the desirability graphs for the ultrasonic bath and probe. In these optimized techniques, both responses (extraction yield and lutein recovery) were transformed into a dimensionless individual desirability function ranging from 0 to 1, with 0 corresponding to the lowest desirability level and 1 to the most desirable condition. Good agreement was observed between the predicted and experimental responses at the optimal conditions, with run 6 being the optimal response for the ultrasound bath and probe experiments (Tables 1 and 2). Taken together, these results suggest that the ultrasonic bath equipment used in this study exhibits good performance for the extraction yield and lutein recovery of *Coccomyxa onubensis* because this green technology positively contributed to both responses.

Subsequently, the predicted conditions obtained using the desirability function were used to evaluate the antimicrobial effects of *C. onubensis* extracts. The ethanolic aliquots were evaporated under a nitrogen stream and resuspended in DMSO. The methanolic extract was obtained using the conventional extraction method (solvent extraction).

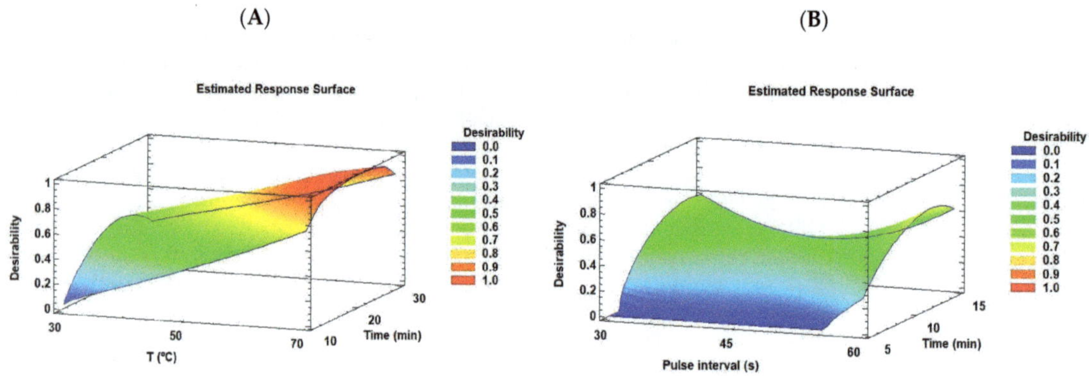

Figure 5. Graphical representation of the desirability functions for the different optimization criteria for (**A**) ultrasonic bath and (**B**) ultrasonic probe.

2.4. Antimicrobial Activity and Anti-Inflammatory Metabolite Identification

Table 3 presents the antimicrobial activity of *C. onubensis* extracted using UAE and conventional extraction against known Gram-negative and Gram-positive bacteria: *Pseudomonas aeruginosa, Escherichia coli, Staphylococcus aureus, Enterococcus hirae,* and *Bacillus subtilis*. The assay was performed in vitro using the broth microdilution method, one of the most used methods to determine the MIC of antimicrobial agents, including antibiotics and other substances that can kill (bactericidal activity) or inhibit the growth (bacteriostatic activity) of bacteria. The methods described here are targeted for testing the susceptibility to antibiotic agents, rather than other antimicrobial biocides such as preservatives and disinfectants. Serial dilutions of *C. onubensis* extracts, ranging from 0.50 µg/mL to 2.20 mg/mL, were assayed.

Table 3. Antimicrobial activity of the optimal *C. onubensis* extracts obtained using UAE compared with conventional extraction.

Bacteria	Biocidal Effect (µg/mL)				
	Gram-Negative		Gram-Positive		
	P. aeruginosa	*E. coli*	*S. aureus*	*E. hirae*	*B. subtilis*
Extraction method					
Conventional	192/96	n.e./n.e.	n.e./n.e.	192/96	192/96
Ultrasonic bath	552/276	552/276	138/69	9/4	9/4
Ultrasonic probe	262/131	262/131	262/131	4/2	131/65

The biocidal effect is described as MBC/MIC referring to minimum bactericide and minimum inhibitory concentration of *C. onubensis* extracts. Positive control with amoxicillin-clavulanic acid resulted in MBC values from 122 to 145 mg/mL for all pathogenic microorganisms tested. Abbreviations: no effect (n.e.).

In all extraction methods using ethanol or methanol as the extractant, relatively low concentrations of the extracts exhibited high efficiency against all pathogens (Table 3).The authors of [37] described the ratings of the antimicrobial efficiency of extracts based on their MIC values as follows: strong inhibitor, MIC < 500 µg/mL; moderate inhibitor, MIC of 600–1500 µg/mL; and weak inhibitor, MIC > 1600 µg/mL. Based on these criteria, *C. onubensis* extracts exhibited strong inhibition against selected pathogens. In particular, UAE (ultrasonic bath and probe) improved the antimicrobial activity of the extracts against *Enterococcus hirae* and *Bacillus subtilis*, followed by *Staphylococcus aureus* and *Escherichia coli* (MIC data range of 2–276 µg/mL of extracts). The antimicrobial activity of the conventional extract against *Pseudomonas aeruginosa* was better than that of *C. onubensis* extracted using UAE. In addition, *C. onubensis* extracts were more effective against Gram-positive bacteria than against Gram-negative bacteria. This effect has been previously described in antibiotics because Gram-positive bacteria have a complex and multilayered cell wall, making it difficult for the active compounds to penetrate the bacteria [38]. The biocidal effects of *C. onubensis* extracts were higher than those of extracts of other microalgae species. Saeed Niazi et al. [39] investigated the antimicrobial potential of the methanol extracts of the green microalgae isolated from the Persian Gulf and highlighted their effects against *Staphylococcus aureus*, *Bacillus Cereus*, *Escherichia coli*, and *Pseudomonas aeruginosa* (the MIC was 0.75, 1.5, 3, and 6 mg/mL and MBC was 1.5, 1.5, and 6 mg/mL, respectively).

The aim of the broth dilution method is to determine the lowest concentration of the assayed antimicrobial agent (MIC) that, under defined test conditions, inhibits visible bacterial growth. MIC values are used to determine the susceptibilities of bacteria to drugs. Furthermore, they are used to evaluate the activity of new antimicrobial agents. In the broth dilution method, often determined in the 96-well microtiter plate format, bacteria are inoculated into a liquid growth medium in the presence of different concentrations of the antimicrobial agent. Growth is assessed after incubation for a defined period (16–20 h) and the MIC value is determined. However, this method only applies to aerobic bacteria and can be completed in 3 days [40]. Notably, DMSO at concentrations of 15% or lower does not affect the growth of the microorganisms tested, as indicated by Navarro et al. [8].

Solvent selection is vital for determining the antimicrobial compounds (including various chemicals) derived from microalgae [41]. For example, a polar solvent such as ethanol (dielectric constant of 24.3) will mainly extract polar compounds such as polar pigments and phenolic compounds; it is widely used owing to its low toxicity and high extraction yields [42].

To identify the compounds present in the most effective antimicrobial extracts (namely, conventional, ultrasonic bath, and ultrasonic probe), GC–MS analysis was performed. Among the extracted compounds, those having been reported to exert antimicrobial and anti-inflammatory properties were identified, though anti-inflammatory activity of the extracts was not analyzed in this study. Table 4 lists the masses obtained, which were compared with the exact masses from different libraries using the NIST MS 2.3 software, to identify the potential compounds and based on the antimicrobial activity in the literature.

Table 4. Tentative identification of the anti-inflammatory and antimicrobial bioactive compounds present in *C. onubensis* extracts obtained using solvent extraction and UAE.

Bioactive Compound	Retention Time (min)	Molecular Ion (*m/z*) M+	Fragments Profile
Neophytadiene	9.444	278	123, 96, 83, 70, 69, 67, 58, 55, 43
Phytol	10.545	296	123, 95, 81, 72, 69, 68, 58, 55, 43, 41
Campesterol	16.875	400	145, 107, 105, 95, 81, 57, 55, 44, 41
Stigmasterol	17.250	440	91, 81, 79, 69, 67, 55, 44, 43, 41

Independently of the extraction method used, there were no differences in the type of compounds identified. As per GC–MS analysis, a common qualitative pattern was observed in the major biomolecular structures found in the extracts, which included phytol residues, fatty acid residues, sterol, and terpenoid compounds. Table 5 lists the name and properties of the antimicrobial molecules present in the microalgal extracts and detected via GC–MS analysis.

The aim of the antimicrobial activity assay was not to detect novel, natural antibiotic compounds. We believe that the path from detecting antibiotic compounds in natural extracts to their formulation as an antibiotic drug for humans is complex; moreover, a potentially useful natural antibiotic should, for instance, display an extremely high specific antibiotic activity against a resistant, pathogenic microorganism in humans. This will probably awaken the interest in the identified molecule as a potential commercial drug. However, the cost of producing microalgal biomass enriched in a given molecule and purifying it until homogeneity remains considerably higher than the production costs of common chemicals with antibiotic properties. Indeed, a molecule such as lutein, for example, can be accumulated by up to several milligrams per biomass gram of *C. onubensis*.

Assuming 1% (dry weight basis) of lutein accumulation in the acidophilic microalgal biomass, 1 kg of biomass would be required to produce 1 g of lutein. The cost for producing 1 kg of biomass can approximately be even more than 10€, depending on the cultivation conditions and procedure. Furthermore, the compound must be extracted and purified to formulate it as a drug for the healthcare industry. Therefore, we suggest using biomass extracts enriched in bioactive compounds, such as anti-inflammatory and antimicrobial bioactive molecules, as a more feasible strategy to formulate products that can have further application in several daily human activities in different fields, including disinfection or nutraceuticals promoting healthy body's normal state, rather than producing natural large molecules as nonspecific anti-inflammatory or antibiotic drugs.

Table 5. Major anti-inflammatory and antimicrobial biomolecules present in *C. onubensis* extracts and their biological functions.

Biomolecule	Chemical Structure	Physiological Role	Bioactivity
Phytol	Diterpenoid	Antioxidant biosynthesis precursor	Anti-inflammatory, antimicrobial
Lutein	Xanthophyll	Light absorption and antioxidant activity against ROS	Anti-inflammatory, antimicrobial, antioxidant
Neophytadiene	Diterpene	Cell defense against stress	Anti-inflammatory, antimicrobial, anxiolytic-like, antidepressant-like, anticonvulsant
Campesterol Stigmasterol	Sterol	Cell defense against oxidative stress, cell membrane fluidity regulation	Anti-inflammatory, antimicrobial, antioxidant, Anticancer

The data are collected by [12,43,44].

A small number of structures that can be responsible for a part of the antimicrobial activity were identified in *C. onubensis* extracts. Phytol was one such structure. It is a diterpene alcohol (Figure 6) and a natural compound that chemically belongs to the class of diterpenes. Furthermore, phytol is part of the chemical structure of the major photosynthetic pigment chlorophyll and can be extracted using ethanol or methanol, among other solvents. Phytol exhibits weak anti-inflammatory and antimicrobial properties; however, it can be converted to other molecules exhibiting antimicrobial properties. Some prominent examples are phytol-derived compounds such as phytol acetate and phytol esters; these compounds possess antimicrobial properties against specific bacterial strains, including *Staphylococcus aureus* and *Escherichia coli*. The stronger antimicrobial properties of phytol derivates can increase their potential as preservative components in antimicrobial coatings and packaging materials.

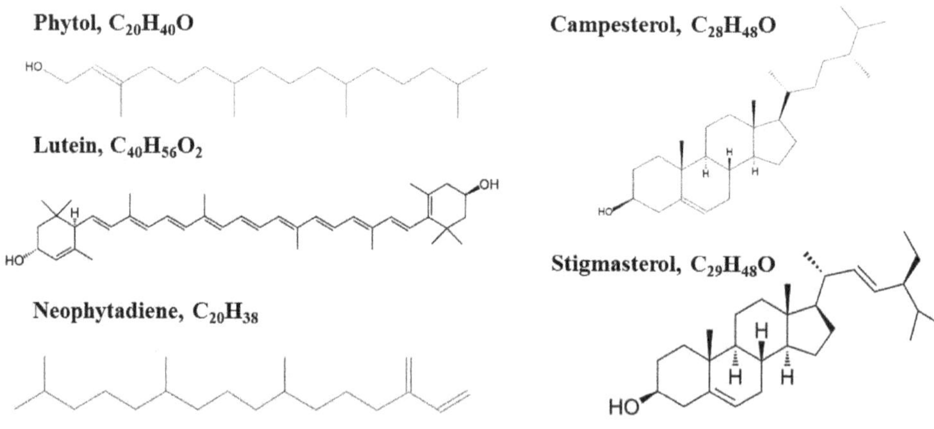

Figure 6. Chemical structures of the compounds displaying antimicrobial activity identified in alcohol solvent-based *C. onubensis* extracts.

Lutein (Figure 6) is a naturally occurring carotenoid pigment found in microalgae, with *C. onubensis* being an outstanding example of a potential producer. Recent studies have revealed its potential anti-inflammatory and antimicrobial properties [44,45]. Furthermore, several studies have reported that lutein exhibits antimicrobial activity against many microorganisms. In particular, lutein inhibits the growth of pathogenic bacteria such as *Staphylococcus aureus* and *Escherichia coli* [45]. Although the mechanisms of action of lutein as an antimicrobial compound remain unelucidated, lutein can interfere with bacterial cell membrane integrity; this is coherent with the linear, long-chain hydrocarbon nature of its chemical structure. The antimicrobial properties of lutein are attributed to its antioxidant and anti-inflammatory effects. Lutein can scavenge reactive oxygen species and decrease oxidative stress, a cellular process frequently associated with microbial infections. Furthermore, its anti-inflammatory properties help to modulate immune responses and contribute to its antimicrobial effects. Neophytadiene is a diterpene (Figure 6) that exerts antimicrobial, antioxidant, anti-inflammatory, and anxiolytic-like activities. However, only a few studies have reported the presence of neophytadiene in microalgae, with no study on its presence in acidophilic microalgae. Neophytadiene has been identified in plants, particularly in conifers. Nevertheless, their recently unveiled pharmacological properties may increase the interest in specific microalgae species as bioresources with the potential for producing neophytadiene. Although the biological activities of neophytadiene have not been extensively studied, it plays a vital role in cell defense mechanisms against oxidative stress. Stigmasterol and campesterol, sterol-derived structures (Figure 6), have been identified in *C. onubensis* extracts. Stigmasterol was originally identified as a typical sterol in fungi; however, it has also been identified in certain microalgae species, including *Chlorella*, *Nannochloropsis*, *Dunaliella*, and *C. onubensis* (the present study). Campesterol is a phytosterol typically found in fruits and vegetables. Nevertheless, the specific biological functions of both sterols in microalgae remain unelucidated. However, their potential roles in membrane fluidity regulation and stress response mechanisms, including UV light and extreme temperature, have been reported [46].

In microalgae, the abovementioned compounds, whose structures are all shown in Figure 6, are biochemically synthesized via the non-mevalonate sterol biosynthesis pathway, which is also called the 1-deoxy-D-xylulose-5-phosphate/2-methyl-D-erythritol-4-phosphate pathway. In microalgae, this pathway is activated under different stress conditions and meets the common features of inducing oxidative stress: high-light irradiance, UV radiation, extreme temperature, or presence of metal ions (a typical chemical scenario in the highly acidic, natural habitat of *C. onubensis*). However, this pathway does not create such oxidative scenarios in microalgal cultures [47,48].

3. Materials and Methods

3.1. Biomass and Chemicals

The microalga used in this study was *C. onubensis* (SAG 2510). The biomass was kindly donated by the research group Algal Biotechnology from the University of Huelva (Spain) and was previously described by Ruiz-Domínguez et al. [33]. The main chemical compound used in UAE was ethanol (99.5%) as a green solvent; it was purchased from VWR Prolabo Chemicals (Barcelona, Spain). Other chemicals used in chromatographic analyses were ethyl acetate, water, acetonitrile, and methanol. All solvents were heavy metal-free based on the specifications supplied by VWR Prolabo. The carotenoid standards used in high-performance liquid chromatography (HPLC) were provided by Sigma-Aldrich (Madrid, Spain). The microorganism strains to test the antimicrobial activity of the extracts were acquired, characterized, and validated by the Spanish Collection of Type Cultures at the University of Valencia (Spain). They are the bacterial species specified in the UNE-EN-13697:2015+A1 Standard because they represent the pathogens of greatest interest for studying sanitizers in the food industry [*Pseudomonas aeruginosa* (NCIMB 8626/ATCC 15442), *Staphylococcus aureus* (NCIMB 9518/ATCC 6538), *Enterococcus hirae* (NCIMB 6459/ATCC 10541), *Escherichia coli* (NCIMB 8545/ATCC 10536), and *Bacillus subtilis* (NCIMB 8054/ATCC 6633)].

3.2. Conventional Extraction

Conventional solvent extraction was performed using methanol. The extraction was performed in a shaker incubator at 30 °C for 24 h. The biomass-to-solvent ratio was 1:100 (g:mL). Cell debris were removed by centrifuging the samples (OHAUS, Frontier 5816R, Parsippany, NJ, USA) at 8000 rpm (9300× *g*) and 15 °C for 5 min. The experiment was performed in triplicate (n = 3, ± SD) and the supernatant was pooled and labeled as the crude extract until use. A part of the extract was used to obtain the extraction yield. The extraction yield was determined gravimetrically after drying the sample under a nitrogen stream to eliminate the solvent (24-Hole Nitrogen Evaporators, MD200–2N, Ollital Technology, Fugian, China). The extraction yield for conventional extraction was 12.31% ± 0.02% *w*/*w* on average and was expressed as mg of extract/g of the dry weight of *C. onubensis* (benchmark extraction).

3.3. Ultrasonic Green Extraction Design

UAE was performed using a ratio of 1:100 (g:mL) biomass and ethanol as the solvent (green extractant). Experiments were conducted using two different ultrasonic equipment: an ultrasonic bath (Ultrasons H-D 3000866 Selecta, Barcelona, Spain) with a power of 330 W (approximately 1–5 W/cm^2 of the acoustic power or ultrasound power intensity delivered into a liquid), a frequency of 50 Hz, and temperature control and an ultrasonic probe (VCX 750, Vibra Cell Sonics, Newtown, CT, USA) with a power of 750 W, frequency of 20 kHz, a 1/8" (3 mm) titanium probe (maximum amplitude: 40% equal to 226 W/cm^2 of the acoustic power), and temperature control (12 °C ± 4 °C). To optimize the operation parameters, different values were tested under two different factorial designs (3 × 2) that will elucidate the effects of 2 factors with 3 levels in 11 runs (plus two central points). For the ultrasonic bath, the factors were extraction time (10–30 min) and temperature (30–70 °C). For the ultrasonic probe, the factors were extraction time (5–15 min) and pulsed duration/interval referring to the "on" time (10 s) and "off" time (30–60 s) of the sonicator. Tables 1 and 2 present the schematics of the ultrasonic experiments. After the extraction, the mixtures were centrifuged at 8000 rpm (9300× *g*) and 15 °C for 5 min to recover the ethanol extracts rich in valuable compounds. They were stored in the dark at −18 ± 2 °C until subsequent analysis.

3.4. Quantification of Carotenoids

The ethanolic ultrasonic extracts were evaporated, resuspended in chromatographic methanol, and filtered (Ø = 0.22 µm filter) together with the conventional extracts (in methanol). Thereafter, they were transferred into a chromatography vial and immediately used to quantify carotenoids via liquid chromatography. HPLC was performed on the Beckman System Gold binary delivery system equipped with a UV–vis photodiode array detector (Beckman Instruments, Fullerton, CA, USA) using a C18 column (150 mm × 4.6 mm i.d., 5 µm, SunFire ^{TM} column; Waters, Milford, MA, USA). The flow rate was maintained at 1 mL/min and injection volume was 40 µL of the algal extracts. Ethyl acetate was used as mobile phase A and acetonitrile/water (9:1 *v*/*v*) was used as mobile phase B. The mobile phase gradient was as follows: 0–16 min, 0–60% solvent A; 16–30 min, 60% A; and 30–35 min, 100% A. The selected carotenoids were detected at a wavelength of 450 nm and by comparing the peak areas obtained from the methanolic *C. onubensis* extracts with those obtained from the injected standards (Sigma-Aldrich, 0–50 ppm, ~R^2 = 0.998). Lutein concentration was referred to as dry biomass or weight of the extract, whereas lutein recovery was calculated using Equation (9):

$$\text{Lutein recovery } (\%\ \text{w/w}) = (\text{Wc}/\text{Wt}) \times 100 \tag{9}$$

where Wc is the mass of lutein (mg) extracted under the ultrasonic conditions described in this study and Wt is the mass of lutein conventionally extracted (using solvent extraction with an average lutein concentration of 3.24 ± 0.11 mg/g, expressed as mg of lutein/g of the dry weight of *C. onubensis*, benchmark extraction).

3.5. Quantification of the Bioactive Extracts Using Gas Chromatography–Mass Spectrometry (GC–MS)

The chemical composition of the optimal and conventional *C. onubensis* extracts was analyzed using GC–MS (Agilent 5977B mass selective detector, Santa Clara, CA, USA) Compound identification was achieved by comparing the mass spectra with the NIST Mass Spectrometry Data Center (MS 2.3 software version) as well as by comparing the retention indices with the literature values. Detection was performed in the electron impact ionization mode (70 eV) under the following conditions: capillary column, HP-5 MS (30 m × 0.25 mm; film thickness 0.25 μm); temperature program, 40 °C (held for 1 min), raised to 300 °C at a rate of 25 °C/min (held for 10 min), and increased to 325 °C at a rate of 10 °C/min (held to 5 min); injector temperature, 250 °C; carrier gas, helium; and flow rate, 1 mL/min.

3.6. Antimicrobial Activity Assay

The extracts subjected to the antimicrobial activity assay were selected from the optimal conditions obtained using the statistical software based on desirability function together with conventional extraction (in methanol). The ultrasonic bath conditions were 70 °C for 22 min, whereas for the ultrasonic probe, the factors were an extraction time of 12 min and pulsed duration/interval of 10/60 s of the sonicator. The biomass-to-solvent ratio was the same (1:100). The ethanolic and methanolic extracts were evaporated under nitrogen flow and resuspended in dimethyl sulfoxide (DMSO) for the antimicrobial assay. The microorganisms selected were the Gram-negative bacteria *Pseudomonas aeruginosa* and *Escherichia coli* and the Gram-positive bacteria *Staphylococcus aureus, Enterococcus hirae,* and *Bacillus subtilis.* The biocidal effects of the extracts were determined using the serial dilution method in 96-well microplates, as described by Navarro et al. [8] and Wiegand et al. [40]. The effectiveness of each biocide against a specific microorganism was tested in triplicate by performing successive dilutions of 50% of the previous well using an automated multichannel pipette. The final volume of each microwell was 200 μL, with 5×10^5 colony-forming units of the tested microorganism. The minimum inhibitory concentration (MIC), which is the lowest concentration of the antimicrobial agent that completely inhibits the visible growth of a microorganism after incubation in the test medium, was calculated. The MIC of each extract was determined by visually inspecting the bottom of the well through an enlarged digital imaging system and determining bacterial growth based on the presence of sediment or defined turbidity. A low MIC value corresponds to more efficient antimicrobial activity. The minimum bactericidal concentration (MBC) of each biocide was determined by inoculating agar plates with the content of the highest dilution wells above the MIC and observing no growth after incubation for 48 h at 37 °C.

3.7. Statistical Analysis and Multiple Response Optimization

Ultrasonic experimental designs and data analysis were performed using RSM with Statgraphics Centurion XVIII software (StatPoint Technologies, Inc., Warrenton, VA, USA). The effect of each factor and its statistical significance on each of the response variables such as extraction yield and lutein recovery from *C. onubensis* were also analyzed using the standardized Pareto chart. Furthermore, it was used for data elaboration and statistical analysis, with a 95% level of significance. In addition, mathematical models were obtained, and the significances were accepted at a *p*-value of ≤ 0.05. All measurements were performed in triplicate ($n = 3$). In the factorial design (3×2) involving two factors X_1 and X_2, the proposed quadratic model (Equation (10)) for each response variable was as follows:

$$Z = \beta_0 + \beta_1 X_1 + \beta_2 X_2 + \beta_{12} X_1 X_2 + \beta_{11} X_1^2 + \beta_{22} X_2^2 \quad (10)$$

where Z = estimated response, β_0 = constant, β_1 and β_2 = linear coefficients, β_{12} = interaction coefficients between the two factors, and β_{11} and β_{22} = quadratic coefficients.

Multiple response optimization was performed using the desirability function described by Del Castillo et al. [49], which provides an overall objective function starting from fitting equations obtained for each response variable (the total desirability, D). Equation (11) given below, D ranges from 0 to 1, represents the geometric mean of the desirable range of each response (di), which also varies from 0 (undesirable value) to 1 (the most desirable value).

$$D = \left(\prod_{i=1}^{n} d_i^{r_i}\right)^{\frac{1}{\sum r_i}} \tag{11}$$

where r_i is the weight assigned by the user for each response variable and d_i is the maximization factor of the response variables [50].

4. Conclusions

Extraction techniques are a relevant step in microalgae biorefinery because their optimization leads to a better recovery of molecules with biotechnological interest. In the present study, we confirm that UAE is an easy-to-use, rapid, and green technology with improved data for *C. onubensis* compared with solvent extraction. In particular, the ultrasonic bath increased the anti-inflammatory terpenoid lutein recovery up to 34% more than that of solvent extraction. Furthermore, *C. onubensis* extracts exhibited biocidal effects against Gram-positive and Gram-negative bacteria. Nevertheless, only some molecules produced by *C. onubensis* in alcohol-based extracts were identified; according to the literature, the referred molecules exhibit anti-inflammatory properties and can also be responsible for the determined antimicrobial activity. Five terpene biosynthesis-derived molecules (two sterols) were identified: phytol, neophytadiene, lutein, stigmasterol, and campesterol. Based on their terpenoid-derived nature, we suggest that oxidative conditions, including high-light irradiance and UV light, induce the production of terpenoids, addressing their accumulation and the subsequent enhancement of the antimicrobial activity of *C. onubensis* extracts. Therefore, *C. onubensis*, as a microorganism isolated from acidic medium, jointly with the green extraction techniques (UAE) can be a good combination to improve valuable molecules recovery applied in the field of biotechnology as well as in other industries. In addition, eventual extraction efficiency improvement through the selection and optimization of solvent use or a mix of them should be targeted for studies at a larger scale.

Supplementary Materials: The following supporting information can be downloaded at https://www.mdpi.com/article/10.3390/md21090471/s1. Figure S1: HPLC profile of main carotenoids present in *C. onubensis* (1) neoxanthin, (2) violaxanthin, (3) lutein, (4) zeaxanthin, and (5) β-carotene; Table S1: Regression coefficients and *p*-value for extraction yield, lutein purity, and lutein recovery in their original units and statistics for the fit obtained by multiple linear regression using ultrasonic bath equipment; Table S2: Regression coefficients and *p*-value for extraction yield lutein recovery in their original units and statistics for the fit obtained by multiple linear regression using ultrasonic probe equipment.

Author Contributions: Conceptualization, M.C.R.-D. and C.V.; methodology, M.C.R.-D., F.N. and Á.B.; software, M.C.R.-D.; validation, M.C.R.-D., F.N. and C.V.; formal analysis, M.C.R.-D., F.N., Á.B. and R.G.; investigation, M.C.R.-D., M.R. and L.M.; resources, M.C.; data curation, M.C.R.-D.; writing—original draft preparation, M.C.R.-D. and C.V.; writing—review and editing, C.V.; supervision, M.C.R.-D. and C.V.; project administration, M.C.R.-D. and M.C.; funding acquisition, M.C.R.-D. and C.V. All authors have read and agreed to the published version of the manuscript.

Funding: This work was partially supported by the Andalusian Research, Development and Innovation Plan (Junta de Andalucía, Spain) with FEDER funds (Project P20_00930) and by the University of Huelva through a grant obtained in the program "Requalitication of the Spanish Universitary System 2021–2023, María Zambrano" (Real Decreto 289/2021, 20 April and Order UNI/551/2021, 26 May) and "IV Convocatoria de Micro-proyectos—Cátedra de la Provincia UHU 2023".

Institutional Review Board Statement: Not applicable.

Informed Consent Statement: Not applicable.

Data Availability Statement: All the data are contained within the article and the Supplementary Materials.

Conflicts of Interest: The authors declare no conflict of interest.

References

1. Evangelista, V.; Barsanti, L.; Frassanito, A.M.; Passarelli, V.; Gualtieri, P. *Algal Toxins: Nature, Occurrence, Effect and Detection*; Springer Science & Business Media: Berlin/Heidelberg, Germany, 2008; pp. 353–391.
2. Tan, J.S.; Lee, S.Y.; Chew, K.W.; Lam, M.K.; Lim, J.W.; Ho, S.H.; Show, P.L. A review on microalgae cultivation and harvesting, and their biomass extraction processing using ionic liquids. *Bioengineered* **2020**, *11*, 116–129. [CrossRef] [PubMed]
3. Canganella, F.; Wiegel, J. Extremophiles: From abyssal to terrestrial ecosystems and possibly beyond. *Naturwissenschaften* **2011**, *98*, 253–279. [CrossRef]
4. Mendes-Silva, T.d.C.D.; da Silva Andrade, R.F.; Ootani, M.A.; Mendes, P.V.D.; da Silva, M.R.F.; Souza, K.S.; dos Santos Correia, M.T.; da Silva, M.V.; de Oliveira, M.B.M. Biotechnological potential of carotenoids produced by extremophilic microorganisms and application prospects for the cosmetics industry. *Adv. Microbiol.* **2020**, *10*, 397–410. [CrossRef]
5. Fuentes, J.-L.; Montero, Z.; Cuaresma, M.; Ruiz-Domínguez, M.-C.; Mogedas, B.; Nores, I.G.; González del Valle, M.; Vílchez, C. Outdoor large-scale cultivation of the acidophilic microalga *Coccomyxa onubensis* in a vertical close photobioreactor for lutein production. *Processes* **2020**, *8*, 324. [CrossRef]
6. Vaquero, I.; Ruiz-Domínguez, M.C.; Márquez, M.; Vílchez, C. Cu-mediated biomass productivity enhancement and lutein enrichment of the novel microalga *Coccomyxa onubensis*. *Process Biochem.* **2012**, *47*, 694–700. [CrossRef]
7. Bermejo, E.; Ruiz-Domínguez, M.C.; Cuaresma, M.; Vaquero, I.; Ramos-Merchante, A.; Vega, J.M.; Vílchez, C.; Garbayo, I. Production of lutein, and polyunsaturated fatty acids by the acidophilic eukaryotic microalga *Coccomyxa onubensis* under abiotic stress by salt or ultraviolet light. *J. Biosci. Bioeng.* **2018**, *125*, 669–675. [CrossRef]
8. Navarro, F.; Forján, E.; Vázquez, M.; Toimil, A.; Montero, Z.; Ruiz-Domínguez, M.d.C.; Garbayo, I.; Castaño, M.Á.; Vílchez, C.; Vega, J.M. Antimicrobial activity of the acidophilic eukaryotic microalga *Coccomyxa onubensis*. *Phycol. Res.* **2017**, *65*, 38–43. [CrossRef]
9. Navarro, F.; Toimil, A.; Ramírez, S.; Montero, Y.; Fuentes, J.L.; Perona, J.S.; Castaño, M.Á.; Pásaro, R.; Vega, J.M.; Vílchez, C. The acidophilic microalga *Coccomyxa onubensis* and atorvastatin equally improve antihyperglycemic and antihyperlipidemic protective effects on rats fed on high-fat diets. *J. Appl. Phycol.* **2020**, *32*, 3923–3931. [CrossRef]
10. Navarro, F.; Forján, E.; Vázquez, M.; Montero, Z.; Bermejo, E.; Castaño, M.Á.; Toimil, A.; Chagüaceda, E.; García-Sevillano, M.Á.; Sánchez, M.; et al. Microalgae as a safe food source for animals: Nutritional characteristics of the acidophilic microalga *Coccomyxa onubensis*. *Food Nutr. Res.* **2016**, *60*, 30472. [CrossRef]
11. Cardozo, K.H.M.; Guaratini, T.; Barros, M.P.; Falcão, V.R.; Tonon, A.P.; Lopes, N.P.; Campos, S.; Torres, M.A.; Souza, A.O.; Colepicolo, P.; et al. Metabolites from algae with economical impact. *Comp. Biochem. Physiol. Toxicol. Pharmacol. CBP* **2007**, *146*, 60–78. [CrossRef]
12. Plaza, M.; Santoyo, S.; Jaime, L.; García-Blairsy Reina, G.; Herrero, M.; Señoráns, F.J.; Ibáñez, E. Screening for bioactive compounds from algae. *J. Pharm. Biomed. Anal.* **2010**, *51*, 450–455. [CrossRef]
13. De Morais, M.G.; Vaz, B.d.S.; de Morais, E.G.; Costa, J.A.V. Biologically Active Metabolites Synthesized by Microalgae. *BioMed Res. Int.* **2015**, *2015*, 835761. [CrossRef]
14. Suganya, T.; Varman, M.; Masjuki, H.; Renganathan, S. Macroalgae and microalgae as a potential source for commercial applications along with biofuels production: A biorefinery approach. *Renew. Sustain. Energy Rev.* **2016**, *55*, 909–941. [CrossRef]
15. Poojary, M.; Barba, F.; Aliakbarian, B.; Donsì, F.; Pataro, G.; Dias, D.; Juliano, P. Innovative alternative technologies to extract carotenoids from microalgae and seaweeds. *Mar. Drugs* **2016**, *14*, 214. [CrossRef]
16. Hu, Y.; Yan, B.; Chen, Z.S.; Wang, L.; Tang, W.; Huang, C. Recent technologies for the extraction and separation of polyphenols in different plants: A review. *J. Renew. Mater.* **2022**, *10*, 1471–1490. [CrossRef]
17. Mason, T.; Chemat, F.; Vinatoru, M. The extraction of natural products using ultrasound or microwaves. *Curr. Org. Chem.* **2011**, *15*, 237–247. [CrossRef]
18. Zheng, S.; Zhang, G.; Wang, H.; Long, Z.; Wei, T.; Li, Q. Progress in ultrasound-assisted extraction of the value-added products from microorganisms. *World J. Microbiol. Biotechnol.* **2021**, *37*, 71. [CrossRef]
19. Santos, H.M.; Capelo, J.L. Trends in ultrasonic-based equipment for analytical sample treatment. *Talanta* **2007**, *73*, 795–802. [CrossRef] [PubMed]
20. Juliano, P.; Augustin, M.A.; Xu, X.Q.; Mawson, R.; Knoerzer, K. Advances in high frequency ultrasound separation of particulates from biomass. *Ultrason. Sonochem.* **2017**, *35*, 577–590. [CrossRef] [PubMed]

21. Roselló-Soto, E.; Galanakis, C.M.; Brnčić, M.; Orlien, V.; Trujillo, F.J.; Mawson, R.; Knoerzer, K.; Tiwari, B.K.; Barba, F.J. Clean recovery of antioxidant compounds from plant foods, by-products and algae assisted by ultrasounds processing. Modeling approaches to optimize processing conditions. *Trends Food Sci. Technol.* **2015**, *42*, 134–149. [CrossRef]
22. Maligan, J.; Widayanti, V.; Zubaidah, E. Production and Identification of Antimicrobial Compounds from Marine Microalgae Tetraselmis chuii Using Ultrasound Assisted Extraction Method. In Proceedings of the 4th International Conference on Food Engineering and Biotechnology (ICFEB 2013), Copenhagen, Denmark, 19–20 May 2013.
23. Kumar, K.; Srivastav, S.; Sharanagat, V.S. Ultrasound assisted extraction (UAE) of bioactive compounds from fruit and vegetable processing by-products: A review. *Ultrason. Sonochem.* **2021**, *70*, 105325. [CrossRef] [PubMed]
24. Sivaramakrishnan, R.; Incharoensakdi, A. Microalgae as feedstock for biodiesel production under ultrasound treatment—A review. *Bioresour. Technol.* **2018**, *250*, 877–887. [CrossRef] [PubMed]
25. Liu, Y.; Liu, X.; Cui, Y.; Yuan, W. Ultrasound for microalgal cell disruption and product extraction: A review. *Ultrason. Sonochem.* **2022**, *87*, 106054. [CrossRef]
26. Pingret, D.; Fabiano-Tixier, A.-S.; Chemat, F. Degradation during application of ultrasound in food processing: A review. *Food Control* **2013**, *31*, 593–606. [CrossRef]
27. Vintila, A.C.N.; Vlaicu, A.; Radu, E.; Ciltea-Udrescu, M.; Enascuta, E.C.; Banu, I.; Oprescu, E.-E. Evaluation of ultrasound assisted extraction of bioactive compounds from microalgae. *J. Food Meas. Charact.* **2022**, *16*, 2518–2526. [CrossRef]
28. Vaquero, I.; Mogedas, B.; Ruiz-Domínguez, M.C.; Vega, J.M.; Vílchez, C. Light-mediated lutein enrichment of an acid environment microalga. *Algal Res.* **2014**, *6*, 70–77. [CrossRef]
29. Bezerra, M.A.; Santelli, R.E.; Oliveira, E.P.; Villar, L.S.; Escaleira, L.A. Response surface methodology (RSM) as a tool for optimization in analytical chemistry. *Talanta* **2008**, *76*, 965–977. [CrossRef]
30. Lou, Z.; Wang, H.; Zhang, M.; Wang, Z. Improved extraction of oil from chickpea under ultrasound in a dynamic system. *J. Food Eng.* **2010**, *98*, 13–18. [CrossRef]
31. Zhang, Z.-S.; Wang, L.-J.; Li, D.; Jiao, S.-S.; Chen, X.D.; Mao, Z.-H. Ultrasound-assisted extraction of oil from flaxseed. *Sep. Purif. Technol.* **2008**, *62*, 192–198. [CrossRef]
32. Carrera, C.; Ruiz-Rodríguez, A.; Palma, M.; Barroso, C.G. Ultrasound assisted extraction of phenolic compounds from grapes. *Anal. Chim. Acta* **2012**, *732*, 100–104. [CrossRef]
33. Ruiz-Domínguez, M.C.; Medina, E.; Salinas, F.; Bugueño, W.; Fuentes, J.-L.; Vílchez, C.; Garbayo, I.; Cerezal-Mezquita, P. Methodological Optimization of Supercritical Fluid Extraction of Valuable Bioactive Compounds from the Acidophilic Microalga *Coccomyxa onubensis*. *Antioxidants* **2022**, *11*, 1248. [CrossRef] [PubMed]
34. Deenu, A.; Naruenartwongsakul, S.; Kim, S.M. Optimization and economic evaluation of ultrasound extraction of lutein from *Chlorella vulgaris*. *Biotechnol. Bioprocess Eng.* **2013**, *18*, 1151–1162. [CrossRef]
35. Sun, Y.; Ma, G.; Ye, X.; Kakuda, Y.; Meng, R. Stability of all-trans-β-carotene under ultrasound treatment in a model system: Effects of different factors, kinetics and newly formed compounds. *Ultrason. Sonochem.* **2010**, *17*, 654–661. [CrossRef] [PubMed]
36. Molina Grima, E.; Belarbi, E.H.; Acién Fernández, F.G.; Robles Medina, A.; Chisti, Y. Recovery of microalgal biomass and metabolites: Process options and economics. *Biotechnol. Adv.* **2003**, *20*, 491–515. [CrossRef] [PubMed]
37. Aligiannis, N.; Kalpoutzakis, E.; Mitaku, S.; Chinou, I.B. Composition and Antimicrobial Activity of the Essential Oils of Two Origanum Species. *J. Agric. Food Chem.* **2001**, *49*, 4168–4170. [CrossRef] [PubMed]
38. Ördög, V.; Stirk, W.A.; Lenobel, R.; Bancířová, M.; Strnad, M.; van Staden, J.; Szigeti, J.; Németh, L. Screening microalgae for some potentially useful agricultural and pharmaceutical secondary metabolites. *J. Appl. Phycol.* **2004**, *16*, 309–314. [CrossRef]
39. Saeed Niazi, V.; Behboudi, H.; Tavakoli, S.; Aminian, F.; Ranjbar, R. Antimicrobial Potential of the Green Microalgae Isolated from the Persian Gulf. *Iran. J. Public Health* **2022**, *51*, 1134–1142. [CrossRef]
40. Wiegand, I.; Hilpert, K.; Hancock, R.E. Agar and broth dilution methods to determine the minimal inhibitory concentration (MIC) of antimicrobial substances. *Nat. Protoc.* **2008**, *3*, 163–175. [CrossRef]
41. Syukriah, A.R.N.; Liza, M.S.; Harisun, Y.; Fadzillah, A.A.M. Effect of solvent extraction on antioxidant and antibacterial activities from *Quercus infectoria* (Manjakani). *Int. Food Res. J.* **2014**, *21*, 1031–1037.
42. Franco, D.; Sineiro, J.; Rubilar, M.; Sánchez, M.; Jerez, M.C.D.; Pinelo, M.; Costoya, N.; Núñez, M.J. Polyphenols from Plant Materials: Extraction and Antioxidant power. *Electron. J. Environ. Agric. Food Chem.* **2008**, *7*, 3210–3216.
43. Luo, X.; Su, P.; Zhang, W. Advances in Microalgae-Derived Phytosterols for Functional Food and Pharmaceutical Applications. *Mar. Drugs* **2015**, *13*, 4231–4254. [CrossRef] [PubMed]
44. Kilic, N.K.; Erdem, K.; Donmez, G. Bioactive compounds produced by *Dunaliella* species, antimicrobial effects and optimization of the efficiency. *Turk. J. Fish. Aquat. Sci.* **2019**, *19*, 923–933. [CrossRef]
45. Kusmiati; Ningsih, E.B.; Ramadhani, I.; Amir, M. Antibacterial and antioxidant activity test of crude lutein extracted from sunflower (*Helianthus annuus* L.). *AIP Conf. Proc.* **2021**, *2331*, 41594. [CrossRef]
46. Lopes, G.; Sousa, C.; Valentão, P.; Andrade, P.B. Sterols in Algae and Health. In *Bioactive Compounds from Marine Foods: Plant and Animal Sources*; Hernández-Ledesma, B., Herrero, M., Eds.; Wiley: Hoboken, NJ, USA, 2013; pp. 173–191. [CrossRef]
47. Sasso, S.; Pohnert, G.; Lohr, M.; Mittag, M.; Hertweck, C. Microalgae in the postgenomic era: A blooming reservoir for new natural products. *FEMS Microbiol. Rev.* **2012**, *36*, 761–785. [CrossRef]

48. Fernandes, T.; Cordeiro, N. Microalgae as Sustainable Biofactories to Produce High-Value Lipids: Biodiversity, Exploitation, and Biotechnological Applications. *Mar. Drugs* **2021**, *19*, 573. [CrossRef] [PubMed]
49. Del Castillo, E.; Montgomery, D.C.; McCarville, D.R. Modified desirability functions for multiple response optimization. *J. Qual. Technol.* **1996**, *28*, 337–345. [CrossRef]
50. Costa, N.R.; Lourenço, J.; Pereira, Z.L. Desirability function approach: A review and performance evaluation in adverse conditions. *Chemom. Intell. Lab. Syst.* **2011**, *107*, 234–244. [CrossRef]

marine drugs

Article

Omega-3 Polyunsaturated Fatty Acid Eicosapentaenoic Acid or Docosahexaenoic Acid Improved Ageing-Associated Cognitive Decline by Regulating Glial Polarization

Juan Xia [1,2,†], Longen Yang [1,2,†], Chengyi Huang [1,2], Shuyi Deng [1,2], Zhiyou Yang [1,2,3], Yongping Zhang [1,2,3], Cai Zhang [1,2] and Cai Song [1,2,3,*]

1 Research Institute for Marine Drugs and Nutrition, College of Food Science and Technology, Guangdong Ocean University, Zhanjiang 524088, China
2 Guangdong Provincial Key Laboratory of Aquatic Product Processing and Safety, College of Food Science and Technology, Guangdong Ocean University, Zhanjiang 524088, China
3 Marine Medicine Research and Development Center, Shenzhen Institutes of Guangdong Ocean University, Shenzhen 518120, China
* Correspondence: cai.song@dal.ca; Tel.: +86-0759-593-4119
† These authors contributed equally to this work.

Abstract: Neuroinflammation induced by microglial and astrocyte polarizations may contribute to neurodegeneration and cognitive impairment. Omega (n)-3 polyunsaturated fatty acids (PUFAs) have anti-inflammatory and neuroprotective effects, but conflicting results were reported after different n-3 PUFA treatments. This study examined both the change in glial polarizations in ageing rats and the differential effects of two omega-3 PUFAs. The results showed that both PUFAs improved spatial memory in ageing rats, with docosahexaenoic acid (DHA) being more effective than eicosapentaenoic acid (EPA). The imbalance between microglial M1/M2 polarizations, such as up-regulating ionized calcium binding adaptor molecule 1 (IBA1) and down-regulating CD206 and arginase-1 (ARG-1) was reversed in the hippocampus by both n-3 PUFAs, while the DHA effect on CD206 was stronger. Astrocyte A1 polarization presented increasing S100B and C3 but decreasing A2 parameter S100A10 in the ageing brain, which were restored by both PUFAs, while DHA was more effective on S100A10 than EPA. Consistent with microglial M1 activation, the concentration of pro-inflammatory cytokines tumor necrosis factor (TNF)-α, interleukin (IL)-1β and IL-6 were significantly increased, which were attenuated by DHA, while EPA only suppressed IL-6. In correlation with astrocyte changes, brain-derived neurotrophic factor precursor was increased in ageing rats, which was more powerfully down-regulated by DHA than EPA. In summary, enhanced microglial M1 and astrocytic A1 polarizations may contribute to increased brain pro-inflammatory cytokines, while DHA was more powerful than EPA to alleviate ageing-associated neuroimmunological changes, thereby better-improving memory impairment.

Keywords: ageing; memory impairment; glial polarization; n-3 PUFAs

Citation: Xia, J.; Yang, L.; Huang, C.; Deng, S.; Yang, Z.; Zhang, Y.; Zhang, C.; Song, C. Omega-3 Polyunsaturated Fatty Acid Eicosapentaenoic Acid or Docosahexaenoic Acid Improved Ageing-Associated Cognitive Decline by Regulating Glial Polarization. *Mar. Drugs* **2023**, *21*, 398. https://doi.org/10.3390/md21070398

Academic Editors: Elena Talero and Javier Ávila-Román

Received: 1 June 2023
Revised: 5 July 2023
Accepted: 6 July 2023
Published: 10 July 2023

1. Introduction

Ageing is a causative factor for the development of neurodegenerative diseases like cognitive decline and Alzheimer's disease (AD) [1]. Neuroanatomical studies reported that hippocampal atrophy contributes to learning and memory impairment, and the inducers could be neuroinflammation and a reduction in neurogenesis and synaptic plasticity [2]. Microglia and astroglia, the vital central immune cells can be activated into inflammatory or neurotoxic phenotypes (M1 microglia and A1 astrocytes) or neuroprotective types (M2 microglia and A2 astrocytes) [3]. Ageing-induced chronic low-grade inflammation promotes microglia into a sensitized, reactive, or primed pro-inflammatory M1 phenotype [4,5], which increased the release of pro-inflammatory factors, such as tumor necrosis factor (TNF)-α, interleukin (IL)-1β, and IL-6, and decreased the expression of anti-inflammatory M2 microglia marker arginase 1(ARG-1) and CD206, as well as anti-inflammatory cytokines IL-4 and IL-10 [6]. As a consequence of microglial M1 activation, these pro-inflammatory cytokines thereafter trigger astrocyte A1 polarization. The increased A1 marker C3 and decreased A2 neurotrophic factor brain-derived neurotrophic factor (BDNF) led to neurotransmitter deficiency and neuronal apoptosis [7]. However, the neuroinflammatory mechanisms mediated by glial polarization imbalance associated with ageing-induced memory impairment were not fully understood.

As mentioned above, glial polarizations may contribute to neurodegeneration and cognitive impairment. The regulation of glial function may provide a novel therapeutic direction to alleviate ageing-related neurodegenerative disease. Omega (n)-3 polyunsaturated fatty acids (PUFAs), indispensable components of cell membranes, play an important role in maintaining the membrane structural integrity and fluidity of immune and neuronal cells. The synthesization of n-3 PUFAs is restricted in human bodies but is mainly obtained from marine sources [8]. n-3 PUFAs reduction has been reported in the physiological processes of ageing, which is associated with brain atrophy and memory impairment [9,10]. We and others have reported that n-3 PUFAs eicosapentaenoic acid (EPA) or docosahexaenoic acid (DHA) supplementation exerted preventative, ameliotropic, and/or neuroprotective effects on psychiatric and neurological disorders, such as depression [11] and Alzheimer's disease [12,13]. The mechanism by which n-3 PUFAs effectively improved these diseases was through inhibiting the expression of microglia M1 phenotype marker CD11b and inducible nitric oxide synthase (iNOS) and downregulating the level of proinflammatory cytokines IL-1β, TNF-α and IL-6, while up-regulating the microglia M2 phenotype marker CD206 and anti-inflammatory cytokines IL-10 [14]. After n-3 PUFAs balance microglia M1 and M2 polarizations, the damage from neuroinflammation on neurons was alleviated, which was mainly involved in nuclear factor kappa-B (NF-κB) pathway inhibition [15]. Similarly, n-3 PUFAs suppress the pro-inflammatory and toxic astroglia A1 phenotype polarizations by decreasing the expression of C3 and S100B [16] but promote astrocyte differentiation from neural stem cells accompanied by an increase in neurotrophic factors BDNF and glial cell line-derived neurotrophic factor (GDNF) [17]. However, DHA was reported to be better in terms of its neuroprotective effect by promoting neuronal survival, synaptic vesicle fusion, neurotransmitter releases and neuronal plasticity [18]. Differently, we and others reported that EPA but not DHA markedly attenuated mental-disorder-associated and inflammation-induced cognitive impairment. However, whether and how EPA and DHA treatments modulate glial polarizations and improve memory impairment in the ageing brain is still unknown. Thus, the present study aimed to determine whether the imbalance between two types of glial polarizations is associated with ageing-induced memory impairment and compare and evaluate the therapeutic effects of n-3 PUFAs DHA and EPA on memory impairment and their underlying neuroimmune mechanisms.

2. Results

2.1. Memory and Locomotor Impairment Occurred in Ageing Rats, Which Was Better Improved by DHA Than EPA

In the Morris water maze (MWM) test, the ageing rats showed increased latency to find the platform (Figure 1A) and decreased time spent on and numbers in the target quadrant ($p < 0.01$) (Figure 1B,C), indicating significant memory impairment in ageing rats when compared to the control group. Even though both EPA and DHA reversed memory impairment in ageing rats by decreasing the latency to the platform ($p < 0.05$) (Figure 1A) and increasing the time spent and the number of entries into the target quadrant ($p < 0.05$) in ageing rats (Figure 1B,C), rats fed DHA showed longer time spent in the target quadrant than the EPA ($p < 0.05$) (Figure 1B). In the open field test (OFT), the total locomotor distance was decreased in ageing rats compared with the control ($p < 0.05$), which was improved by EPA ($p < 0.05$) or DHA ($p < 0.05$) treatment (Figure 1D). EPA treatment was more powerful than DHA in the increase in central entries compared with ageing rats (Figure 1F). However, there were no differences between controls and ageing rats in the number of rearing and center entries (Figure 1E,F).

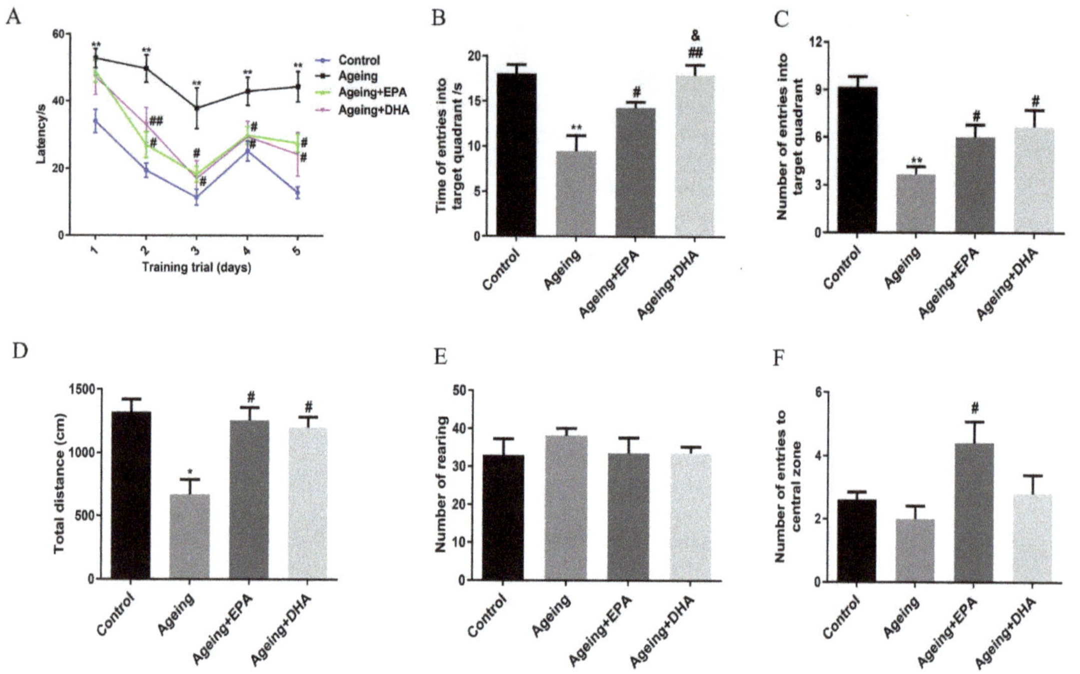

Figure 1. Memory and locomotor impairment that occurred in ageing rats was better improved by DHA than by EPA. The latency time (**A**), time (**B**) and number (**C**) of entries into the target quadrant in MWM behavioral test. The total distance (**D**), number of rearing (**E**) and entries into the central zone (**F**) in the OFT. The data were expressed as mean ± SEM (n = 12 in control group, n = 9 in other three groups). * $p < 0.05$, ** $p < 0.01$ versus control group; # $p < 0.05$, ## $p < 0.01$ versus ageing group. & $p < 0.05$ versus ageing + EPA group.

2.2. n-3 and n-6 PUFAs Imbalance Were Both Improved by DHA and EPA

Compared with the control group, significantly reduced contents of n-3 PUFAs EPA ($p < 0.05$), DPA ($p < 0.05$), and DHA ($p < 0.01$) (Figure 2A–C) were found in ageing rat brains, without changing n-6 PUFAs cis-8,11,14-eicosatrienoic acid and arachidonic acid (AA) concentrations (Figure 2D,E). The ratio of n-6/n-3 PUFAs was markedly increased in ageing rats compared with the control group ($p < 0.01$) (Figure 2F). Supplementation of EPA or DHA elevated the level of EPA ($p < 0.001$) or DHA ($p < 0.05$), respectively (Figure 2A,C), and both EPA and DHA treatments significantly reduced the ratio of n-6/n-3 PUFAs in ageing rats ($p < 0.05$) without an effect on DPA and n-6 PUFA contents (Figure 2B,D–F).

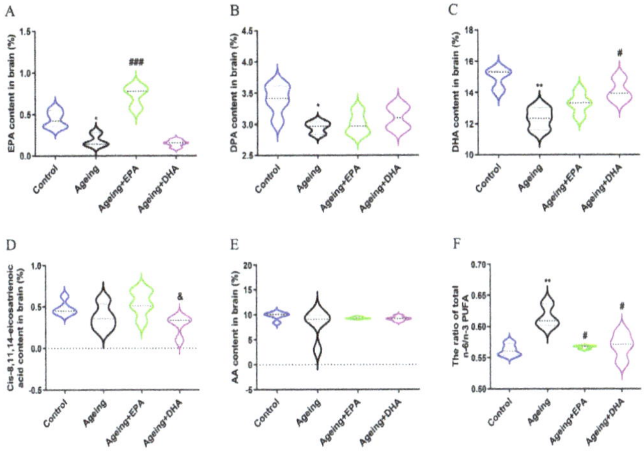

Figure 2. The increase in n-6/n-3 PUFAs ratio in ageing rats was effectively reversed by EPA and DHA. The contents of n-3 PUFAs EPA (**A**), DPA (**B**), and DHA (**C**) and contents of n-6 PUFAs cis-8,11,14-eicosatrienoic acid (**D**) and AA (**E**) as well as n-6/n-3 PUFAs ratio (**F**). The data were expressed as mean ± SEM (n = 5). * $p < 0.05$, ** $p < 0.01$ versus control group; # $p < 0.05$, ### $p < 0.001$ versus ageing group. & $p < 0.05$ versus ageing + EPA group.

2.3. Abnormal Microglial M1 and M2 Polarizations in the Hippocampus of Ageing Rats Was Better Ameliorated by DHA Than EPA through Upregulating CD206

Compared with the control group, mRNA expression of M1 microglia markers ionized calcium binding adaptor molecule 1 (IBA1) ($p < 0.05$) was upregulated, while the mRNA expression of M2 microglia marker CD206 ($p < 0.05$) and ARG-1 ($p < 0.05$) was downregulated in the hippocampus of ageing rats (Figure 3A,C,D). Either EPA or DHA treatment significantly inhibited M1 microglial and restored M2 microglial polarization (Figure 3A–D). Nevertheless, DHA was better in the upregulation of CD206 mRNA expression than EPA (Figure 3D). Increased protein expression of iNOS ($p < 0.01$) was found in the ageing hippocampus, which was attenuated by either EPA or DHA ($p < 0.05$) (Figure 3E,F). DHA rather than EPA treatment significantly upregulated CD206 protein expression compared with ageing rats ($p < 0.01$) (Figure 3G,H).

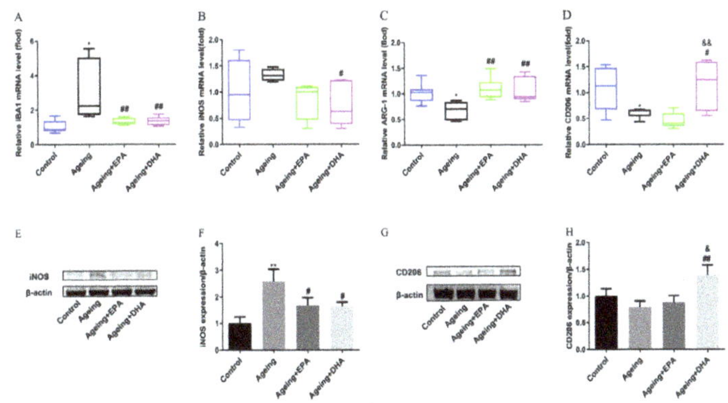

Figure 3. Imbalance between microglial M1 and M2 polarization that occurred in ageing rats was attenuated by EPA and DHA. The mRNA expression of IBA1 (**A**), iNOS (**B**), ARG-1 (**C**), and CD206 (**D**). The protein expression of iNOS (**E**,**F**) and CD206 (**G**,**H**). The data were expressed as mean ± SEM (n = 6). * $p < 0.05$, ** $p < 0.01$ versus control group; # $p < 0.05$, ## $p < 0.01$ versus ageing group. & $p < 0.05$, && $p < 0.01$ versus ageing+ EPA group.

2.4. Neuroinflammation in Ageing Rats Were Both Inhibited by DHA and EPA

Compared with the control group, the concentration of pro-inflammatory cytokines IL-1β, IL-6, and TNF-α were significantly increased in the hippocampus of ageing rats (IL-1β, $p < 0.001$; IL-6, $p < 0.01$; TNF-α, $p < 0.05$) (Figure 4A–C), which was equally reversed by EPA or DHA treatment ($p < 0.05$) (Figure 4A–C). Meanwhile, the anti-inflammatory cytokine IL-10 was significantly upregulated ($p < 0.01$), without changing the IL-4 concentrations (Figure 4D,E). Neither EPA nor DHA had a significant effect on these anti-inflammatory cytokine concentrations in ageing rats (Figure 4D,E).

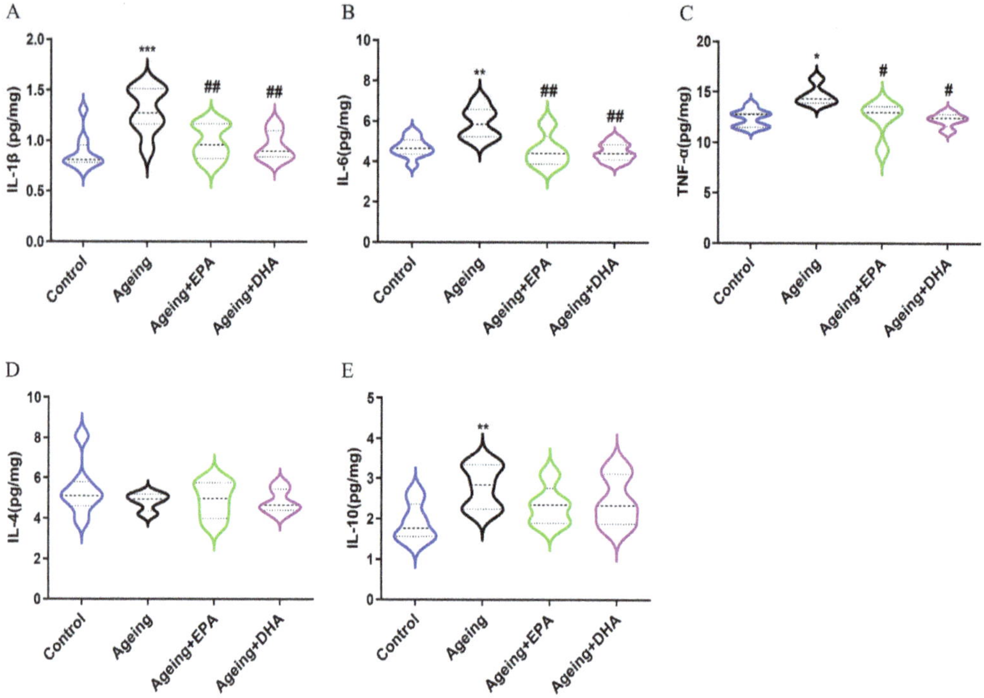

Figure 4. Neuroinflammatory responses upregulated during normal ageing were inhibited by EPA and DHA. The concentrations of pro-inflammatory cytokines IL-1β (**A**), IL-6 (**B**), and TNF-α (**C**). The concentrations of anti-inflammatory cytokines IL-4 (**D**) and IL-10 (**E**). * $p < 0.05$, ** $p < 0.01$, *** $p < 0.001$ versus control group; # $p < 0.05$, ## $p < 0.01$ versus ageing group.

2.5. Abnormal Astroglia A1/A2 Phenotypic Polarizations in the Hippocampus of Ageing Rats Were Both Attenuated by EPA and DHA

The mRNA expressions of activated astroglia marker glial fibrillary acidic protein (GFAP) ($p < 0.01$), A1 marker S100B ($p < 0.01$) and C3 ($p < 0.001$) were significantly upregulated, while A2 marker S100A10 ($p < 0.05$) was markedly downregulated (Figure 5A–D) in the ageing hippocampus. Both GFAP and C3 changes were confirmed at the level of protein expression. EPA and DHA equally inhibited the mRNA expression of GFAP ($p < 0.05$), S100B ($p < 0.05$) and C3 ($p < 0.01$) in the hippocampus of ageing rats (Figure 5A–C). Surprisingly, the decrease in mRNA expression of astroglia A2 marker S100A10 and the increase in protein expression of GFAP were significantly attenuated by DHA treatment, but not EPA ($p < 0.05$) (Figure 5D–F).

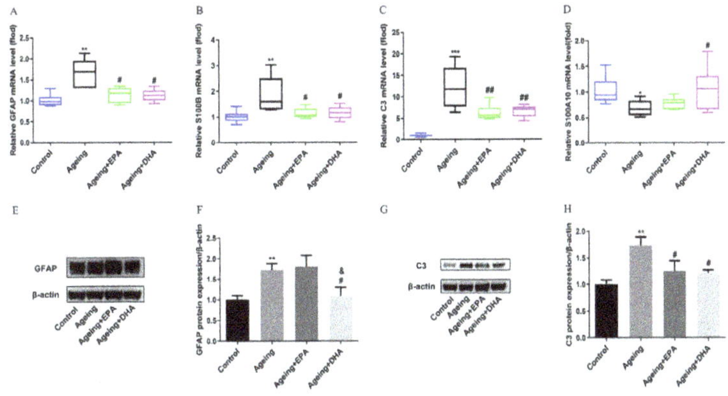

Figure 5. Astroglia polarization imbalance induced in the ageing rat hippocampus was reversed by EPA and DHA. The mRNA expressions of GFAP (**A**), S100B (**B**), C3 (**C**), and S100A10 (**D**). The WB bands of marker proteins (**E**,**G**), and the relative protein expression of GFAP (**F**) and C3 (**H**). The data were expressed as mean ± SEM (n = 6). * $p < 0.05$, ** $p < 0.01$, *** $p < 0.001$ versus control group; # $p < 0.05$, ## $p < 0.01$ versus ageing group; & $p < 0.05$ versus ageing + EPA group.

2.6. Activation of proBDNF-p75(NTR) in the Hippocampus of Ageing Rats Was Inhibited by EPA and DHA

Compared with the control group, mRNA expression of BDNF and its high-affinity receptor tropomyosin-related kinase B (TrkB) were not changed, while p75 receptor mRNA expression ($p < 0.05$) was significantly increased in ageing rats (Figure 6A–C). EPA and DHA had no significant effects on BDNF and TrkB mRNA expression; however, both DHA and EPA indiscriminately reduced the mRNA expression of p75 ($p < 0.05$) (Figure 6A–C). Importantly, the expression of BDNF precursor protein proBDNF ($p < 0.01$) and its receptor p75 ($p < 0.01$) were significantly upregulated in the hippocampus of ageing rats (Figure 6D–F). Both DHA or EPA inhibited the expression of proBDNF ($p < 0.05$) or p75 receptor ($p < 0.05$) (Figure 6D–F). In concert with mRNA expression, no change in BDNF protein expression occurred between control and ageing animals, while TrkB protein was upregulated in ageing rats compared with controls ($p < 0.05$) (Figure 6G–I). EPA and DHA did not affect BDNF/TrkB, but normalized proBDNF/p75 protein expressions in ageing rats.

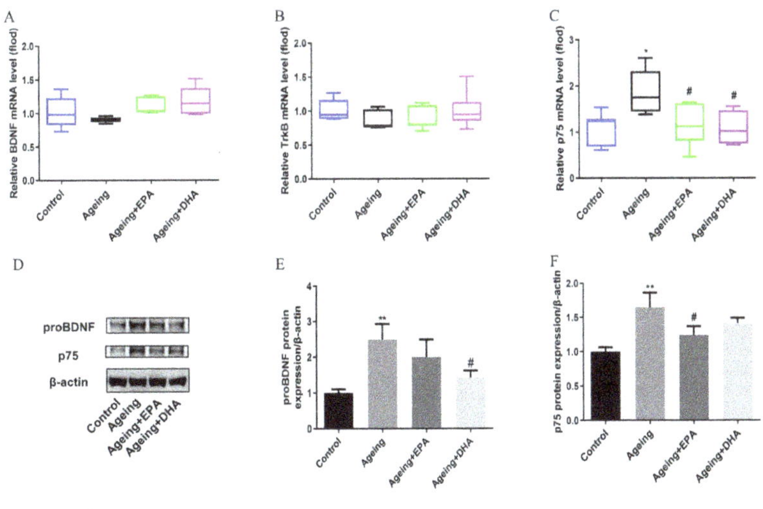

Figure 6. *Cont.*

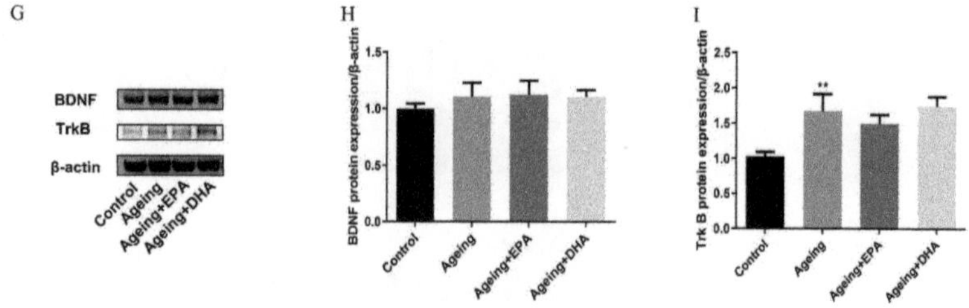

Figure 6. The dysregulation of neurotrophic factors and receptors that occurred in ageing rats was improved by EPA and DHA. The mRNA expression of BDNF (**A**), TrkB (**B**), and p75 (**C**). The WB bands of protein (**D**,**G**), and the relative protein expression of proBDNF (**E**), p75 (**F**), BDNF (**H**), and TrkB (**I**). The data were expressed as mean ± SEM (n = 6). * $p < 0.05$, ** $p < 0.01$ versus control group; # $p < 0.05$ versus ageing group.

3. Discussion

In this study, the possible mechanism by which the imbalance between different glial cell polarizations may contribute to ageing-associated cognitive decline was first explored. Growing evidence has suggested different actions of EPA and DHA in the treatment of psychiatric and neurodegenerative diseases. Then, the effects of different n-3 PUFAs on cognitive impairment and related glial polarizations in the ageing brain were studied. The current study distinguished different effects between EPA and DHA and revealed some possible mechanisms in the treatment of ageing-associated cognitive decline. Several new findings were discussed as follows:

First, the results showed that both spatial learning and memory in MWM and exploration in OFT were impaired in normal ageing rats compared to adult rats. Meanwhile, the concentration of brain n-3 PUFAs, such as EPA, DPA and DHA, were decreased in ageing brains. A longitudinal cohort study of elderly Japanese with cognitive decline over 10 years displayed decreased serum DHA, which was positively correlated with cognitive decline [19]. Thus, DHA supplementation was reported to improve cognitive decline in older adults with mild cognitive impairment [20,21] or animal models of AD [22]. Most studies were based on supplementing a mixture of EPA and DHA, whereas the present study compared the effects of EPA and DHA on memory impairment in ageing animals. This study demonstrated that both EPA and DHA could improve the ageing-related memory impairment in the MWM test, while DHA effects were better than EPA in terms of increasing the time of entries into the target quadrant. However, a previous study reported that both EPA and DHA indiscriminately exhibited anti-ageing effects by antioxidation and reducing ageing-related proteins [23]. Since EPA supplementation or its derivatives could significantly increase the amount of EPA and DHA in the brain [24,25], EPA's effect may partly depend on DHA function. The PUFAs profile in our study showed that both EPA and DHA treatments respectively reversed the decrease of EPA or DHA in ageing rats' brains, which was more conducive to revealing their commonalities and specific mechanisms for improving ageing-related memory impairment.

Secondly, ageing-associated chronic low-grade inflammation is one of the contributors to cognitive deficits, which was mainly triggered by activated microglia [26]. Numerous studies have reported the activation of the microglia M1 phenotype in normal ageing, which upregulated feature genes IBA1 and iNOS and produced pro-inflammatory cytokines IL-1β, IL-6, and TNF-α [27–29]. The alternative M2 phenotype exhibits anti-inflammatory activity with the expression of representative markers ARG-1 and CD206, along with secreted anti-inflammatory cytokines IL-4 and IL-10 [30]. Thus, shifting M1 polarization to M2 might be a potential therapy for neurodegenerative diseases. The anti-neuroinflammatory activities of EPA and DHA attributed to the balance between M1 and M2 polarization had been confirmed to improve depression [31], Alzheimer's disease [32], Parkinson's disease [33] and other neuropsychiatric disorders by our and others' group, which are consistent with the findings in the current study. Both EPA and DHA reversed the increased M1 marker IBA1 mRNA and iNOS protein expression and decreased M2 marker ARG-1 mRNA expression in the hippocampus of ageing rats, while DHA showed stronger effects than EPA in terms of increasing the expression of CD206. An important finding that supports our results was that DHA promoted the expression of CD206, M2 phenotype in human CHME3 microglia cells, but not EPA [32]. The promotion of M2 macrophage polarization by DHA primarily relied on the modulation of the p38 MAPK signaling pathway and autophagy [34]. CD206 plays an important role in the procedure of anti-neuroinflammation as an M2 phenotype marker of microglia. Previous studies have demonstrated that the therapeutic effects of DHA on traumatic brain injury depend on increasing CD206-positive phagocytic microglia [35]. Thus, CD206 might play a critical neuroprotective role in DHA-related cognitive improvement, which deserves further research. Furthermore, both DHA and EPA equally inhibited the production of pro-inflammatory cytokines IL-1β, IL-6, and TNF-α in the hippocampus of ageing rats, without changing anti-inflammatory cytokines IL-4 and IL-10.

Similar to microglia, astroglia depolarize to two phenotypes, A1 and A2. A2 plays supporting and maintaining roles for brain homeostasis in normal conditions, which could be damaged by neuroinflammation [36]. Transcriptional analysis for astroglia in ageing mice showed that the increase in astroglia A1 pro-inflammatory or cytotoxic response phenotype could drive an ageing-related cognitive decline [7]. The present study further demonstrated ageing-dependent remodeling of astroglia, with increased GFAP and A1 marker genes S100B and C3 complement and decreased protective A2 phenotype marker gene S100A10. Both EPA and DHA inhibited C3 expression in the ageing brain, while DHA, interestingly, was more effective than EPA in reducing GFAP but upregulating A2 marker S100A10 transcription. GFAP, a common marker of astroglia, was generally elevated in rodents and humans during ageing [37], which indicates a potential marker of cognitive impairment induced by ageing [38], AD [39], heart failure [40] and type 2 diabetes [41]. Previous studies showed that DHA regulated glial cell polarization by activating PPARγ in AD [42] and enhanced the expression of GFAP by the up-regulation of both PI3K/AKT-dependent FABP7–PPARγ interaction and MKP3 in astrocytes in young rat brains [43]. Conversely, and excitingly, the present study showed, for the first time, the specific inhibition of DHA on GFAP in ageing rats, which was supported by a similar effect observed in a chronic constriction injury [44].

Thirdly, BDNF, a neurotrophin produced by neurons and astroglia, plays a crucial role in maintaining neurogenesis and neuronal plasticity, the alteration of which was related to cognitive deficits in ageing and AD [45]. There are two BDNF receptors; the first is the high-affinity TrkB receptor, which promotes cell survival, neuronal growth, synaptic plasticity and neurotransmitter release. Exogenous BDNF inhibits ageing-activated microglia through the TrkB-Erk-CREB pathway [46]. The other is p75 receptor, which promotes cell apoptosis and inhibits cell survival through binding to proBDNF. In contrast, some studies reported no relationship between ageing-associated neurodegeneration and mRNA expression of BDNF or TrkB receptors in the hippocampus [47]. Even though the present study confirmed that no significant changes were observed in mRNA and protein expression of BDNF between the ageing and control rats, TrkB was increased in ageing rats. Unexpectedly, EPA and DHA had no effects on BDNF and TrkB expression in ageing rats. However, as reported previously, proBDNF, the precursor protein of BDNF, with an opposite effect on neuronal plasticity through binding to the p75 receptor, was increased in the hippocampus, which mainly contributes to ageing-related memory impairments [48]. Consistently, the present study showed that protein expression of proBDNF and its receptor p75 were elevated in ageing rats, while both EPA and DHA decreased the mRNA expression of p75. More interesting, the present study, for the first time, found that EPA and DHA exhibited different and specific effects on the protein expression of these factors as DHA downregulated proBDNF protein, while EPA downregulated p75 protein.

In summary, the present study demonstrated that cognitive decline occurred during the progress of normal ageing, the mechanisms of which mainly involved an imbalance between microglial M1 and M2, and astroglia A1and A2 phenotypic polarizations, as well as the upregulation of proBDNF and p75 signaling. In addition, the decrease in endogenous n-3 PUFAs in ageing progression and improvement of n-3 PUFAs on memory deficit, undoubtedly, provided a new chance for the prevention or treatment of ageing-related cognitive disorders. Both EPA and DHA improved the ageing-related learning and memory impairment, while DHA effects were more pronounced than EPA. The common mechanisms of EPA or DHA were as follows: (1) both EPA and DHA restored the ratio of the n-3/n-6 PUFAs profile; (2) EPA and DHA equally inhibited the ageing-related neuroinflammation induced by M1 microglia excessive activation; (3) EPA and DHA both inhibited astroglia A1 phenotype by decreasing C3 and S100B expression; (4) both DHA and EPA effectively suppressed the proBDNF/p75 pathway without changing BDNF-TrkB signaling. Furthermore, differences between EPA and DHA were revealed: (1) DHA was superior to EPA in promoting M2 microglial polarization; (2) DHA was more able than EPA to block GFAP expression while promoting A2 polarization. Collectively, compared with EPA, DHA is preferable for treating ageing-related cognitive and memory impairments.

4. Materials and Methods

4.1. Animals and Experimental Procedure

Twelve adult male (3-month old) and twenty-seven male ageing (24-month old) Sprague–Dawley (SD) rats were housed as two per cage in a standard environment with 23 ± 1 °C and relative humidity 50 ± 10% under a regular 12 h light/dark cycle (light on 7:00 A.M.). The experimental process was performed according to the National Institutes of Health Guide for the Care and Use of Laboratory Animals and approved by the Animal Ethics Committee of Guangdong Ocean University.

Adult rats were used as a control while the ageing rats were divided into three groups: ageing (saline), ageing + EPA (500 mg/kg/day EPA) and ageing + DHA (500 mg/kg/day DHA). Rats were orally administered with saline, EPA or DHA for 8 weeks, with the dosage of EPA or DHA as described previously [49–51], and then the MWM and OFT were performed. One day after behavioral tests, animals were sacrificed, and hippocampal samples were collected on ice and quick-frozen in liquid nitrogen, then stored in a −80 °C refrigerator for further experiments.

4.2. Morris Water Maze

To evaluate spatial and learning memory, a tank (150 cm in diameter) was filled with water at 25 ± 1 °C. A 10 cm-diameter platform was submerged 2 cm underwater. On the first trial day, the start positions varied randomly in four quadrants without repetition. Four trials were arranged for each day to allow the rats to reach the platform. If the rats failed to find the platform within 1 min, they were gently guided to the platform to stay for 15 s. The latency to reach the platform was recorded for each animal. On the fourth day, the platform was relocated to a different quadrant in the maze and the latency to locate the platform was recorded. On the last day, the platform was removed, and the rats were allowed to swim freely for 1 min. The latency to the platform, the time and the number of entries into the target quadrant were recorded by video track. The data was processed by the SuperMaze behavior analysis system (Shanghai Xinruan Information Technology Co., Ltd., Shanghai, China).

4.3. Open Field Test

The OFT was popularly used for rodent locomotor and exploratory behavior in a novel environment [52]. Round open field apparatus (diameter 100 cm, high 50 cm) with a white painted wall and floor was used, in which the animal total distance was recorded by a digital camera within 3 min.

4.4. Brain n-3/n-6 PUFA Analysis by Gas Chromatography-Mass Spectrometry

Three n-3 PUFAs, including EPA, docosapentaenoic acid (DPA) and DHA, and two n-6 PUFAs including cis-8,11,14-eicosatrienoic acid and AA, were profiled by GC-MS (Agilent, Santa Clara, CA, USA). The lipid extraction method in this study was referred to by Folch [53]. Briefly, the brain tissues were homogenized in chloroform/methanol solution (2/1), vortexed with 0.9% NaCl solution, and centrifuged at 500 rpm for 20 min. The underlayer phase was collected and dried with a nitrogen blower. To make the fatty acid methyl derivatization, the precipitates were incubated with 1.5% sulfuric acid methanol and methylene chloride solution at 100 °C for 1 h. Hexane and saturated sodium chloride solution were added and centrifuged at 500 rpm for 2 min after cooling. Finally, the hexane phase was separated and then dried by a nitrogen blower for gas chromatography–mass spectrometry (GC-MS) analysis. The GC-MS analysis was performed on InterCap Pure-WAX Capillary columns (30 m × 0.25 mm, 0.25 μm) utilizing helium as a carrier gas at a flow rate of 1.0 mL/min. The mass spectrometry was recorded at a proton energy of 70 eV.

4.5. Cytokine Detections

The pro-inflammatory cytokines (IL-1β, IL-6, and TNF-α) and anti-inflammatory factors (IL-4 and IL-10) in the hippocampus were quantified by the enzyme-linked immunosorbent assay (ELISA) kits according to the manufacturer's instructions. The hippocampal tissues were homogenized in 1 × PBS solution (pH 7.4) at a ratio of 1:9 and centrifuged at 2000 rpm for 15 min at 4 °C. The supernatant was collected for quantification.

4.6. Real-Time Quantitative Polymerase Chain Reaction Analysis

The total RNA extraction was performed according to the manufacturer's instructions. Briefly, the hippocampus was homogenized with 1 mL TRIzol reagent (15596026, Invitrogen, Carlsbad, CA, USA) and incubated for 10 min at room temperature (RT). Chloroform (200 μL) was added and mixed by vigorous shaking for 30 s and standing for 10 min at RT. The RNA precipitation was obtained after centrifugation (12,000 rpm for 15 min) and washed once with 75% ethanol. The precipitates were finally dissolved in 20 μL diethylpyrocarbonate (DEPC) water. A measure of 1 μg of total RNA was used to perform reverse transcription by the QuantiTect Reverse Transcription Kit (Vazyme, Nanjing, China), and real-time PCR was performed with AceQ Universal SYBR qPCR Master Mix (Vazyme, China) on an ABI7500. The amplification reaction condition was set as follows: denaturation at 95 °C for 10 min, followed by 40 cycles of denaturation at 95 °C for 10 s and annealing/extension at 60 °C for 30 s. The expression of fold changes for target genes was calculated by the $2^{-\Delta\Delta ct}$ method. The primer sequences were listed in Table 1.

Table 1. The primer sequences of target genes.

Genes	Forward Primer	Reverse Primer
BDNF	F:5′-CAAAAGGCCAACTGAAGC	R:5′-CGCCAGCCAATTCTCTTT
TrkB	F:5′-CACACACAGGGCTCCTTA	R:5′-AGTGGTGGTCTGAGGTTGG
p75	F:5′-TGCTCCATTTCCATCTCAG	R:5′-GATAGGTCCGTAATCCTCTTC
iNOS	F:5′-TGGAGCGAGTTGTGGATTGT	R:5′-GTAGTGATGTCCAGGAAGTAGGT
CD206	F:5′-GTTTCCATCGAGACTGCTGC	R:5′-GCCACTTTCCTTCAACATTTCG
ARG-1	F:5′-GGTAGAGAAAGGTCCCGCAG	R:5′-CAGACCGTGGGTTCTTCACA
C3	F:5′-TGTGGGTGGATGTGAAGGAC	R:5′-CTTGTCCACAGCCACTAGCC
IBA1	F:5′-CAACAAGCACTTCCTCGATGATC	R:5′-TGAAGGCCTCCAGTTTGGACT
GFAP	F:5′-CCAAGATGAAACCAACCT	R:5′-CGCTGTGAGGTCTGGCTT
S100B	F:5′-CTCTGTCTACCCTCCTAGTCC	R:5′-GACATCAATGAGGGCAACCAT
S100A10	F:5′-TATCACTAGTGGCGGGGCTC	R:5′-ATCAAGGTGTGGGTACCAGG
β-actin	F:5′-ACGGTCAGGTCATCACTATCG	R:5′-GTTTCATGGATGCCACAGGATT

4.7. Western Blotting Analysis

The total hippocampus protein was extracted by RIPA lysis buffer containing 1 mM phenylmethanesulfonyl fluoride (PMSF) on ice and quantified using a BCA protein assay kit. Each sample (30 μg total protein) was separated by 12% sodium dodecyl sulfate-polyacrylamide gel electrophoresis (SDS-PAGE) and transferred onto polyvinylidene fluoride (PVDF) membranes. The membranes were blocked with Tris-buffered saline containing 5% skim milk for 2 h at RT and incubated with primary antibodies at 4 °C overnight. Subsequently, the membranes were washed three times with TBST (0.1% Tween-20) and incubated with HRP-labeled secondary antibodies for 2 h at RT. The bands were detected using an enhanced chemiluminescence (ECL) system and analyzed by Image J Software. The primary antibodies used were as follows: iNOS (1:500, sc-7271, Santa Cruz, CA, USA), CD206 (1:500, sc-70586, Santa Cruz, CA, USA), GFAP (1:500, sc-33673, Santa Cruz, CA, USA), C3 (1:500, sc-28294, Santa Cruz, CA, USA), proBDNF (1:500, sc-65514, Santa Cruz, CA, USA), TrKB (1:500, sc-377218, Santa Cruz, CA, USA), BDNF (1:1000, 28205-1-AP, proteintech, Wuhan, China), p75 (1:500, sc-271708, Santa Cruz, CA, USA) and β-actin (1:1000, #4967, CST, Danvers, MA, USA). The secondary antibodies were HRP-conjugated anti-mouse (1:1000, #7076, CST, MA, USA) and HRP-conjugated rabbit (1:1000, #7074, CST, MA, USA).

4.8. Statistical Analysis

The data were expressed as the mean ± standard error (SEM) and analyzed by GraphPad Prism (GraphPad Software 9.0.0, San Diego, CA, USA). Statistical comparisons were performed using one-way analysis of variance (ANOVA) followed by Fisher's LSD or Tukey's post hoc test. Statistical significance was considered as $p < 0.05$.

Author Contributions: Conceptualization, J.X., L.Y. and C.S.; formal analysis, J.X., L.Y., C.H., S.D. and C.S.; methodology, J.X., L.Y., Z.Y., Y.Z., C.Z. and C.S.; investigation, J.X., L.Y. and C.S.; data curation, J.X., L.Y., C.H. and S.D.; writing—original draft preparation, J.X. and C.S.; writing—review and editing, J.X., L.Y. and C.S.; supervision, Z.Y., Y.Z. and C.S.; project administration and funding acquisition, C.S. All authors have read and agreed to the published version of the manuscript.

Funding: This research was funded by Guangdong Provincial Natural Science Fund Project, grant number 2021A1515011579; Shenzhen Science and Technology plan (International Cooperation Research) project, grant number GJHZ20190823111414825, and Zhanjiang Science and Technology Project, grant number 2021A05046.

Institutional Review Board Statement: The animal study protocol was approved by the Ethics Committee of Guangdong Ocean University (Protocol 2017-16—approval 12 September 2017).

Data Availability Statement: All data are contained within this article.

Conflicts of Interest: The authors declare no conflict of interest.

References

1. Hou, Y.; Dan, X.; Babbar, M.; Wei, Y.; Hasselbalch, S.G.; Croteau, D.L.; Bohr, V.A. Ageing as a risk factor for neurodegenerative disease. *Nat. Rev. Neurol.* **2019**, *15*, 565–581. [CrossRef] [PubMed]
2. Bettio, L.E.B.; Rajendran, L.; Gil-Mohapel, J. The effects of aging in the hippocampus and cognitive decline. *Neurosci. Biobehav. Rev.* **2017**, *79*, 66–86. [CrossRef] [PubMed]
3. Kwon, H.S.; Koh, S.H. Neuroinflammation in neurodegenerative disorders: The roles of microglia and astrocytes. *Transl. Neurodegener.* **2020**, *9*, 42. [CrossRef] [PubMed]
4. Ward, R.J.; Dexter, D.T.; Crichton, R.R. Ageing, neuroinflammation and neurodegeneration. *Front. Biosci. Sch. Ed.* **2015**, *7*, 189–204. [CrossRef]
5. Cornell, J.; Salinas, S.; Huang, H.Y.; Zhou, M. Microglia regulation of synaptic plasticity and learning and memory. *Neural Regen. Res.* **2022**, *17*, 705–716.
6. Yao, K.; Zu, H.B. Microglial polarization: Novel therapeutic mechanism against Alzheimer's disease. *Inflammopharmacology* **2020**, *28*, 95–110. [CrossRef]
7. Clarke, L.E.; Liddelow, S.A.; Chakraborty, C.; Münch, A.E.; Heiman, M.; Barres, B.A. Normal aging induces A1-like astrocyte reactivity. *Proc. Natl. Acad. Sci. USA* **2018**, *115*, E1896–E1905. [CrossRef]

8. Swanson, D.; Block, R.; Mousa, S.A. Omega-3 fatty acids EPA and DHA: Health benefits throughout life. *Adv. Nutr.* **2012**, *3*, 1–7. [CrossRef]
9. Cutuli, D. Functional and Structural Benefits Induced by Omega-3 Polyunsaturated Fatty Acids during Aging. *Curr. Neuropharmacol.* **2017**, *15*, 534–542. [CrossRef]
10. Chiu, C.C.; Frangou, S.; Chang, C.J.; Chiu, W.C.; Liu, H.C.; Sun, I.W.; Liu, S.I.; Lu, M.L.; Chen, C.H.; Huang, S.Y.; et al. Associations between n-3 PUFA concentrations and cognitive function after recovery from late-life depression. *Am. J. Clin. Nutr.* **2012**, *95*, 420–427. [CrossRef]
11. Yan, L.; Gu, M.Q.; Yang, Z.Y.; Xia, J.; Li, P.; Vasar, E.; Tian, L.; Song, C. Endogenous n-3 PUFAs attenuated olfactory bulbectomy-induced behavioral and metabolomic abnormalities in Fat-1 mice. *Brain Behav. Immun.* **2021**, *96*, 143–153. [CrossRef]
12. Zhang, Y.P.; Brown, R.E.; Zhang, P.C.; Zhao, Y.T.; Ju, X.H.; Song, C. DHA, EPA and their combination at various ratios differently modulated Aβ-induced neurotoxicity in SH-SY5Y cells. *Prostaglandins Leukot. Essent. Fatty Acids* **2018**, *136*, 85–94. [CrossRef] [PubMed]
13. Song, C.; Shieh, C.H.; Wu, Y.S.; Kalueff, A.; Gaikwad, S.; Su, K.P. The role of omega-3 polyunsaturated fatty acids eicosapentaenoic and docosahexaenoic acids in the treatment of major depression and Alzheimer's disease: Acting separately or synergistically? *Prog. Lipid Res.* **2016**, *62*, 41–54. [CrossRef]
14. Joffre, C.; Rey, C.; Layé, S. N-3 Polyunsaturated Fatty Acids and the Resolution of Neuroinflammation. *Front. Pharmacol.* **2019**, *10*, 1022. [CrossRef] [PubMed]
15. Liu, B.P.; Zhang, Y.P.; Yang, Z.Y.; Liu, M.J.; Zhang, C.; Zhao, Y.T.; Song, C. ω-3 DPA Protected Neurons from Neuroinflammation by Balancing Microglia M1/M2 Polarizations through Inhibiting NF-κB/MAPK p38 Signaling and Activating Neuron-BDNF-PI3K/AKT Pathways. *Mar. Drugs* **2021**, *19*, 587. [CrossRef]
16. Cao, J.; Dong, L.J.; Luo, J.L.; Zeng, F.N.; Hong, Z.X.; Liu, Y.Z.; Zhao, Y.B.; Xia, Z.Y.; Zuo, D.M.; Xu, L.; et al. Supplemental N-3 Polyunsaturated Fatty Acids Limit A1-Specific Astrocyte Polarization via Attenuating Mitochondrial Dysfunction in Ischemic Stroke in Mice. *Oxid. Med. Cell. Longev.* **2021**, *2021*, 5524705. [CrossRef]
17. Yu, J.Z.; Wang, J.; Sheridan, S.D.; Perlis, R.H.; Rasenick, M.M. N-3 polyunsaturated fatty acids promote astrocyte differentiation and neurotrophin production independent of cAMP in patient-derived neural stem cells. *Mol. Psychiatry* **2021**, *26*, 4605–4615. [CrossRef] [PubMed]
18. Walczewska, A.; Stępień, T.; Bewicz-Binkowska, D.; Zgórzyńska, E. The role of docosahexaenoic acid in neuronal function. *Postep. Hig. Med. Dosw.* **2011**, *65*, 314–327. [CrossRef] [PubMed]
19. Otsuka, R.; Tange, C.; Nishita, Y.; Kato, Y.; Imai, T.; Ando, F.; Shimokata, H. Serum docosahexaenoic and eicosapentaenoic acid and risk of cognitive decline over 10 years among elderly Japanese. *Eur. J. Clin. Nutr.* **2014**, *68*, 503–509. [CrossRef]
20. Li, M.Y.; Li, W.; Gao, Y.M.; Chen, Y.J.; Bai, D.; Weng, J.X.; Du, Y.; Ma, F.; Wang, X.Y.; Liu, H.; et al. Effect of folic acid combined with docosahexaenoic acid intervention on mild cognitive impairment in elderly: A randomized double-blind, placebo-controlled trial. *Eur. J. Nutr.* **2021**, *60*, 1795–1808. [CrossRef]
21. Bai, D.; Fan, J.T.; Li, M.Y.; Dong, C.X.; Gao, Y.M.; Fu, M.; Huang, G.W.; Liu, H. Effects of Folic Acid Combined with DHA Supplementation on Cognitive Function and Amyloid-β-Related Biomarkers in Older Adults with Mild Cognitive Impairment by a Randomized, Double Blind, Placebo-Controlled Trial. *J. Alzheimer's Dis.* **2021**, *81*, 155–167. [CrossRef]
22. Zussy, C.; John, R.; Urgin, T.; Otaegui, L.; Vigor, C.; Acar, N.; Canet, G.; Vitalis, M.; Morin, F.; Planel, E.; et al. Intranasal Administration of Nanovectorized Docosahexaenoic Acid (DHA) Improves Cognitive Function in Two Complementary Mouse Models of Alzheimer's Disease. *Antioxidants* **2022**, *11*, 838. [CrossRef]
23. Hsu, Y.M.; Yin, M.C. EPA or DHA enhanced oxidative stress and aging protein expression in brain of d-galactose treated mice. *BioMedicine* **2016**, *6*, 17. [CrossRef]
24. Fu, S.S.; Wen, M.; Zhao, Y.C.; Shi, H.H.; Wang, Y.M.; Xue, C.H.; Wei, Z.H.; Zhang, T.T. Short-term supplementation of EPA-enriched ethanolamine plasmalogen increases the level of DHA in the brain and liver of n-3 PUFA deficient mice in early life after weaning. *Food Funct.* **2022**, *13*, 1906–1920. [CrossRef] [PubMed]
25. Yalagala, P.C.R.; Sugasini, D.; Dasarathi, S.; Pahan, K.; Subbaiah, P.V. Dietary lysophosphatidylcholine-EPA enriches both EPA and DHA in the brain: Potential treatment for depression. *J. Lipid Res.* **2019**, *60*, 566–578. [CrossRef] [PubMed]
26. Kohman, R.A. Aging microglia: Relevance to cognition and neural plasticity. *Methods Mol. Biol.* **2012**, *934*, 193–218. [PubMed]
27. Yao, K.; Zhao, Y.F. Aging modulates microglia phenotypes in neuroinflammation of MPTP-PD mice. *Exp. Gerontol.* **2018**, *111*, 86–93. [CrossRef]
28. Wang, Q.Q.; Yao, H.M.; Liu, W.Y.; Ya, B.L.; Cheng, H.J.; Xing, Z.K.; Wu, Y.L. Microglia Polarization in Alzheimer's Disease: Mechanisms and a Potential Therapeutic Target. *Front. Aging Neurosci.* **2021**, *13*, 772717. [CrossRef]
29. D'avila, J.C.; Siqueira, L.D.; Mazeraud, A.; Azevedo, E.P.; Foguel, D.; Castro-Faria-Neto, H.C.; Sharshar, T.; Chrétien, F.; Bozza, F.A. Age-related cognitive impairment is associated with long-term neuroinflammation and oxidative stress in a mouse model of episodic systemic inflammation. *J. Neuroinflamm.* **2018**, *15*, 28. [CrossRef]
30. Guo, S.R.; Wang, H.; Yin, Y.F. Microglia Polarization from M1 to M2 in Neurodegenerative Diseases. *Front. Aging Neurosci.* **2022**, *14*, 815347. [CrossRef]
31. Gu, M.Q.; Li, Y.Y.; Tang, H.T.; Zhang, C.; Li, W.D.; Zhang, Y.P.; Li, Y.J.; Zhao, Y.T.; Song, C. Endogenous Omega (n)-3 Fatty Acids in Fat-1 Mice Attenuated Depression-Like Behavior, Imbalance between Microglial M1 and M2 Phenotypes, and Dysfunction of Neurotrophins Induced by Lipopolysaccharide Administration. *Nutrients* **2018**, *10*, 1351. [CrossRef]

32. Hjorth, E.; Zhu, M.; Toro, V.C.; Vedin, I.; Palmblad, J.; Cederholm, T.; Freund-Levi, Y.; Faxen-Irving, G.; Wahlund, L.O.; Basun, H.; et al. Omega-3 fatty acids enhance phagocytosis of Alzheimer's disease-related amyloid-β42 by human microglia and decrease inflammatory markers. *J. Alzheimer's Dis.* **2013**, *35*, 697–713. [CrossRef] [PubMed]
33. Li, P.; Song, C. Potential treatment of Parkinson's disease with omega-3 polyunsaturated fatty acids. *Nutr. Neurosci.* **2022**, *25*, 180–191. [CrossRef] [PubMed]
34. Kawano, A.; Ariyoshi, W.; Yoshioka, Y.; Hikiji, H.; Nishihara, T.; Okinaga, T. Docosahexaenoic acid enhances M2 macrophage polarization via the p38 signaling pathway and autophagy. *J. Cell. Biochem.* **2019**, *120*, 12604–12617. [CrossRef] [PubMed]
35. Harvey, L.D.; Yin, Y.; Attarwala, I.Y.; Begum, G.; Deng, J.; Yan, H.Q.; Dixon, C.E.; Sun, D. Administration of DHA Reduces Endoplasmic Reticulum Stress-Associated Inflammation and Alters Microglial or Macrophage Activation in Traumatic Brain Injury. *ASN Neuro* **2015**, *7*, 1759091415618969. [CrossRef]
36. Liu, L.R.; Liu, J.C.; Bao, J.S.; Bai, Q.Q.; Wang, G.Q. Interaction of Microglia and Astrocytes in the Neurovascular Unit. *Front. Immunol.* **2020**, *11*, 1024. [CrossRef]
37. Nichols, N.R.; Day, J.R.; Laping, N.J.; Johnson, S.A.; Finch, C.E. GFAP mRNA increases with age in rat and human brain. *Neurobiol. Aging* **1993**, *14*, 421–429. [CrossRef]
38. Wruck, W.; Adjaye, J. Meta-analysis of human prefrontal cortex reveals activation of GFAP and decline of synaptic transmission in the aging brain. *Acta Neuropathol. Commun.* **2020**, *8*, 26. [CrossRef]
39. Chatterjee, P.; Pedrini, S.; Stoops, E.; Goozee, K.; Villemagne, V.L.; Asih, P.R.; Verberk, I.M.W.; Dave, P.; Taddei, K.; Sohrabi, H.R.; et al. Plasma glial fibrillary acidic protein is elevated in cognitively normal older adults at risk of Alzheimer's disease. *Transl. Psychiatry* **2021**, *11*, 27. [CrossRef]
40. Traub, J.; Otto, M.; Sell, R.; Homola, G.A.; Steinacker, P.; Oeckl, P.; Morbach, C.; Frantz, S.; Pham, M.; Störk, S.; et al. Serum glial fibrillary acidic protein indicates memory impairment in patients with chronic heart failure. *Heart Fail.* **2022**, *9*, 2626–2634. [CrossRef]
41. Ayala-Guerrero, L.; García-delaTorre, P.; Sánchez-García, S.; Guzmán-Ramos, K. Serum Levels of Glial Fibrillary Acidic Protein Association with Cognitive Impairment and Type 2 Diabetes. *Arch. Med. Res.* **2022**, *53*, 501–507. [CrossRef] [PubMed]
42. Heras-Sandoval, D.; Pedraza-Chaverri, J.; Pérez-Rojas, J.M. Role of docosahexaenoic acid in the modulation of glial cells in Alzheimer's disease. *J. Neuroinflamm.* **2016**, *13*, 61. [CrossRef] [PubMed]
43. Tripathi, S.; Kushwaha, R.; Mishra, J.; Gupta, M.K.; Kumar, H.; Sanyal, S.; Singh, D.; Sanyal, S.; Sahasrabuddhe, A.A.; Kamthan, M.; et al. Docosahexaenoic acid up-regulates both PI3K/AKT-dependent FABP7–PPARγ interaction and MKP3 that enhance GFAP in developing rat brain astrocytes. *J. Neurochem.* **2017**, *140*, 96–113. [CrossRef] [PubMed]
44. Manzhulo, I.V.; Ogurtsova, O.S.; Kipryushina, Y.O.; Latyshev, N.A.; Kasyanov, S.P.; Dyuizen, I.V.; Tyrtyshnaia, A.A. Neuron-astrocyte interactions in spinal cord dorsal horn in neuropathic pain development and docosahexaenoic acid therapy. *J. Neuroimmunol.* **2016**, *15*, 90–97. [CrossRef]
45. Oh, H.; Lewis, D.A.; Sibille, E. The Role of BDNF in Age-Dependent Changes of Excitatory and Inhibitory Synaptic Markers in the Human Prefrontal Cortex. *Neuropsychopharmacology* **2016**, *41*, 3080–3091. [CrossRef]
46. Wu, S.Y.; Pan, B.S.; Tsai, S.F.; Chiang, Y.T.; Huang, B.M.; Mo, F.E.; Kuo, Y.M. BDNF reverses aging-related microglial activation. *J. Neuroinflamm.* **2020**, *17*, 210. [CrossRef]
47. Lapchak, P.A.; Araujo, D.M.; Beck, K.D.; Finch, C.E.; Johnson, S.A.; Hefti, F. BDNF and trkB mRNA expression in the hippocampal formation of aging rats. *Neurobiol. Aging* **1993**, *14*, 121–126. [CrossRef]
48. Buhusi, M.; Etheredge, C.; Granholm, A.C.; Buhusi, C.V. Increased Hippocampal ProBDNF Contributes to Memory Impairments in Aged Mice. *Front. Aging Neurosci.* **2017**, *9*, 284. [CrossRef]
49. Minogue, A.M.; Lynch, A.M.; Loane, D.J.; Herron, C.E.; Lynch, M.A. Modulation of amyloid-beta-induced and age-associated changes in rat hippocampus by eicosapentaenoic acid. *J. Neurochem.* **2007**, *103*, 914–926. [CrossRef]
50. Trofimiuk, E.; Braszko, J.J. Long-term administration of cod liver oil ameliorates cognitive impairment induced by chronic stress in rats. *Lipids* **2011**, *46*, 417–423. [CrossRef]
51. Zhu, W.; Chi, N.; Zou, P.; Chen, H.; Tang, G.; Zhao, W. Effect of docosahexaenoic acid on traumatic brain injury in rats. *Exp. Ther. Med.* **2017**, *14*, 4411–4416. [CrossRef] [PubMed]
52. Bao, X.H.; Qi, C.X.; Liu, T.T.; Zheng, X.Y. Information transmission in mPFC-BLA network during exploratory behavior in the open field. *Behav. Brain Res.* **2021**, *414*, 113483. [CrossRef] [PubMed]
53. Folch, J.; Lees, M.; Sloane Stanley, G.H. A simple method for the isolation and purification of total lipides from animal tissues. *J. Biol. Chem.* **1957**, *226*, 497–509. [CrossRef] [PubMed]

marine drugs

Article

Novel Antioxidant Peptides Identified from *Arthrospira platensis* Hydrolysates Prepared by a Marine Bacterium *Pseudoalteromonas* sp. JS4-1 Extracellular Protease

Congling Liu [1,†], Gong Chen [1,†], Hailian Rao [1], Xun Xiao [1], Yidan Chen [1], Cuiling Wu [2], Fei Bian [3,*] and Hailun He [1,*]

1 State Key Laboratory of Medical Genetics, School of Life Sciences, Central South University, Changsha 410013, China
2 Department of Biochemistry, Changzhi Medical College, Changzhi 046000, China
3 Institute of Crop Germplasm Resources, Shandong Academy of Agricultural Sciences, Jinan 250100, China
* Correspondence: bxf.9@163.com (F.B.); helenhe@csu.edu.cn (H.H.); Tel.: +86-531-6665-9499 (F.B.); +86-0731-8265-0230 (H.H.)
† These authors contributed equally to this work.

Abstract: Crude enzymes produced by a marine bacterium *Pseudoalteromonas* sp. JS4-1 were used to hydrolyze phycobiliprotein. Enzymatic productions showed good performance on DPPH radical and hydroxyl radical scavenging activities (45.14 ± 0.43% and 65.11 ± 2.64%, respectively), especially small peptides with MWCO <3 kDa. Small peptides were fractioned to four fractions using size-exclusion chromatography and the second fraction (F2) had the highest activity in hydroxyl radical scavenging ability (62.61 ± 5.80%). The fraction F1 and F2 both exhibited good antioxidant activities in oxidative stress models in HUVECs and HaCaT cells. Among them, F2 could upregulate the activities of SOD and GSH-Px and reduce the lipid peroxidation degree to scavenge the ROS to protect *Caenorhabditis elegans* under adversity. Then, 25 peptides total were identified from F2 by LC-MS/MS, and the peptide with the new sequence of INSSDVQGKY as the most significant component was synthetized and the ORAC assay and cellular ROS scavenging assay both illustrated its excellent antioxidant property.

Keywords: phycobiliprotein; antioxidant peptide; bacterial extracellular protease; enzymatic hydrolyzation; *C. elegans*

Citation: Liu, C.; Chen, G.; Rao, H.; Xiao, X.; Chen, Y.; Wu, C.; Bian, F.; He, H. Novel Antioxidant Peptides Identified from *Arthrospira platensis* Hydrolysates Prepared by a Marine Bacterium *Pseudoalteromonas* sp. JS4-1 Extracellular Protease. *Mar. Drugs* **2023**, *21*, 133. https://doi.org/10.3390/md21020133

Academic Editors: Elena Talero and Javier Ávila-Román

Received: 30 December 2022
Revised: 16 February 2023
Accepted: 16 February 2023
Published: 20 February 2023

1. Introduction

The microalgae *Arthrospira* belongs to the cyanobacteria division, the cyanophycean class, and the Oscillatoriaceae family. *Arthrospira* sp. is a photosynthetic cyanobacterium that has recently grown in value as a nutritional supplement and food additive due to its high protein content of around 60–70% in their biomass and various valuable natural compounds, such as pigments, β-carotenes, polysaccharides and peptides. In addition, *Arthrospira* sp. has a wide range of uses in the pharmaceutical, cosmetic, and nutritional industries. Because of its high protein content and nutritive value of amino acid composition, *Arthrospira* sp. was evaluated as a potential meat substitute and has received increased attention [1]. Phycobiliprotein is a class of proteins with antioxidant, antitumor and anti-inflammatory properties, but the heat sensitivity and allergenicity limit its wider applications. Phycobiliprotein is highly abundant in *Arthrospira*, accounting for about 20% of its dry weight [2]. Using proteases to hydrolyze the phycobiliprotein to produce antioxidant peptides is an efficient way to reach higher bioactive potential and promote high value utilization of marine proteins while avoiding its natural limitations. Moreover, the proteins of microalgae can be converted by enzymatic hydrolysis into value-added products with better functional properties such as antioxidant peptides (APs), anti-hypertension peptides, and antibacterial peptides, etc. Therefore, more and more researchers are interested in obtaining effective bioactive peptides (BPs) from *Arthrospira* sp. hydrolysates.

Enzymatic hydrolysis of the dietary proteins generates BPs, which are short chains of 2–15 amino acids residues, can enhance nutritional value, safety, bioactive function and reduce the allergenicity [3,4]. Numerous studies have demonstrated that BPs with antioxidant activity produced by proteolysis can achieve antioxidation by scavenging free radicals, chelating transition metals, and enhancing the activity of endogenous antioxidant enzymes such as catalase (CAT), superoxide dismutase (SOD), and glutathione peroxidase (GSH-Px). Redox homeostasis (balance) is an important cellular process that plays a vital role in maintaining the normal physiological steady state of the human body. A disturbance of balance between oxidants and antioxidants results in oxidative stress. Reactive oxygen species (ROS) are molecules derived from aerobic cellular metabolism; in the oxidative stress condition excessive production of ROS can disrupt the intracellular redox balance and cause numerous diseases such as cancer, diabetes, atherosclerosis, cardiovascular and neurodegenerative [4]. A growing number of studies have demonstrated that APs from protease hydrolysates can effectively scavenge free radicals or inhibit the production of ROS. Meanwhile, food-derived APs also performed nontoxicity, are low cost, absorb efficiency advantages, and were considered as potential substitutes for commercial synthetic antioxidants [5,6].

Since different proteases have specific hydrolyzed performance on protein substrate and can produce diverse BPs with different length, sequence, and composition, these hydrolyzes may exhibit various hydrophobicity and functionalities [7,8]. Therefore, it is necessary to find a new enzymes resource to hydrolyze *Arthrospira platensis* (*A. platensis*) and produce novel APs with high antioxidant activity and bioactive properties. Our previous study suggested using extracellular proteases secreted by a marine bacterium *Pseudoalteromonas* sp. JS4-1 to hydrolyze *chlorella* can obtain hydrolysates with high antioxidant activity [9]. In this study, proteases from *Pseudoalteromonas* sp. JS4-1 were used to hydrolyze *A. platensis* proteins and produced diverse Aps; a series of biochemical assays, cell models, and animal models were applied for testing the bioactivity of the APs.

2. Results

2.1. Screening of Proteases for Enzymatic Hydrolysis

Pseudoalteromonas is a strictly marine genus and able to produce extracellular enzymes, which have good advantages in degrading natural marine proteins [10]. In order to evaluate which enzyme is more suitable for hydrolyzing *Arthrospira* sp. proteins, the extracellular enzymes from eight strains named *Pseudoalteromonas* sp. B27-3, *Pseudoalteromonas* sp. B47-6, *Pseudoalteromonas* sp. B62-3, *Pseudoalteromonas* sp. WH05-1, *Pseudoalteromonas* sp. WH06-2, *Pseudoalteromonas* sp. WH16-2, *Pseudoalteromonas* sp. JS4-1 and *Pseudoalteromonas* sp. ZB23-2 were extracted and screened by zymography. The enzymatic hydrolysates were identified by SDS-PAGE and their DPPH RSA and OH RSA activities were also detected. The zymography showed with the exception of *Pseudoalteromonas* sp. WH06-2 and *Pseudoalteromonas* sp. WH16-2, all the other strains can secrete a variety of proteases, indicating that most marine bacteria could produce extracellular proteases with a variety of cleavage sites, and suitable for enzymatic hydrolysis (Figure 1A). Proteases from *Pseudoalteromonas* sp. B27-3 and *Pseudoalteromonas* sp. JS4-1 have stronger hydrolysis activity than other bacterial proteases toward *Arthrospira* sp. proteins. SDS-PAGE showed that the hydrolysates existed in smaller fragments compared with the control protein, and almost no protein band left on the gel, while the hydrolysates produced by other bacterial proteases still left visible bands on the gel (Figure 1B). The hydrolysates of *Pseudoalteromonas* sp. JS4-1 proteases have the highest DPPH RSA (45.14 ± 0.43%) and OH RSA (65.11 ± 2.64%) than other productions (Figure 1C,D). Therefore, proteases from *Pseudoalteromonas* sp. JS4-1 were selected as a candidate for *Arthrospira* sp. proteins hydrolyzing.

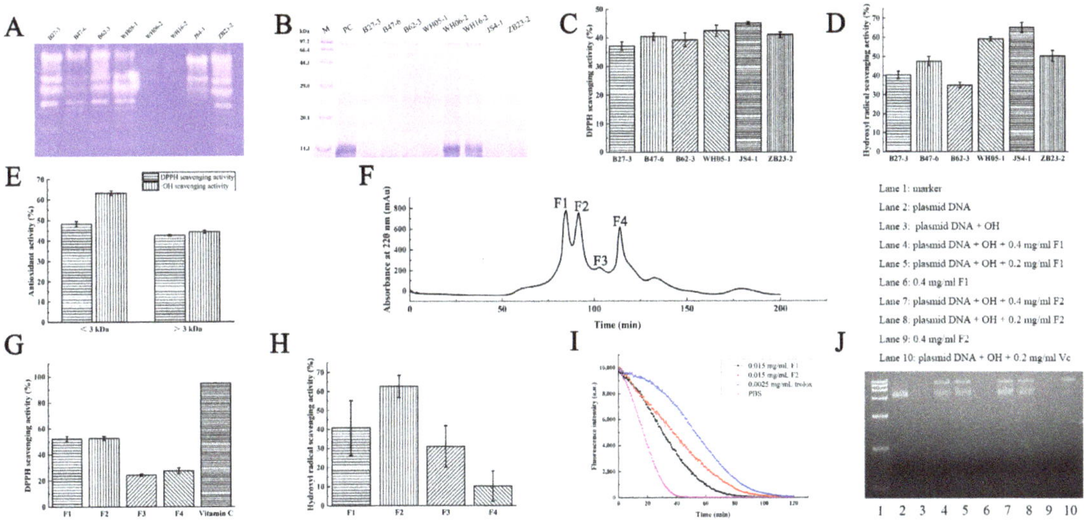

Figure 1. Antioxidant activities of phycobiliprotein hydrolysis products with JS4-1. (**A**) Comparison of catalytic ability on casein of proteases from *Pseudoalteromonas* sp. B27-3, *Pseudoalteromonas* sp. B47-6, *Pseudoalteromonas* sp. B62-3, *Pseudoalteromonas* sp. WH05-1, *Pseudoalteromonas* sp. WH06-2, *Pseudoalteromonas* sp. WH16-2, *Pseudoalteromonas* sp. JS4-1 and *Pseudoalteromonas* sp. ZB23-2. (**B**) SDS-PAGE electrophoresis of phycobiliprotein and hydrolysates of different proteases (Line M: protein marker, Line PC: phycobiliprotein). (**C**) DPPH radical and (**D**) hydroxyl radical scavenging activity of phycobiliprotein hydrolysates of different proteases. (**E**) Antioxidant activities of the ultrafiltration fraction of phycobiliprotein hydrolysis products treated with optimized hydrolysis condition. Data were expressed as mean ± SD (n = 3). (**F**) Size-exclusion liquid chromatography of the <3 kDa ultrafiltration fractions on a Sephadex LH-20 column. The F1–F4 represent purified fractions. (**G**) DPPH radical scavenging activity of F1-F4 compared with Vitamin C. (**H**) Hydroxyl radical scavenging activity of F1–F4. (**I**) Peroxyl radical scavenging activity (ORAC) of F1 and F2 compared with PBS and Trolox. (**J**) Protective effects of F1 and F2 on hydroxyl radical-induced oxidation of plasmid DNA compared with Vitamin C. Data were expressed as mean ± SD (n = 3).

2.2. Optimization of Enzymatic Hydrolyzing Parameters

Enzymatic hydrolyzation was optimized using the one-variable-at-a-time (OVAT) approach by changing one parameter while keeping other parameters constant. First, the hydrolyzing was performed at 50 °C from 1 h to 7 h with 1 h at intervals, at a E: S ratio of 1:10, to estimate the optimal hydrolyzing time. Data showed that the hydrolysis degree increased along with the hydrolyzing process, especially in the initial three hours. After five hours hydrolyzation, the hydrolysis degree was not improved significantly. DPPH RSA as an evaluation index suggested that when the E/S ratio ranged from 1:6 to 1:10, the DPPH RSA of the hydrolytes exhibited no significant difference. While the E/S ratio adjusted to 1:12 and 1:14, the DPPH RSA of the hydrolytes dropped 6.72% and 15.95%, respectively, indicating that the substrate exceeded the catalytic capacity of the enzymes. When the hydrolyzing temperature increased, the DPPH RSA improved and reached the maximal activity at 50 °C, but when the temperature exceeded >50 °C the DPPH RSA decreased, probably due to the enzymes being inactivated at higher temperatures. After OVAT optimization, when E:S ratio was 1:10 and enzymatic hydrolyzing performed at 50 °C for 4 h, the DPPH RSA and the OH RSA of the hydrolysates achieved the maximal level of 48.51 ± 2.42% and 67.16 ± 3.21%, respectively. The ninhydrin colorimetry was used to determine the hydrolysis degree, and data showed that about 0.67 ± 0.02 mM of •NH_2 was released after hydrolyzing reaction.

In order to evaluate the free RSA of small peptides, the hydrolysates were separated into 2 parts by ultrafiltration. The DPPH RSA of < 3 kDa and > 3 kDa parts were 48.26 ± 1.34% and 42.60 ± 0.25%, respectively. The OH RSA of < 3 kDa and > 3 kDa parts were 63.30 ± 1.11% and 44.30 ± 0.68%, respectively (Figure 1E). It has been proven that the small peptides (<3 kDa) contained more antioxidant components and were more easy to exert biological effects [11]. The bioactive peptides with small molecular weight have broad application potential in food, medicine, cosmetics and other industries due to their good absorbability and fewer side effects besides their higher bioactivity [12].

2.3. Antioxidant Activities of Size-Exclusion Liquid Chromatography Fractions of Phycobiliprotein Hydrolysis Products

In order to investigate the antioxidant components of the small peptides, the peptides with MWCO <3 kDa were further separated using size-exclusion liquid chromatography, and in total four fractions defined as F1 to F4 were obtained (Figure 1F). By measuring the DPPH RSA (F1 to F4 were 52.06 ± 2.13%, 52.56 ± 1.38%, 24.43 ± 0.85% and 27.58 ± 2.24%, respectively; Figure 1G) and the OH RSA (F1 to F4 were 40.75 ± 14.35%, 62.61 ± 5.80%, 30.86 ± 10.90%, and 10.21 ± 7.91%, respectively; Figure 1H), F1 and F2 showed higher antioxidant activity than F3 and F4. We could obtain about 40 mg of F1 and 40 mg of F2 lyophilized powder after the size-exclusion liquid chromatography from 1 g of *A. Platensis* powder.

Then, the ORAC assay was performed to compare the antioxidant activity of F1 and F2. At peptide concentration of 0.015 mg/mL, both F1 and F2 could significantly slow down the decrease rate of the fluorescence decay compared with PBS. The Trolox equivalent antioxidant capacity of F1 and F2 were 0.27 ± 0.02 mmol TE/g and 0.41 ± 0.04 mmol TE/g, respectively (Figure 1I). F2 exhibited higher antioxidant activity than F1.

The hydroxyl radicals could open the circular of supercoiled DNA (SC DNA) and make it become partly opened circular DNA (OC DNA), which can slow down the DNA mobility on agarose gel. Comparing with the no damaged plasmid DNA pET-22b (Figure 1J, lane 2), when the plasmid DNA has no protection, it can be completely degraded by hydroxyl radicals into small fragments and no band observed on the gel (Figure 1J, lane 3). An amount of 0.2 mg/mL of Vitamin C can protect the plasmid DNA from oxidative damage, but the protection is limited and the OC DNA is the major form retained on the gel (Figure 1J, lane 10). F1 and F2 concentrations range from 0.2 mg/mL to 0.4 mg/mL; all can significantly protect plasmid DNA from oxidative damage and these protections were far better than Vitamin C, which make the SC DNA and OC DNA co-existing on the gel (Figure 1J, lane 4, 5, 7, 8). So, this study proved that both F1 and F2 have strong antioxidant activity to protect DNA from oxidative stress.

2.4. Antioxidant Activities of F1 and F2 at Cellular Levels

The intracellular oxidized cell model is a common method that is used to assess the antioxidative capacity of compounds. Vascular endothelial cell injury is associated with several factors, among them oxidative stress is an important cause of cardiovascular disease. APs are able to protect vascular endothelial cell function which is a key point in the prevention and treatment for cardiovascular diseases. The skin is the first immune defense of the body and susceptible to a variety of physical and chemical stimuli. Protecting skin cells could be an effective strategy to prevent and treat dermatosis. Therefore, we chose HUVECs and HaCaT cells as cellular models to investigate the intracellular antioxidant effects of F1 and F2 [13–16].

MTT assay indicated that both F1 and F2 had no significant cytotoxicity on HUVECs and HaCaT cells; moreover, they can obviously increase the cell viability at a concentration range of 100–500 μg/mL (Figure 2A,B and Figure 3A,B).

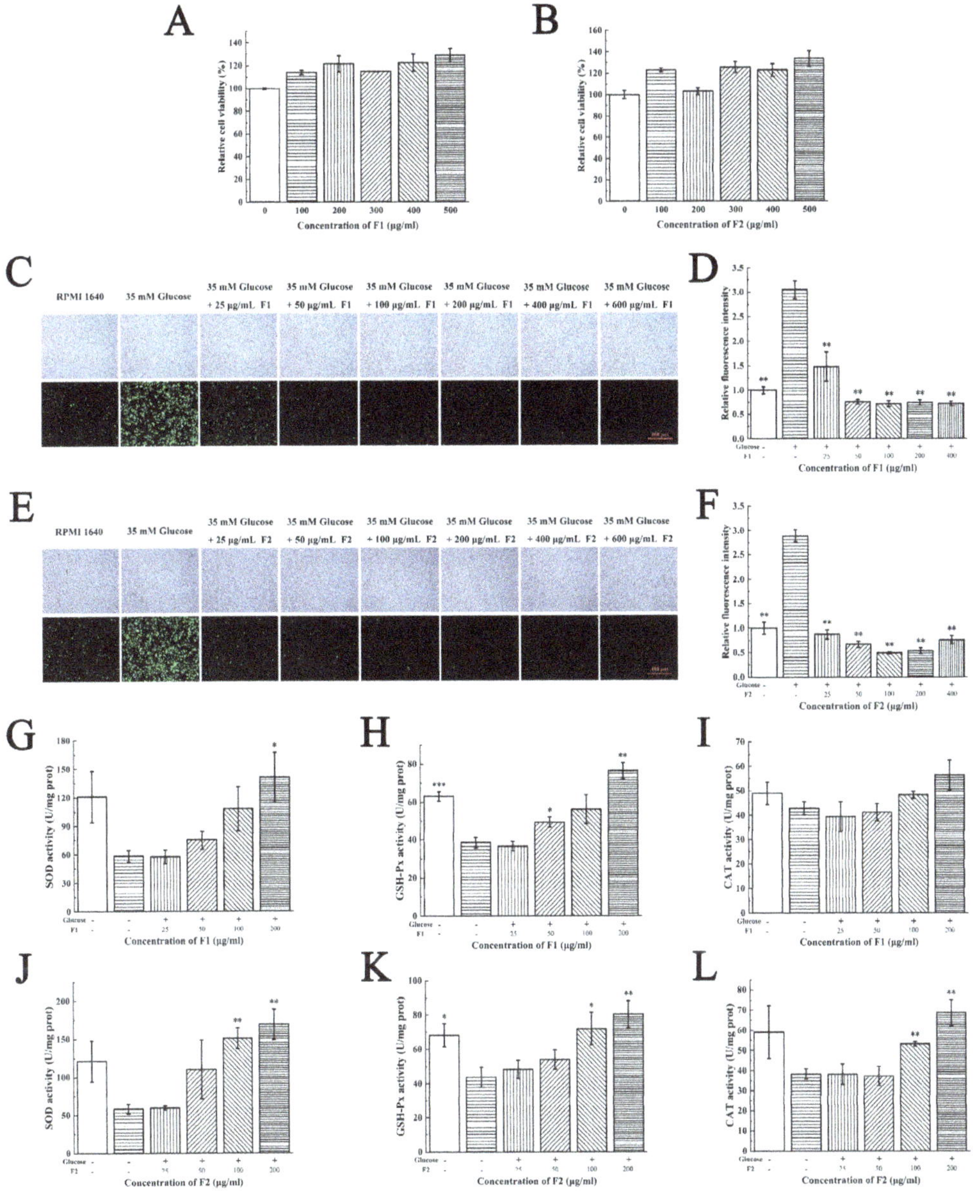

Figure 2. Antioxidant activities of F1 and F2 on HUVEC cells. The effects of different concentration of (**A**) F1 and (**B**) F2 on cell viability were measured by MTT assay. ROS scavenging capacities of (**C**) F1 and (**E**) F2 on HUVEC cells indicated as a green DCFH-DA fluorescence. Images were taken using fluorescence microscope (magnification, 10×). (**D**,**F**) Statistical analysis was performed to quantify relative fluorescence density accordingly (Student *t*-test). (**G**,**J**) Superoxide dismutase (SOD), (**H**,**K**) glutathione peroxidase (GSH-Px) and (**I**,**L**) catalase (CAT) activities in HUVEC cells were detected after treatment with F1 and F2, respectively. The Normality of data was analyzed by Shapiro–Wilk test. Data were expressed as mean ± SD (n = 3). * represents $p < 0.05$, ** represents $p < 0.01$ and *** represents $p < 0.001$, compared with glucose group (Student *t*-test).

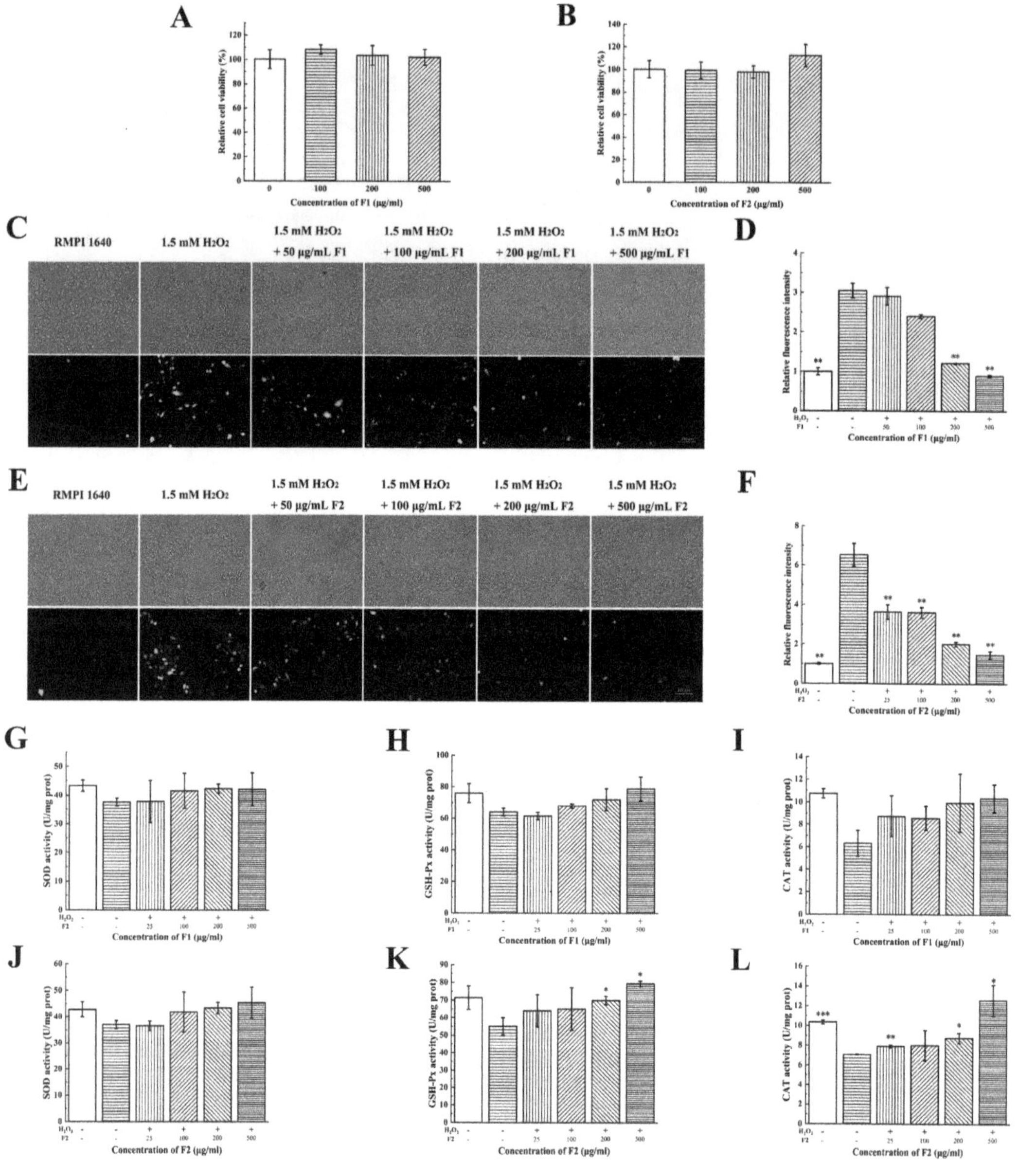

Figure 3. Antioxidant activities of F1 and F2 on HaCaT cells. The effects of different concentration of (**A**) F1 and (**B**) F2 on cell viability were measured by MTT assay. ROS scavenging capacities of (**C**) F1 and (**E**) F2 on HaCaT cells indicated as green DCFH-DA fluorescence. Images were taken using fluorescence microscope (magnification, 10×). (**D**,**F**) Statistical analysis was performed to quantify relative fluorescence density accordingly (Student *t*-test). (**G**,**J**) Superoxide dismutase (SOD), (**H**,**K**) glutathione peroxidase (GSH-Px) and (**I**,**L**) catalase (CAT) activities in HaCaT cells were detected after treatment with F1 and F2, respectively. The Normality of data was analyzed by Shapiro–Wilk test. Data were expressed as mean ± SD (n = 3). * represents $p < 0.05$, ** represents $p < 0.01$ and *** represents $p < 0.001$, compared with H_2O_2 group (**G**,**J** Mann–Whitney test; **H**,**I**,**K**,**L** Student *t*-test).

An amount of 35 mM of glucose was added to the HUVECs and 1.5 mM H_2O_2 was used in the HaCaT cells. Cells treated with glucose and H_2O_2 exhibited higher fluorescence intensity than control groups, indicating the oxidative stress models have been successfully established (Figure 2C,E and Figure 3C,E). The intracellular ROS were labelled by DCFH-DA probe. Cells treated with low concentration of F1 and F2 (25 μg/mL) displayed less fluorescence intensity than the glucose group in HUVECs, suggesting the fractions were able to scavenge intracellular ROS to protect HUVECs from hyperglycemia-induced oxidative stress (Figure 2D,F). Cells treated with a high concentration of F1 (100 μg/mL) and a low concentration of F2 (50 μg/mL) displayed less fluorescence intensity than the H_2O_2 group in HaCaT cells, suggesting that the fractions were able to scavenge intracellular ROS to protect HaCaT cells from H_2O_2 induced oxidative stress (Figure 3D,F). The antioxidant activities of peptides on a cellular level in this study were close to those reported extracts of plants [17–19], which suggested the peptides in our study may also have great potential applications.

To further investigate the underlying mechanisms of fractions scavenging intracellular ROS, we detected the activities of antioxidant enzymes such as SOD, GSH-Px and CAT after being treated with fractions due to their role in scavenging ROS and suppressing oxidative stress [20,21]. The glucose and H_2O_2 reduced the activities of SOD, GSH-Px and CAT in HUVECs and HaCaT cells. The administration of F1 and F2 dose-dependently increased the activities of these enzymes at a concentration range of 25–200 μg/mL (Figure 2G–L and Figure 3G–L).

2.5. *Effects of F2 on Resistance to Oxidative Stress in C. elegans*

Caenorhabditis elegans (*C. elegans*) have many highly conserved genes and signaling pathways involved in the regulation of aging and oxidative stress [22,23]. The genes and pathways in aging, oxidative stress and inflammation between *C. elegans* and human are highly homologous [24]. *C. elegans* as an animal model is widely used for anti-aging and anti-oxidative studies.

In this study, 0.5 mg/mL of F1 and F2 had no toxicity and no significant life-extending effect on N2 (Figure 4A). The in vivo ROS assay showed that 0.5 mg/mL of F1 had no significant ROS scavenging effects, while F2 could scavenge ROS at low concentration (0.2 mg/mL, Figure 4E,F). The fluorescence accumulation assay of 80 nematodes showed similar results (Figure 4G).

Then, we speculate whether the APs can enhance the stress resistance of N2. Oxidizing agent H_2O_2 (10 mM), pesticide paraquat (10 mM) and heat (35 °C) were chosen as stress conditions. Considering F2 has higher anti-oxidative activity than F1, in this study N2 were treated with F2 and then exposed to stress conditions. Data showed that the F2 can extend the stressed N2 survival time in all stress conditions, and the survival rate was positively correlated with the addition of F2. An amount of 0.2 mg/mL of F2 increased 35% survival rate after 7 h of H_2O_2 stress, 50% after 120 h of paraquat stress and 24% under heat stress (Figure 4B–D). Although F2 did not help the *C. elegans* live longer, it really improved the resistant ability of *C. elegans* obviously to the oxidation, pesticide and heat stress.

We then investigated the MDA levels and activities of antioxidant enzymes in nematodes. The MDA level is an important biomarker of lipid peroxidation [25]. F2 could significantly reduce the MDA levels in nematodes (Figure 4H). As shown in Figure 4I–K, F2 could have dose-dependently increased the activities of SOD and GSH-Px, while it had no effects on CAT. These results suggested that F2 could upregulate the activities of SOD and GSH-Px and reduce the lipid peroxidation degree to scavenge the ROS in the organisms. Activating these antioxidant enzymes contributes to maintaining the redox balance, delaying aging process and prolonging lifetime [26]. Antioxidant enzymes with high activity are able to reduce ROS levels, maintain normal physiological steady state and prevent numerous diseases such as cancer, diabetes, neurodegenerative diseases and so on [4].

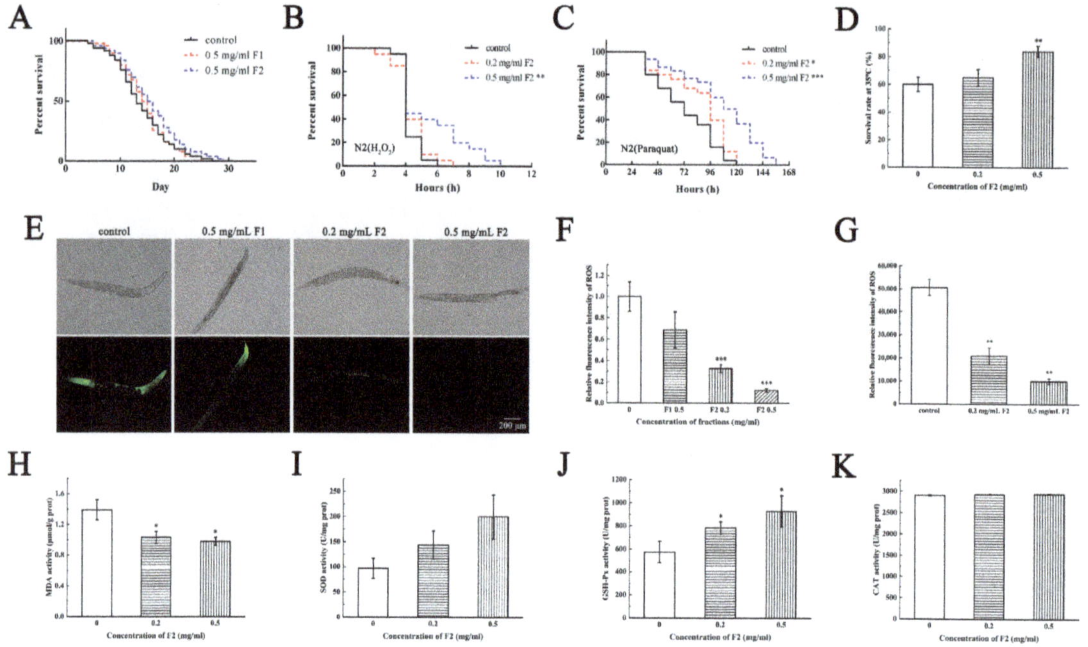

Figure 4. Effects of F2 on resistance to oxidative stress in Caenorhabditis elegans. (**A**) Survival curves of C. elegans treated with F1 and F2. Survival curves of C. elegans pre-treated with F2 at given concentrations for 3 days and then exposed to (**B**) 10 mM H_2O_2 and (**C**) 10 mM paraquat. * represents $p < 0.05$, ** represents $p < 0.01$ and *** represents $p < 0.001$, compared with H_2O_2 and paraquat group (Student *t*-test). (**D**) Survival rate of C. elegans pre-treated with F2 at given concentrations for 3 days and then cultured at 35 °C for 10 h. (**E**) Antioxidant activities of F2 on *C. elegans* were detected by DCFH-DA probe. Images were taken using fluorescence microscope (magnification, 5×). (**F**) Statistical analysis was performed to quantify relative fluorescence density (Student *t*-test). (**G**) The ROS content in *C. elegans* was determined based on the average integrated area of the fluorescene intensity of probes during 6 h intervals. (**H**) MDA levels, (**I**) superoxide dismutase (SOD), (**J**) glutathione peroxidase (GSH-Px) and (**K**) catalase (CAT) activities in C. elegans were detected after treated with F2 for 3 days. The Normality of data was analyzed by Shapiro–Wilk test. Data were expressed as mean ± SD (n = 3). * represents $p < 0.05$, ** represents $p < 0.01$ and *** represents $p < 0.001$, compared with blank control group (Student *t*-test).

2.6. Antioxidant Activities of Synthesized Peptide INSSDVQGKY

Since F2 has the highest antioxidant activity in all fractions, its components were investigated using LC-MS/MS. The total ion chromatography of F2 was shown in Figure S1. A total of 25 peptides with relative abundance >1% were identified, and their sequence and abundance were listed in Table 1. Among them, one peptide with the sequence of INSSDVQGKY accounting for 17.83% was considered as the major component of F2. The MS/MS of INSSDVQGKY was shown in Figure S2.

Table 1. Peptides identified through liquid chromatography mass spectrum (> 1%).

Amino Acid Sequence	Mass (Da)	Source (Position)	Counts (%)
INSSDVQGKY	1109.535	Allophycocyanin beta-subunit [*Arthrospira platensis* qy3] (9–18)	17.83
VGGSVPREY	962.482	Elongation factor G [*Arthrospira platensis*] (531–539)	9.12
SSYPNRPVP	1015.509	Photosystem II lipoprotein Psb27 [*Arthrospira platensis* C1] (102–110)	8.81
GIFDTMNH	933.401	ATP-dependent Clp protease proteolytic subunit [*Arthrospira platensis* C1] (74–81)	6.91
YTPDYTPK	983.460	Ribulose bisphosphate carboxylase large chain [*Arthrospira platensis* C1] (26–33)	3.34
IDPSHGTGF	929.424	Phospho-2-dehydro-3-deoxyheptonate aldolase [*Arthrospira platensis*] (277–285)	3.33
TPEPKPEPKPEPKPEP	1795.936	CAAD domain-containing protein [*Arthrospira platensis* C1] (52–67)	3.25
YEQMPEPKY	1183.522	NAD(P)H-quinone oxidoreductase subunit K [*Arthrospira platensis* C1] (105–113)	3.14
DVDWSDYQKQ	1282.547	Ferredoxin-NADP reductase [*Arthrospira platensis*] (380–389)	2.78
VGGSVPKEY	934.476	Elongation factor G [*Arthrospira platensis* C1] (496–504)	2.76
TDVPANHPY	1012.461	S-layer domain protein [*Arthrospira platensis* C1] (283–291)	2.42
QGTLEKYI	950.507	Putative adenylate cyclase [*Arthrospira platensis* C1] (375–382)	2.40
AGIDEINRT	987.499	Phycocyanin alpha-subunit [*Arthrospira platensis* qy3] (113–121)	2.32
SESPNLILMDIQMP	1586.768	Multi-sensor hybrid histidine kinase [*Arthrospira platensis* C1] (1700–1713)	2.32
APYDESEVVFH	1291.572	Rod linker polypeptide CpcI [*Arthrospira platensis*] (107–117)	2.20
IDTIPTGGK	900.492	50S ribosomal protein L13 [*Arthrospira platensis* C1] (144–152)	1.96
LIAGGTGPM	815.421	Phycocyanin alpha-subunit [*Arthrospira platensis* qy3] (99–107)	1.92
DVDWSDYQK	1154.488	Ferredoxin-NADP reductase [*Arthrospira platensis*] (380–388)	1.90
LPEEPMTGK	1000.490	Transcriptional regulator AbrB family [*Arthrospira platensis* C1] (9–17)	1.80
VINSSDVQGKY	1208.604	Allophycocyanin beta-subunit [*Arthrospira platensis* qy3] (8–18)	1.68
LVTQQPLGGKAQ	1238.698	DNA-directed RNA polymerase subunit beta [*Arthrospira platensis*] (992–1003)	1.52
KELTKKSPNSP	1227.682	NYN domain-containing protein [*Arthrospira platensis*] (398–408)	1.37
IETKEIPVPT	1125.628	Assimilatory sulfite reductase (ferredoxin) [*Arthrospira platensis* C1] (592–601)	1.34
VDYQEQPREY	1325.589	Major membrane protein I [*Arthrospira platensis* C1] (81–90)	1.30
GANYEDEWK	1110.462	Transketolase [*Arthrospira platensis*] (308–316)	1.30

Sequence alignment showed that the peptide INSSDVQGKY was a part of a beta-subunit of allophycocyanin. This was a novel AP first be sequenced and proved with antioxidant activity. The structure of INSSDVQGKY was predicted and it was a linear peptide with a little helix (Figure 5A). In order to validate the antioxidant function of this peptide, it was synthetized. ORAC assay of the synthetized peptide was 0.32 ± 0.03 mmol TE/g and slightly lower than F2 (0.41 ± 0.04 mmol TE/g) (Figure 5B). The peptide had no significant cytotoxicity on HaCaT cells within 400 μg/mL (Figure 5C). The peptide dose-dependently reduced the intracellular ROS levels at a concentration range from 100 μg/mL to 400 μg/mL (Figure 5D,E). This peptide is therefore hypothesized to be the major antioxidant component of F2. The antioxidant activity of INSSDVQGKY is close to peptide PNN which was obtained from hydrolysis of phycobiliprotein by protease K [27] and may have potential applications in further study.

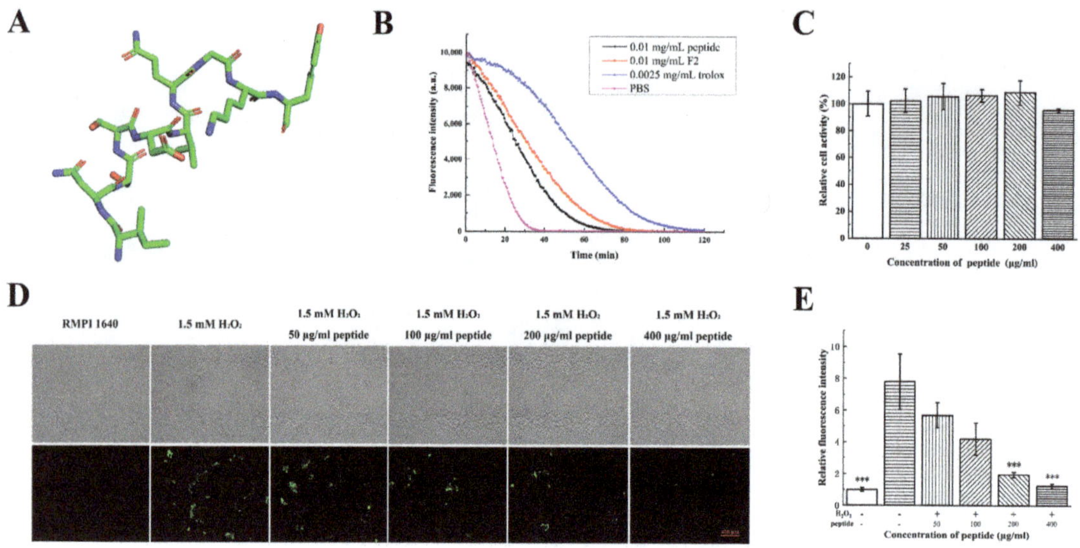

Figure 5. Antioxidant activities of synthesized peptide INSSDVQGKY. (**A**) Predicted structure of peptide INSSDVQGKY. (**B**) ORAC assay of synthesized peptide INSSDVQGKY compared with Trolox and PBS. (**C**) The effects of different concentration of synthesized peptide INSSDVQGKY on HaCaT cell viability were measured by MTT assay. (**D**) Antioxidant activities of synthesized peptide INSSDVQGKY on HaCaT cells were detected by DCFH-DA probe. Images were taken using fluorescence microscope (magnification, 10×). (**E**) Statistical analysis was performed to quantify relative fluorescence density. The Normality of data was analyzed by Shapiro–Wilk test. Data were expressed as mean ± SD (*n* = 3). *** represents $p < 0.001$, compared with H_2O_2 group (Student *t*-test).

3. Discussion

Phycobiliprotein is an ideal marine protein resource with antioxidant properties widely spread in Rhodophyta and Cyanophyta, which could be used in bioactive peptides preparation [28]. This study aimed to apply the bacterial extracellular protease in preparing antioxidant peptides from phycobiliprotein. To utilize phycobiliprotein more comprehensively and avoid its natural limitations, we reported bacterial extracellular protease of *Pseudoalteromonas* sp. JS4–1 hydrolyzed phycobiliprotein to obtain antioxidant peptides in a single process. The bioactive fractions exhibited antioxidant activities both in vitro and in vivo. The biochemical mechanisms of the antioxidant activities of the fractions at cellular and *C. elegans* levels may be through activating the activities of antioxidant enzymes and scavenging the oxidative stress in the organisms. The molecular mechanism underlying this biological progress still need to be further studied in the future.

In this study, extracellular proteases from marine bacteria were used to hydrolyze phycobiliprotein. Marine proteases can degrade organics in the ocean naturally; thus, they have extraordinary advantages in digesting marine-sourced proteins such as fish and plants compared with land proteases [9,29,30]. Compared with commercial proteases, marine proteases have higher catalyzing efficiency and shorter hydrolyzing time. The unique cleavage sites of marine proteases also produced peptides with different amino acid sequences which is helpful to select more bioactive peptides and develop various applications of phycobiliprotein [31]. The identified peptide INSSDVQGKY exhibited antioxidant activities at a concentration of 100–200 μg/mL which was close to some newly found antioxidant peptides derived from other resources [32,33]. The cellular ROS scavenging assay is strong proof of the outstanding antioxidant activity of the peptide during biological process.

ROS are important mediators of biological process. The increased formation of ROS could cause oxidative stress which is associated with oxidative damage to biomolecules [34].

Oxidative stress happens in various pathophysiological processes and is an important pathophysiological mechanism underlying many diseases. Thus, decreasing the ROS levels to inhibit the oxidative stress has become an efficient and promising way in medicine [35]. Activating endogenous antioxidant enzymes such as SOD, CAT and GSH-Px by small molecules is an effective and protective approach in diseases [36]. The enzymatic hydrolysates and newly identified peptide in this study were able to activate the antioxidant enzymes to scavenge ROS under exogenous stress such as high glucose, and H_2O_2. Li G et al. identified a novel peptide EDEQKFWGK with DPPH RSA 26.76% and the OH RSA 32.19% from porcine plasma hydrolysate, and this peptide could increase SOD, CAT, and GSH-Px activities in HepG2 cells [37]. Park YR et al. identified a novel peptide NCWPFQGVPLGFQAPP with DPPH RSA 50% from clam worms and exhibited good antioxidant and anti-inflammation effects in RAW264 7 cells [38]. Compared with antioxidant peptides from other resources, phycobiliprotein hydrolysates also showed good antioxidant activities (DPPH RSA: 45.14 ± 0.43% and OH RSA: 65.11 ± 2.64%). Further, phycobiliprotein can be easily obtained and produced massively in the industry. So, exploring bioactive peptides from phycobiliprotein is of great economic value.

Conclusively, the enzymatic hydrolysis of phycobiliprotein is a promising and efficient method to produce antioxidant peptides. The peptides hydrolyzed from phycobiliprotein by marine proteases exhibited antioxidant activities both at cellular levels and in *C. elegans*.

4. Materials and Methods

4.1. Strains and Reagents

The *A. platensis* powder was purchased from a local supplier in Shanghai. Strains used in this study were isolated from the sediment of the South China Sea and stored in the lab: *Pseudoalteromonas* sp. JS4-1 (GenBank accession number: MT116988), *Pseudoalteromonas* sp. B27-3, *Pseudoalteromonas* sp. B47-6, *Pseudoalteromonas* sp. B62-3, *Pseudoalteromonas* sp. WH05-1, *Pseudoalteromonas* sp. WH06-2, and *Pseudoalteromonas* sp. ZB23-2. Immortal human umbilical vein endothelial cell line (HUVECs) and human immortal keratinocyte line (HaCaT) were obtained from the National Collection of Authenticated Cell Cultures (Shanghai, China). Wild type *C. elegans* (N2) were obtained from Caenorhabditis Genetics Center (CGC).

Ninhydrin, penicillin-streptomycin solution (cell culture grade), trypsin-EDTA solution, dimethyl sulfoxide (DMSO), thiazolyl blue tetrazolium bromide (MTT), and phosphate buffered saline (PBS) were purchased from Sangon Biotech (Shanghai, China). Trisodium citrate dihydrate, $SnCl_2$, n-propanol, $FeSO_4$, 30% H_2O_2, cholesterol and glucose were purchased from Sinopharm Chemical Reagent Co., Ltd. (Shanghai, China). In addition, 1,1-diphenyl-2-picrylhydrazyl (DPPH), 1,10-phenanthroline monohydrate (OP), fluorescein sodium salt, 2,20-azobis (2-amidinopropane) dihydrochloride (AAPH), 6-hydroxy-2,5,7,8-tetramethylchroman-2-carboxylic acid (Trolox), L-leucine, antifade mounting medium and NaN_3 were purchased from Sigma-Aldrich, Ltd. (St. Louis, MO, USA). Amicon™ Ultra-15 centrifugal filter units were purchased from Millipore® (Billerica, MA, USA). Sephadex LH-20 was purchased from GE Healthcare Life Sciences (Uppsala, Sweden). Fetal bovine serum (FBS) was purchased from PAN-BiotechGibco company (Aidenbach, Germany). RPMI-1640 medium was purchased from Gibco® ThermoFisher Scientific Company (Waltham, MA, USA). Methyl viologen dichloride was purchased from Aladdin® (Shanghai, China). BCA assay kit, cell lysis buffer, ROS assay kit, catalase (CAT) assay kit, total superoxide dismutase (SOD) assay kit with NBT, cellular glutathione peroxidase (GSH-Px) assay kit with NADPH and lipid peroxidation MDA assay kit were purchased from Beyotime Biotechnology (Shanghai, China).

4.2. Extraction of Phycobiliprotein and Screening of Proteases

4.2.1. Extraction of Phycobiliprotein from *A. platensis*

One gram of *A. platensis* powder was stirred in 100 mL of sterilized ultrapure water, then frozen in −20 °C and thawed in 37 °C-water bath repeatedly 3–4 times to help proteins released. The protein content was measured by BCA method [39]. The soluble phycobiliprotein (crude extract) was obtained by centrifugation at 12,000× *g* for 20 min, and stored at −20 °C for future use. We could obtain 100 mL of 3.75 mg/mL protein solution from 1 g of *A. platensis* powder and the yield rate was 37.5%.

4.2.2. Preparation of Crude Bacterial Extracellular Protease

The activated seed inoculum was prepared by adding bacterial strains preserved in glycerin into 5 mL 2216E liquid medium (tryptone 5 g/L, yeast extract 1 g/L, $Fe_2(SO_4)_3$ 0.01 g/L, dissolved in artificial sea water, pH 7.8), and incubating at 16 °C for 16–24 h with 180 rpm shaking. Then, 2% of seed inoculum was transformed to fresh fermentation medium (bean powder 20 g/L, corn powder 20 g/L, wheat bran 10 g/L, $Na_2HPO_4 \cdot 12H_2O$ 1 g/L, KH_2PO_4 0.3 g/L, $CaCl_2$ 1 g/L, Na_2CO_3 1 g/L, dissolved in artificial sea water, pH 7.8) and continuously culturing at the same condition. After 5 days cultivation, the crude enzyme was obtained by collecting the supernatant of the fermentation broth at 12,000× *g* centrifugation for 10 min at 4 °C, and stored at −20 °C prior to use.

4.2.3. Substrate Immersing Zymography

Substrate immersing zymography was conducted to determine the enzymatic activities against casein according to a previous study, with minor modifications [40]. The SDS-PAGE gel was immersed in the pre-warmed 0.5% (*w/v*) casein solution and incubated at 37 °C for 1 h with 75 rpm shaking. The gel was then stained with 0.1% of Coomassie Brilliant Blue dye solution for 4 h, and then decolored with 30% of ethanol and 70% of acetic acid (*v/v*) until the bands were clearly visible.

4.3. Antioxidant Activities of Hydrolysates

4.3.1. Optimization of Hydrolysis Conditions

First, the enzyme/substrate (E: S) ratio (*v*/*w*, mL/mg) was designed as 1:6, 1:8, 1:10, 1:12 and 1:14 with 1 mg/mL enzyme at 50 °C, to optimize the E/S ratio during the hydrolyzing. Then, the hydrolysis reaction was carried out at 35 to 60 °C, with 5 °C at intervals, to estimate the optimal temperature for hydrolyzing. Finally, hydrolyzing was taken from 1 h to 7 h at hourly intervals at 50 °C to optimize the reaction time. The reaction was terminated by heating at 95 °C for 10 min. The reaction buffer was ultrapure water. The hydrolysate was centrifuged at 12,000× *g* and 4 °C for 20 min and the hydrolysis degree was measured by ninhydrin coloration method [41]. The $\bullet NH_2$ in hydrolysate could react with ninhydrin and have absorption peak at 570 nm. The standard curve was determined using L-leucine at a concentration of 0, 1, 1.5, 2, 2.5 and 3 mM.

4.3.2. Purification of Hydrolysates

After digestion, the hydrolysates were added into the upper casing of the Amicon™ Ultra-15 centrifugal filter units and centrifuged at 5000× *g* for 30 min to separate peptides with MWCO (molecular weight cut-off) > 3 kDa (in the upper casing) and MWCO < 3 kDa (in the lower casing).

The peptides with MWCO < 3 kDa was further fractionated by NGC Scout Plus fast protein liquid chromatography (FPLC) system (Bio-Rad, USA) and a XK16/70 column (1.6 × 60 cm) equipped with Sephadex LH-20 gel. One mL of sample was injected into the loading loop and the column was continuously eluted with filtered ultrapure water at a flow rate of 0.75 mL/min. The protein absorbance was monitored at 220 nm with a UV detection all the time. Fractions were lyophilized and redissolved in ultrapure water at the same concentration for further assay.

4.3.3. Determination of Antioxidant Activities

To evaluate antioxidant activities of hydrolytes in vitro, DPPH free radical scavenging activity (DPPH RSA) assay, hydroxyl free radical (OH) scavenging activity (OH RSA) assay and oxygen radical absorbance capacity (ORAC) assay were conducted according to previous studies, with minor modifications [42–44]. The DPPH and OH radical scavenging rate were calculated.

The ORAC was defined as Trolox equivalents (mmol TE/g sample or mmol TE/mmol sample) according to the area under the curve (AUC) and calculated using the following formula:

$$ORAC = (AUC_{sample} - AUC_{control})/(AUC_{Trolox} - AUC_{control}) \times (M_{Trolox}/M_{sample}) \quad (1)$$

where AUC_{sample}, $AUC_{control}$ and AUC_{Trolox} were the integral areas under the fluorescence decay curve of the sample, PBS and Trolox, and M_{Tolox} and M_{sample} were the concentrations of the Trolox and sample, respectively.

4.3.4. Protective Effects on Oxidative Damage of Plasmid DNA

To evaluate the protective effects of peptides on the oxidative damage of biological macromolecules (proteins and DNA), the protection effect assay was conducted according to a previous method, with minor modifications [15]. An amount of 8 μL of pET-22b DNA, 2 μL of $FeSO_4$ (2 mM), 8 μL of sample and 2 μL of 0.1% H_2O_2 were mixed in a sterilized tube and incubated at 37 °C for 10 min. For the blank, 12 μL of sterilized water replaced other reagents incubating with the DNA. For the negative control, sample was replaced with sterilized water. For the positive control, the sample was replaced with Vitamin C (200 μg/mL). The reaction solution was then analyzed by 1% agarose gel electrophoresis under 120 V for 30 min.

4.3.5. Determination of Intracellular ROS Levels and the Activities of Intracellular Oxidative Enzymes

HUVECs and HaCaT cells were cultured in RPMI-1640 medium with 10% FBS, 100 U/mL penicillin and 100 μg/mL streptomycin (complete medium) at 37 °C in a humidified atmosphere of 5% CO_2. Cell viability was measured by MTT assay.

The 2′,7′-dichloro-fluorescence diacetate (DCFH-DA) probe can be oxidized by ROS to produce green fluorescent materials which can be observed under fluorescence microscope. An amount of 1×10^5 of cells were plated in a 24-well plate and incubated overnight. For the HUVECs cells, the medium was replaced with RPMI-1640 medium (without FBS instead, added 35 mM of glucose) for 24 h. For the HaCaT cells, the medium was replaced with RPMI-1640 medium (without FBS, but added 1.5 mM of H_2O_2) for 4 h. For the blank, cells were treated with RPMI-1640 medium (without FBS). An amount of 10 μM DCFH-DA probe was added to each well and incubated for 1 h. Cells were washed with PBS for 3 times and re-immersed in 500 μL of PBS. Images were taken using an inverted fluorescence microscope (Axio vert. A1, Carl Zeiss, Oberkochen, Germany).

Next, 2×10^6 of cells were plated in a 6-well plate and treated with the same condition. After incubating for 24 h, cells were washed with PBS for 3 times, then incubating with 150 μL of cell lysis buffer for 30 min on ice. The suspension was centrifuged at 12,000× *g* for 10 min at 4 °C. The protein concentration was measured by BCA method. The CAT, SOD and GSH-Px activities were determined according to the manufacturer's instructions.

4.3.6. Antioxidant Activities Evaluation Using *C. elegans* Model

Maintenance of *C. elegans* and Lifespan Assay

N2 was cultured on nematode growth medium (NGM) plates supplied with *E. coli* OP50 as food at 20 °C. The age-synchronized *C. elegans* were prepared using the alkaline hypochlorite method according to Phaniendra et al. [45].

Fresh NGM plates were coated with *E. coli* OP50 containing a different concentration of APs (0.2 and 0.5 mg/mL) and incubated at 37 °C for 48 h to prepare the sample plates. Age-synchronized nematodes (L4 stage) were plated in blank plates and sample plates (80 worms on each plate), as Day 0 for lifespan assay. Nematodes on each plate were transferred to a new plate to avoid new-born worms and counted every day until all nematodes were dead. The survival rate was calculated.

Determination of ROS Levels and Activities of Antioxidant Enzymes in *C. elegans*

After 3 days of treatment on the sample plates, nematodes were collected into a M9 buffer (added 100 μM DCFH-DA probe) and incubated at 20 °C for 30 min without light. Nematodes were washed with M9 buffer 3 times. For inverted fluorescence microscope observation, nematodes were anesthetized with 1 μL of NaN_3 (25 mM) and 20 μL of suspension was dropped onto the center of the slide, and 5 μL of antifade mounting medium was added for observation.

The ROS accumulation was conducted according to previous studies, with minor modifications [46]. The fluorescence intensity was measured at 15 min intervals for 6 h with excitation 485 nm and emission 530 nm at 25 °C.

For intracellular proteins extraction, nematodes were ultrasonicated on ice and the suspension was centrifuged at 12,000× *g* for 10 min at 4 °C. The protein concentration was measured by BCA method. The CAT, SOD and GSH-Px activities and MDA levels in the supernatant were determined according to the manufacturer's instructions.

Oxidative Stress Experiment

After 3 days of treatment, nematodes were transferred onto NGM plates supplied with 10 mM H_2O_2 or 10 mM paraquat, and each plate with 80 worms. The number of survival nematodes was counted per 30 min intervals (the H_2O_2 plates) and per 12 h intervals (the paraquat plates) until all the nematodes were dead [47]. The survival rate was calculated.

After 3 days of treatment, nematodes were transferred onto the fresh NGM plates (80 worms on each plate) and incubated at 35 °C for 10 h [48]. Numbers of survival nematodes were counted after the heat shock and the survival rate was calculated.

4.3.7. Identification and Solid Phase Synthesis of APs

The sequence and composition of peptides in the active fractions were analyzed by liquid chromatography-mass spectrometry (LC-MS/MS) in Sangon Biotech Co., Ltd. (Shanghai, China). The LC-MS/MS was developed on an UltiMate 3000 RSLCnano system (Thermo, Waltham, MA, USA) coupled with a Q Exactive Plus HPLC-Mass Spectrometer (Thermal, Riverside County, CA, USA). The chromatographic separation was achieved using a C18 column (3 μm, 120 Å, 100 μm × 20 mm). The analysis was performed on a C18 analytical column (2 μm, 120 Å, 750 μm × 150 mm). Mobile phase A was 3% DMSO, 0.1% formic acid, and 97% H_2O and mobile phase B was 3% DMSO, 0.1% formic acid, and 97% ACN. The flow rate was set as 300 nL/min. The MS data was processed with MaxQuant (V1.6.2.10) to identify detected peptides in *Arthrospira platensis* proteomics database. Peptide with maximal abundance was synthesized by solid phase synthesis in ChinaPeptides Co., Ltd. (Shanghai, China) according to the sequence information given by LC-MS/MS.

The structure of peptide was predicted by Phyre2 (http://www.sbg.bio.ic.ac.uk/phyre2 (accessed on 20 December 2022)).

4.4. Statistical Analysis

All the experiments and analysis were carried out in triplicate. The Normality of experimental data was analyzed by Shapiro–Wilk test. Normal experimental data were analyzed with by Student *t*-test. Non-parametric data were analyzed by Mann–Whitney test. Data were presented as the mean ± standard deviation (SD) and $P < 0.05$ was considered as a statistically significant difference.

5. Conclusions

Algal protein resources are considered to be a huge treasury of bioactive peptides. More and more researchers are attempting to prepare novel bioactive peptides from algal protein through protease hydrolysis. In addition, novel proteases from marine bacteria also possess great potential in antioxidant peptide preparation, due to their high efficiency and low cost. In this study, phycobiliprotein was hydrolyzed by extracellular proteases from

Pseudoalteromonas sp. JS4-1 to produce antioxidant peptides. A novel antioxidant peptide INSSDVQGKY was separated and identified from the hydrolysates. The hydrolysates and peptide INSSDVQGKY exhibited excellent antioxidant activities both at cellular levels and in *C. elegans*.

Supplementary Materials: The following supporting information can be downloaded at: https://www.mdpi.com/article/10.3390/md21020133/s1, Figure S1: Total ion chromatography of F2; Figure S2: MS/MS of INSSDVQGKY.

Author Contributions: C.L.: conceptualization, methodology and validation. G.C.: writing—original draft and visualization. H.R., X.X., Y.C., C.W.: investigation. F.B.: writing—review and editing. H.H.: supervision and funding acquisition. All authors have read and agreed to the published version of the manuscript.

Funding: This work was supported by the National Natural Science Foundation of China (31900038 and 31500048), the Natural Science Foundation of Hunan Province, China (2021JJ30029), the Research Fund of The State Key Laboratory of Coal Resources and Safe Mining (SKLCRSM22KF020), The Independent Exploration and Innovation project for postgraduates of Hunan Province (CX20220357, CX20220316), the Independent Exploration and Innovation project for postgraduates of Central South University (2022ZZTS0996), and the Open Sharing Fund for the Large-scale Instruments and Equipments of Central South University (CSUZC202244).

Institutional Review Board Statement: Not applicable.

Informed Consent Statement: Not applicable.

Data Availability Statement: The data presented in this study are available on request from the corresponding author. The data are not publicly available due to privacy.

Conflicts of Interest: The authors declare no conflict of interest.

References

1. Lupatini, A.L.; Colla, L.M.; Canan, C.; Colla, E. Potential application of microalga *Spirulina platensis* as a protein source. *J. Sci. Food Agric.* **2017**, *97*, 724–732. [CrossRef] [PubMed]
2. Romay, C.; González, R.; Ledón, N.; Remirez, D.; Rimbau, V. C-phycocyanin: A biliprotein with antioxidant, anti-inflammatory and neuroprotective effects. *Curr. Protein Pept. Sci.* **2003**, *4*, 207–216. [CrossRef] [PubMed]
3. Marthandam Asokan, S.; Wang, T.; Su, W.T.; Lin, W.T. Antidiabetic Effects of a Short Peptide of Potato Protein Hydrolysate in STZ-Induced Diabetic Mice. *Nutrients* **2019**, *11*, 779. [CrossRef] [PubMed]
4. Qiao, Q.; Chen, L.; Li, X.; Lu, X.; Xu, Q. Roles of Dietary Bioactive Peptides in Redox Balance and Metabolic Disorders. *Oxid. Med. Cell Longev.* **2021**, *2021*, 5582245. [CrossRef] [PubMed]
5. Amigo, L.; Hernández-Ledesma, B. Current Evidence on the Bioavailability of Food Bioactive Peptides. *Molecules* **2020**, *25*, 4479. [CrossRef]
6. Wang, K.; Han, L.; Hong, H.; Pan, J.; Liu, H.; Luo, Y. Purification and identification of novel antioxidant peptides from silver carp muscle hydrolysate after simulated gastrointestinal digestion and transepithelial transport. *Food Chem.* **2021**, *342*, 128275. [CrossRef]
7. Chalamaiah, M.; Dinesh Kumar, B.; Hemalatha, R.; Jyothirmayi, T. Fish protein hydrolysates: Proximate composition, amino acid composition, antioxidant activities and applications: A review. *Food Chem.* **2012**, *135*, 3020–3038. [CrossRef]
8. Lu, X.; Zhang, L.; Sun, Q.; Song, G.; Huang, J. Extraction, identification and structure-activity relationship of antioxidant peptides from sesame (*Sesamum indicum* L.) protein hydrolysate. *Food Res. Int.* **2019**, *116*, 707–716. [CrossRef]
9. Olena, Z.; Yang, Y.; TingTing, Y.; XiaoTao, Y.; HaiLian, R.; Xun, X.; Dong, X.; CuiLing, W.; HaiLun, H. Simultaneous preparation of antioxidant peptides and lipids from microalgae by pretreatment with bacterial proteases. *Bioresour. Technol.* **2022**, *348*, 126759. [CrossRef]
10. Xu, F.; Cha, Q.Q.; Zhang, Y.Z.; Chen, X.L. Degradation and Utilization of Alginate by Marine Pseudoalteromonas: A Review. *Appl. Environ. Microbiol.* **2021**, *87*, e0036821. [CrossRef]
11. Kou, X.; Gao, J.; Xue, Z.; Zhang, Z.; Wang, H.; Wang, X. Purification and identification of antioxidant peptides from chickpea (*Cicer arietinum* L.) albumin hydrolysates. *LWT—Food Sci. Technol.* **2013**, *50*, 591–598. [CrossRef]
12. Hou, H.; Wang, J.; Wang, J.; Tang, W.; Shaikh, A.S.; Li, Y.; Fu, J.; Lu, L.; Wang, F.; Sun, F.; et al. A Review of Bioactive Peptides: Chemical Modification, Structural Characterization and Therapeutic Applications. *J. Biomed. Nanotechnol.* **2020**, *16*, 1687–1718. [CrossRef] [PubMed]
13. Cai, S.; Lu, C.; Liu, Z.; Wang, W.; Lu, S.; Sun, Z.; Wang, G. Derivatives of gecko cathelicidin-related antioxidant peptide facilitate skin wound healing. *Eur. J. Pharmacol.* **2021**, *890*, 173649. [CrossRef] [PubMed]

14. Hseu, Y.C.; Vudhya Gowrisankar, Y.; Wang, L.W.; Zhang, Y.Z.; Chen, X.Z.; Huang, P.J.; Yen, H.R.; Yang, H.L. The in vitro and in vivo depigmenting activity of pterostilbene through induction of autophagy in melanocytes and inhibition of UVA-irradiated α-MSH in keratinocytes via Nrf2-mediated antioxidant pathways. *Redox Biol.* **2021**, *44*, 102007. [CrossRef]
15. Wu, R.; Wu, C.; Liu, D.; Yang, X.; Huang, J.; Zhang, J.; Liao, B.; He, H. Antioxidant and anti-freezing peptides from salmon collagen hydrolysate prepared by bacterial extracellular protease. *Food Chem.* **2018**, *248*, 346–352. [CrossRef]
16. Zou, J.; Fei, Q.; Xiao, H.; Wang, H.; Liu, K.; Liu, M.; Zhang, H.; Xiao, X.; Wang, K.; Wang, N. VEGF-A promotes angiogenesis after acute myocardial infarction through increasing ROS production and enhancing ER stress-mediated autophagy. *J. Cell Physiol.* **2019**, *234*, 17690–17703. [CrossRef]
17. Nensat, C.; Songjang, W.; Tohtong, R.; Suthiphongchai, T.; Phimsen, S.; Rattanasinganchan, P.; Metheenukul, P.; Kumphune, S.; Jiraviriyakul, A. Porcine placenta extract improves high-glucose-induced angiogenesis impairment. *BMC Complement. Med. Ther.* **2021**, *21*, 66. [CrossRef]
18. Rezabakhsh, A.; Rahbarghazi, R.; Malekinejad, H.; Fathi, F.; Montaseri, A.; Garjani, A. Quercetin alleviates high glucose-induced damage on human umbilical vein endothelial cells by promoting autophagy. *Phytomedicine* **2019**, *56*, 183–193. [CrossRef]
19. Son, D.H.; Yang, D.J.; Sun, J.S.; Kim, S.K.; Kang, N.; Kang, J.Y.; Choi, Y.H.; Lee, J.H.; Moh, S.H.; Shin, D.M.; et al. A Novel Peptide, Nicotinyl-Isoleucine-Valine-Histidine (NA-IVH), Promotes Antioxidant Gene Expression and Wound Healing in HaCaT Cells. *Mar. Drugs* **2018**, *16*, 262. [CrossRef]
20. Dong, H.; Liu, M.; Wang, L.; Liu, Y.; Lu, X.; Stagos, D.; Lin, X.; Liu, M. Bromophenol Bis (2,3,6-Tribromo-4,5-dihydroxybenzyl) Ether Protects HaCaT Skin Cells from Oxidative Damage via Nrf2-Mediated Pathways. *Antioxidants* **2021**, *10*, 1436. [CrossRef]
21. Wang, G.; Hao, M.; Liu, Q.; Jiang, Y.; Huang, H.; Yang, G.; Wang, C. Protective effect of recombinant Lactobacillus plantarum against H(2)O(2)-induced oxidative stress in HUVEC cells. *J. Zhejiang Univ. Sci. B* **2021**, *22*, 348–365. [CrossRef] [PubMed]
22. Lin, C.; Xiao, J.; Xi, Y.; Zhang, X.; Zhong, Q.; Zheng, H.; Cao, Y.; Chen, Y. Rosmarinic acid improved antioxidant properties and healthspan via the IIS and MAPK pathways in *Caenorhabditis elegans*. *Biofactors* **2019**, *45*, 774–787. [CrossRef] [PubMed]
23. Nakagawa, H.; Shiozaki, T.; Kobatake, E.; Hosoya, T.; Moriya, T.; Sakai, F.; Taru, H.; Miyazaki, T. Effects and mechanisms of prolongevity induced by *Lactobacillus gasseri* SBT2055 in *Caenorhabditis elegans*. *Aging Cell* **2016**, *15*, 227–236. [CrossRef]
24. Ayuda-Durán, B.; González-Manzano, S.; González-Paramás, A.M.; Santos-Buelga, C. Caernohabditis elegans as a Model Organism to Evaluate the Antioxidant Effects of Phytochemicals. *Molecules* **2020**, *25*, 3194. [CrossRef] [PubMed]
25. Tsikas, D. Assessment of lipid peroxidation by measuring malondialdehyde (MDA) and relatives in biological samples: Analytical and biological challenges. *Anal. Biochem.* **2017**, *524*, 13–30. [CrossRef] [PubMed]
26. Shi, Y.; Meng, X.; Zhang, J. Multi- and trans-generational effects of N-butylpyridium chloride on reproduction, lifespan, and pro/antioxidant status in *Caenorhabditis elegans*. *Sci. Total Environ.* **2021**, *778*, 146371. [CrossRef] [PubMed]
27. Yu, J.; Hu, Y.; Xue, M.; Dun, Y.; Li, S.; Peng, N.; Liang, Y.; Zhao, S. Purification and Identification of Antioxidant Peptides from Enzymatic Hydrolysate of *Spirulina platensis*. *J. Microbiol. Biotechnol.* **2016**, *26*, 1216–1223. [CrossRef]
28. Pagels, F.; Guedes, A.C.; Amaro, H.M.; Kijjoa, A.; Vasconcelos, V. Phycobiliproteins from cyanobacteria: Chemistry and biotechnological applications. *Biotechnol. Adv.* **2019**, *37*, 422–443. [CrossRef]
29. Gómez-Estaca, J.; Bravo, L.; Gómez-Guillén, M.C.; Alemán, A.; Montero, P. Antioxidant properties of tuna-skin and bovine-hide gelatin films induced by the addition of oregano and rosemary extracts. *Food Chem.* **2009**, *112*, 18–25. [CrossRef]
30. Kim, S.Y.; Je, J.Y.; Kim, S.K. Purification and characterization of antioxidant peptide from hoki (*Johnius belengerii*) frame protein by gastrointestinal digestion. *J. Nutr. Biochem.* **2007**, *18*, 31–38. [CrossRef]
31. Wu, R.; Chen, L.; Liu, D.; Huang, J.; Zhang, J.; Xiao, X.; Lei, M.; Chen, Y.; He, H. Preparation of Antioxidant Peptides from Salmon Byproducts with Bacterial Extracellular Proteases. *Mar. Drugs* **2017**, *15*, 4. [CrossRef] [PubMed]
32. Fontoura, R.; Daroit, D.J.; Corrêa, A.P.F.; Moresco, K.S.; Santi, L.; Beys-da-Silva, W.O.; Yates, J.R., 3rd; Moreira, J.C.F.; Brandelli, A. Characterization of a novel antioxidant peptide from feather keratin hydrolysates. *New Biotechnol.* **2019**, *49*, 71–76. [CrossRef] [PubMed]
33. Ma, Y.; Huang, K.; Wu, Y. In Vivo/In Vitro Properties of Novel Antioxidant Peptide from *Pinctada fucata*. *J. Microbiol. Biotechnol.* **2021**, *31*, 33–42. [CrossRef] [PubMed]
34. Li, R.; Jia, Z.; Trush, M.A. Defining ROS in Biology and Medicine. *React. Oxyg Species* **2016**, *1*, 9–21. [CrossRef] [PubMed]
35. He, L.; He, T.; Farrar, S.; Ji, L.; Liu, T.; Ma, X. Antioxidants Maintain Cellular Redox Homeostasis by Elimination of Reactive Oxygen Species. *Cell Physiol. Biochem.* **2017**, *44*, 532–553. [CrossRef] [PubMed]
36. Christofidou-Solomidou, M.; Muzykantov, V.R. Antioxidant strategies in respiratory medicine. *Treat Respir. Med.* **2006**, *5*, 47–78. [CrossRef] [PubMed]
37. Li, G.; Zhan, J.; Hu, L.; Yuan, C.; Ying, X.; Hu, Y. Identification of novel antioxidant peptide from porcine plasma hydrolysate and its effect in in vitro digestion/HepG2 cells model. *J. Food Biochem.* **2022**, *46*, e13853. [CrossRef]
38. Park, Y.R.; Park, C.I.; Soh, Y. Antioxidant and Anti-Inflammatory Effects of NCW Peptide from Clam Worm (*Marphysa sanguinea*). *J. Microbiol. Biotechnol.* **2020**, *30*, 1387–1394. [CrossRef]
39. Walker, J.M. The bicinchoninic acid (BCA) assay for protein quantitation. *Methods Mol. Biol.* **1994**, *32*, 5–8.
40. Liu, D.; Yang, X.; Huang, J.; Wu, R.; Wu, C.; He, H.; Li, H. In situ demonstration and characteristic analysis of the protease components from marine bacteria using substrate immersing zymography. *Appl. Biochem. Biotechnol.* **2015**, *175*, 489–501. [CrossRef]

41. Leach, A.A. The determination of the substitution achieved at the alpha-amino, epsilon-amino and imidazole groups of proteins with special reference to derivatives of gelatin. *Biochem. J.* **1966**, *98*, 506–512. [CrossRef] [PubMed]
42. Dávalos, A.; Gómez-Cordovés, C.; Bartolomé, B. Extending applicability of the oxygen radical absorbance capacity (ORAC-fluorescein) assay. *J. Agric. Food Chem.* **2004**, *52*, 48–54. [CrossRef] [PubMed]
43. Shimada, K.; Fujikawa, K.; Yahara, K.; Nakamura, T. Antioxidative properties of xanthan on the autoxidation of soybean oil in cyclodextrin emulsion. *J. Agric. Food Chem.* **1992**, *40*, 945–948. [CrossRef]
44. Wang, B.; Li, Z.R.; Chi, C.F.; Zhang, Q.H.; Luo, H.Y. Preparation and evaluation of antioxidant peptides from ethanol-soluble proteins hydrolysate of Sphyrna lewini muscle. *Peptides* **2012**, *36*, 240–250. [CrossRef]
45. Alugoju, P.; Janardhanshetty, S.S.; Subaramanian, S.; Periyasamy, L.; Dyavaiah, M. Quercetin Protects Yeast Saccharomyces cerevisiae pep4 Mutant from Oxidative and Apoptotic Stress and Extends Chronological Lifespan. *Curr. Microbiol.* **2018**, *75*, 519–530. [CrossRef]
46. Lin, C.; Su, Z.; Luo, J.; Jiang, L.; Shen, S.; Zheng, W.; Gu, W.; Cao, Y.; Chen, Y. Polysaccharide extracted from the leaves of *Cyclocarya paliurus* (Batal.) Iljinskaja enhanced stress resistance in *Caenorhabditis elegans* via skn-1 and hsf-1. *Int. J. Biol. Macromol.* **2020**, *143*, 243–254. [CrossRef]
47. Sakamoto, T.; Imai, H. Hydrogen peroxide produced by superoxide dismutase SOD-2 activates sperm in *Caenorhabditis elegans*. *J. Biol. Chem.* **2017**, *292*, 14804–14813. [CrossRef]
48. Yasuda, K.; Kubo, Y.; Murata, H.; Sakamoto, K. Cortisol promotes stress tolerance via DAF-16 in *Caenorhabditis elegans*. *Biochem. Biophys. Rep.* **2021**, *26*, 100961. [CrossRef]

marine drugs

Article

Probiotic Potential of the Marine Isolate *Enterococcus faecium* EA9 and In Vivo Evaluation of Its Antisepsis Action in Rats

Eman H. Zaghloul [1,†], Hatem M. Abuohashish [2,†], Amany S. El Sharkawy [1,*], Eman M. Abbas [1], Mohammed M. Ahmed [3] and Salim S. Al-Rejaie [3]

1 National Institute of Oceanography and Fisheries (NIOF), Cairo 11516, Egypt
2 Department of Biomedical Dental Sciences, College of Dentistry, Imam Abdulrahman Bin Faisal University, P.O. Box 1982, Dammam 31441, Saudi Arabia
3 Department of Pharmacology and Toxicology, College of Pharmacy, King Saud University, Riyadh 11451, Saudi Arabia
* Correspondence: as.sharkawy@niof.sci.eg
† These authors contributed equally to this work.

Abstract: This study aims to obtain a novel probiotic strain adapted to marine habitats and to assess its antisepsis properties using a cecal ligation and puncture (CLP) model in rodents. The marine *Enterococcus faecium* EA9 was isolated from marine shrimp samples and evaluated for probiotic potential after phenotypical and molecular identification. In septic animals, hepatic and renal tissues were histologically and biochemically evaluated for inflammation and oxidative stress following the probiotic treatment. Moreover, gene expressions of multiple signaling cascades were determined using RT-PCR. EA9 was identified and genotyped as *Enterococcus faecium* with a 99.88% identity. EA9 did not exhibit any signs of hemolysis and survived at low pH and elevated concentrations of bile salts. Moreover, EA9 isolate had antibacterial activity against different pathogenic bacteria and could thrive in 6.5% NaCl. Septic animals treated with EA9 had improved liver and kidney functions, lower inflammatory and lipid peroxidation biomarkers, and enhanced antioxidant enzymes. The CLP-induced necrotic histological changes and altered gene expressions of IL-10, IL-1β, INF-γ, COX-2, SOD-1, SOD-2, HO-1, AKT, mTOR, iNOS, and STAT-3 were abolished by the EA9 probiotic in septic animals. The isolate *Enterococcus faecium* EA9 represents a promising marine probiotic. The in vivo antisepsis testing of EA9 highlighted its potential and effective therapeutic approach.

Keywords: marine; probiotic; *Enterococcus faecium*; sepsis; inflammation; oxidative stress

Citation: Zaghloul, E.H.; Abuohashish, H.M.; El Sharkawy, A.S.; Abbas, E.M.; Ahmed, M.M.; Al-Rejaie, S.S. Probiotic Potential of the Marine Isolate *Enterococcus faecium* EA9 and In Vivo Evaluation of Its Antisepsis Action in Rats. *Mar. Drugs* **2023**, *21*, 45. https://doi.org/10.3390/md21010045

Academic Editors: Elena Talero and Javier Ávila-Román

Received: 4 November 2022
Revised: 3 January 2023
Accepted: 4 January 2023
Published: 7 January 2023

1. Introduction

Lactic acid bacteria (LAB) are widespread microorganisms that exist as normal microflora in humans and animals [1]. Some members of this ecosystem possess probiotic properties that confer a variety of health benefits, including colon cancer prevention, intestinal constipation alleviation [2], serum cholesterol reduction [3], and immunity boosting [4]. Furthermore, they enhance fatty acid absorption in the gut and boost health by producing antioxidants, vitamins, enzymes, organic acids (lactic and acetic), bacteriocins, and other antimicrobial substances [5].

The selection of a probiotic strain is critical since it must be safe with positive impacts on the host. The probiotic strain must be capable of withstanding digestive tract circumstances, such as acidic pH and high concentrations of bile salts and digestive enzymes [6]. Probiotics have been typically studied using culture-dependent methods that rely on isolation, growth, and identification in the lab using standard morphological and biochemical assays. The results of these in vitro tests must be encountered to evaluate and optimize the feeding dose and survivability of the probiotic strain [7] and to determine whether the candidate strain is suitable to be tested in vivo or not. Probiotic strains must be further examined in vivo to assess their ability to control or prevent diseases and disorders [8].

Enterococci species are facultative, gram-positive, and anaerobic cocci that can be isolated from multiple environments [9]. The therapeutic utilization of this well-known LAB as a probiotic has been documented in several clinical and preclinical studies [9]. In a recent prospective randomized comparative clinical study, the use of *Enterococcus faecium* isolate as a probiotic showed potential therapeutic effects in 62 patients with post-infectious irritable bowel syndrome [10]. Another *Enterococcus faecium* isolate revealed anti-inflammatory and immunomodulatory effects through the regulation of nuclear factor kappa B (NF-κB) and the c-Jun N-terminal kinase (JNK) pathway [11]. Moreover, the preventive effects of *Enterococcus faecium* probiotic were reported in broiler chickens with necrotic enteritis-induced intestinal barrier injury [12]. *Enterococcus faecium* also potentiated the anti-inflammatory and antiarthritic effects of methotrexate in a rat model of arthritis [13]. In Divyashri et al. study, the probiotic containing *Enterococcus faecium* tended to protect against neurological inflammation and oxidative stress [14].

In the medical field, probiotics have established their effective and potential protective effects against several pathological conditions [15] owing to less frequent antimicrobial resistance in comparison with conventional antimicrobial therapies. The life-threatening septic shock is one pathological condition that was targeted by multiple probiotics due to its high prevalence even with the presence of advanced critical care medical services [16,17]. Clinical data suggested that probiotics might delay the disease onset and reduce inflammatory mediators, particularly in children [18]. Likewise, experimental data demonstrated the ability of probiotics to alleviate sepsis-induced pathological response and inflammatory injuries in several organs [19]. Among the medically introduced probiotics, probiotics isolated from the marine environment have multiple therapeutic applications [20,21]. Their industrial, pharmaceutical, and environmental perspectives were documented [22]. Studies suggested that these marine microorganisms could be a selective source for secondary and bioactive compounds, which could be considered an advantage over the non-aquatic microbes [23]. Accordingly, it could be speculated that the present marine-derived *Enterococcus faecium* EA9 isolate could have potential beneficial properties in sepsis conditions.

Therefore, the current study aimed to isolate and identify a promising LAB from marine samples and assess its probiotic potential. The second aim of the present study was to explore whether the isolated potential probiotic could provide in vivo protective effects in septic Wistar albino rats using the cecal ligation and puncture (CLP) model.

2. Results

2.1. Assessment of Probiotic Potential of Isolate EA9

2.1.1. Phenotypic and Biochemical Characterization of Isolate EA9

The isolate EA9 was characterized as Gram-positive, catalase negative, non-spore forming, and spherical in shape (Figure 1A,B). The VITEK 2 system was also used to characterize it biochemically, as shown in Table 1. The biochemical features of EA9 VITEK 2 revealed certain biochemical traits, such as lactose fermentation, resistance to four antibiotics (Polymixin, Bacitracin, Novobiocin, and Optochin), and the capacity to survive in 6.5% NaCl. The isolate EA9 was biochemically identified as *Enterococcus faecium* with a 95% likelihood.

2.1.2. Molecular Identification

The identity of the marine isolate EA9 was subsequently validated by molecular identification using 16S rRNA gene sequencing analysis and sequence homology searches using the NCBI's BLAST software (NCBI, Bethesda, MD, USA). With a 99.88% similarity rate, EA9 was identified as *Enterococcus faecium*. The sequence was deposited in GenBank with the accession number MW218438. Figure 2 depicts the evolutionary relationship between the marine isolate EA9 and its close relatives in the NCBI database.

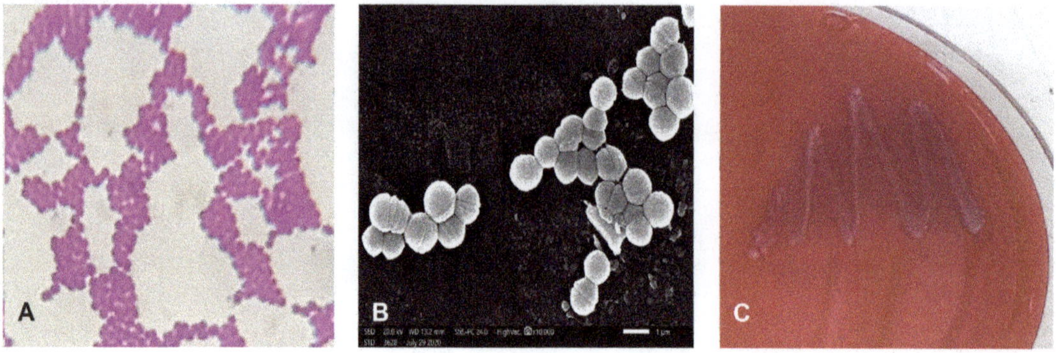

Figure 1. (**A**) Gram staining of isolate EA9 showing the Gram reaction of the isolate. (**B**) EA9 bacterial cells imaged with a scanning electron microscope (SEM) show the EA9 isolate's unique spherical cell shape and clustering. (**C**) Blood hemolysis test for isolate EA9 after 24 h at 37 °C indicating non-hemolytic activity.

Table 1. Biochemical characterization of the marine isolate EA9 using the VITEK 2 microbial identification system (BioMérieux, Craponne, France).

Test	Result	Test	Result
Alpha-mannosidase (AMAN)	−	D-mannosE (dMNE)	+
Phosphatase (PHOS)	−	Methyl-b-d-glucopyranoside (MBdG)	+
Leucine arylamidase (LeuA)	−	Pullulan (PUL)	−
L-proline arylamidase (ProA)	−	D-raffinose (dRAF)	−
Tyrosine arylamidase (TyrA)	−	Optochin resistance (OPTO)	+
D-sorbitol (dSOR)	−	Urease (URE)	−
Polymixin b resistance (POLYB)	+	D-mannitol (dMAN)	−
D-amygdalin (AMY)	−	D-galactose (dGAL)	+
Phosphatidylinositol phospholipase (PIPLC)	−	D-ribose (dRIB)	+
D-xylose (dXYL)	−	L-lactate alkalinization (ILATK)	−
Arginine dihydrolase 1 (ADH1)	+	Lactose (LAC)	+
Beta-galactosidase (BGAL)	+	N-acetyl-D-glucosamine (NAG)	+
Alpha-glucosidase (AGLU)	−	D-maltose (dMAL)	+
Beta glucuronidase (BGURr)	−	O/129 resistance (comp. vibrio.) (O129R)	+
Alpha-galactosidase (AGAL)	−	Salicin (SAL)	+
L-pyrrolydonyl-arylamidase (PyrA)	+	Saccharose/Sucrose (SAC)	−
Beta-glucuronidase (BGUR)	−	D-trehalose (dTRE)	+
Alanine arylamidase (AlaA)	−	Arginine dihydrolase 2 (ADH2s)	+
Ala-phe-pro arylamidase (APPA)	−	Bacitracin resistance (BACI)	+
Cyclodextrin (CDEX)	+	Novobiocin resistance (NOVO)	+
L-aspartate arylamidase (AspA)	−	Growth in 6.5% NaCl (NC6.5)	+
Beta galactopyranosidase (BGAR)	−		

"+" indicates positive test results and "−" indicates negative test results according to the manufacturer.

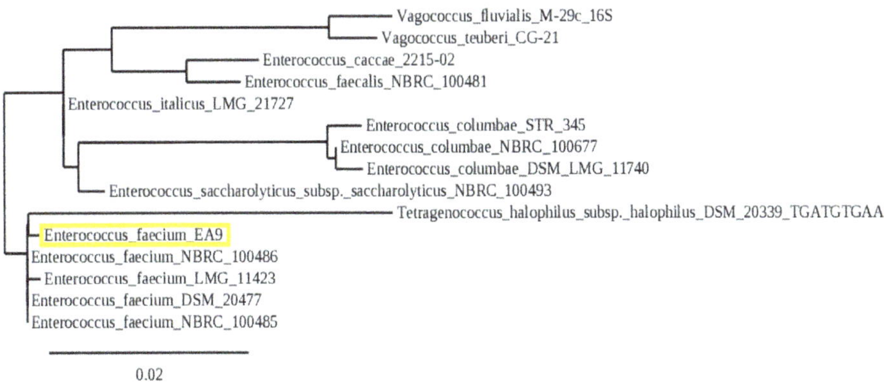

Figure 2. Phylogenetic tree of the marine isolate EA9 (yellow square) based on the 16S rRNA partial gene sequence.

2.1.3. Blood Hemolysis

One of the most important safety factors for evaluating potential probiotic strains is hemolytic activity. EA9 showed no clear or greenish zones around their colonies grown on a blood agar plate, indicating non-hemolytic activity (Figure 1C).

2.1.4. Antibiotic Resistance

The disc diffusion method was used to investigate the antibiotic resistance of the bacterial strain EA9. Out of 11 clinically significant antibiotics examined, isolate EA9 was resistant to 5 types of antibiotics namely: Ampicillin, Ceftriaxone, Piperacillin/Tazobactam, Nalidixic Acid, and Erythromycin (Table 2), whereas it was susceptible to the other 6 antibiotics.

Table 2. Antibiotic resistance of isolate EA9.

Antibiotic	Result
Nalidixic Acid (30 mcg)	R
Ampicillin (10 mcg)	R
Erythromycin (15 mcg)	R
Tetracycline (30 mcg)	S
Piperacillin/Tazobactam (110 mcg)	R
Vancomycin (30 mcg)	S
Oxacillin (1 mcg)	S
Ceftriaxone (30 mcg)	R
Amoxicillin (25 mcg)	S
Ofloxacin (5 mcg)	S
Cephradine (5 mcg)	S

"R" resistant and "S" sensitive.

2.1.5. Resistance to pH

As shown in Figure 3A, the isolate EA9 was able to withstand pH 4 and 3 for 4 h, as shown by an increase in absorbance and survival rates of 98% and 83%, respectively. Similarly, isolate EA9 was able to live and grow at pH 2 for 4 h, although at a slower pace, with a survival rate of 69%, and it continued to develop as reported after 24 h.

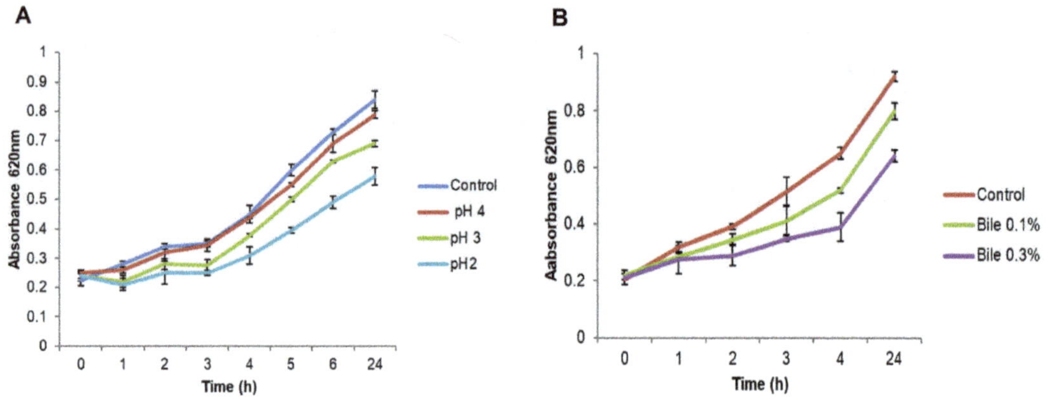

Figure 3. Growth of the marine isolate EA9 at diverse (**A**) pH values and (**B**) concentrations of bile salts for 24 h at 37 °C.

2.1.6. Bile Salts Tolerance

Data in Figure 3B shows that the isolate EA9 could withstand 0.1% and 0.3% bile concentrations, as demonstrated by an increase in absorbance and survival rates of 79.7% and 60%, respectively. Furthermore, EA9 continued to develop as observed after 24 h.

2.1.7. Antimicrobial Activity

The antibacterial activity of EA9 cell free supernatant (CFS) was tested against 7 pathogenic bacteria using the agar well-cut diffusion technique. According to the findings, EA9 CFS exhibited potent antimicrobial activity against all pathogens examined, with inhibition zones ranging from 15 to 21 mm in diameter (Table 3). The EA9 CFS showed the highest activity against the Gram-positive bacteria *Staphylococcus aureus* ATCC 25923 (21 mm) while the lowest activity was recorded against the Gram-negative bacteria *Escherichia coli* ATCC 8739 and *Klebsiella pneumonia* ATCC 13883 (15 mm).

Table 3. Antimicrobial activity of isolate EA9 cell free supernatant against different indicator pathogens.

Indicator Pathogen	Inhibition Zone Diameter (mm)
Pseudomonas fluorescens ATCC 13525	18 ± 0.5
Streptococcus agalactiae ATCC 13813	19 ± 0.44
Aeromonas hydrophila ATCC 13037	18 ± 0.23
Staphylococcus aureus ATCC 25923	21 ± 0.115
Escherichia coli ATCC 8739	15 ± 0.05
Enterococcus faecalis ATCC 29212	18 ± 0.15
Klebsiella pneumonia ATCC 13883	15 ± 0.23

2.2. *In Vivo Testing of the Antisepsis Action of the Isolated EA9 Probiotic*

2.2.1. EA9 Effects on Liver and Kidney Functions Tests

Serum levels of liver functions tests, including AST and ALT, as well as the kidney functions tests, including BUN and creatinine, were significantly ($p \leq 0.01$ and $p \leq 0.05$; respectively) augmented in CLP groups as compared to the SHAM and SHAM + AE9 groups. Treatment of CLP animals with *Enterococcus faecium* EA9 significantly ($p \leq 0.05$) lowered the liver and kidney functions tests as compared to the CLP group (Table 4).

Table 4. Assessment of serum biochemistry following treatment with *Enterococcus faecium* EA9 probiotic in septic Wistar rats.

	SHAM	SHAM + EA9	CLP	CLP + EA9
AST	143.83 ± 13.79 **	141.16 ± 20.06 **	211 ± 25.49	164.5 ± 41.92 *
ALT	57.33 ± 8.94 **	55.33 ± 7.89 **	83.5 ± 13.14	63 ± 12.17 *
BUN	72.16 ± 14.77 *	72.51 ± 16.00 *	104 ± 28.09	74.66 ± 7.89 *
Creatinine	0.71 ± 0.09 *	0.71 ± 0.08 *	1.1 ± 0.33	0.73 ± 0.15 *

Data are presented as mean ± SD (n = 6) and statistically analyzed by one way ANOVA followed by Tukey-Kramer post hoc test. Significance was considered when * $p \leq 0.05$ or ** $p \leq 0.01$ vs. CLP.

2.2.2. EA9 Effects on Liver and Kidney Inflammation

The dry/wet ratio in the liver samples from CLP group was significantly lower than SHAM ($p \leq 0.05$), SHAM + EA9 ($p \leq 0.01$), and CLP + EA9 ($p \leq 0.05$) groups while the dry/wet ratio in the renal samples from CLP group was significantly lower than SHAM ($p \leq 0.01$) and SHAM + EA9 ($p \leq 0.01$) groups (Figure 4A). The levels of inflammatory cytokines, such as TNF-α, IL-1β, and IL-6, were significantly ($p \leq 0.05$ and $p \leq 0.01$) higher in liver and kidney samples from CLP group as compared to SHAM and SHAM + EA9 groups. CLP + EA9 showed markedly ($p \leq 0.05$) lower hepatic levels of TNF-α, IL-1β, and IL-6 as well as renal levels of TNF-α and IL-1β as compared to CLP group (Figure 4B–D).

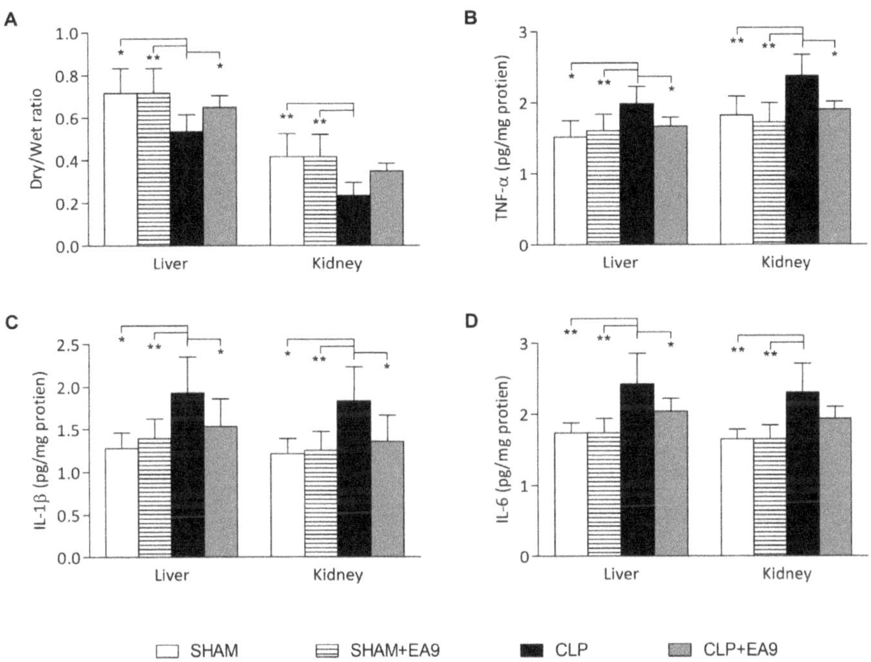

Figure 4. Assessment of inflammation in hepatic and renal tissues following treatment with *Enterococcus faecium* EA9 probiotic in septic Wistar rats. (**A**) degree of edema and fluid infiltration expressed as dry/wet ratio of hepatic and renal tissues, (**B**) hepatic and renal levels of tumor necrosis factor alpha (TNF-α), (**C**) hepatic and renal levels of interleukin-1 beta (IL-1β), and (**D**) hepatic and renal levels of interleukin-6 (IL-6). Data are presented as mean ± SD (n = 6) and statistically analyzed by one way ANOVA followed by Tukey-Kramer post hoc test. Significance was considered when * $p \leq 0.05$ or ** $p \leq 0.01$.

2.2.3. EA9 Effects on Liver and Kidney Oxidative Stress

In the liver samples, TBARS ($p \leq 0.01$) level was increased while the levels of GSH ($p \leq 0.05$) and the enzymatic activities of CAT ($p \leq 0.01$), SOD ($p \leq 0.01$), GPx ($p \leq 0.01$), and GR ($p \leq 0.05$) were significantly decreased in the CLP group as compared to SHAM and SHAM + EA9 groups. Similarly in the kidney samples, TBARS ($p \leq 0.001$) was heightened while GSH ($p \leq 0.05$) levels and the enzymatic activities of CAT ($p \leq 0.05$), SOD ($p \leq 0.01$), GPx ($p \leq 0.01$), and GR ($p \leq 0.01$) were significantly lowered in the CLP group as compared to SHAM and SHAM + EA9 groups. *Enterococcus faecium* EA9 probiotic treatment to the CLP animals significantly lowered the hepatic and renal TBARS ($p \leq 0.05$) level and significantly increased the hepatic GSH ($p \leq 0.05$), hepatic and renal CAT ($p \leq 0.05$) activity, hepatic SOD ($p \leq 0.05$) activity, and hepatic and renal GPx ($p \leq 0.05$) activity (Figure 5A–F).

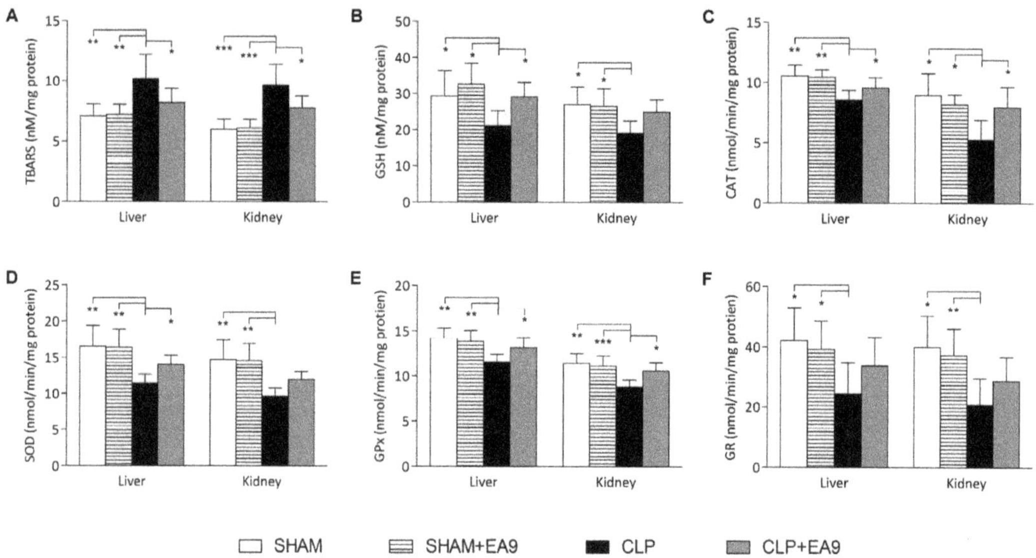

Figure 5. Assessment of oxidative damage in hepatic and renal tissues following treatment with *Enterococcus faecium* EA9 probiotic in septic Wistar rats. (**A**) hepatic and renal levels TBARS as a marker for lipid peroxidation, (**B**) hepatic and renal levels of GSH as an endogenous antioxidant, (**C**) hepatic and renal enzymatic activities of catalase (CAT), (**D**) hepatic and renal enzymatic activities of superoxide dismutase (SOD), (**E**) hepatic and renal enzymatic activities of glutathione peroxidase (GPx), and (**F**) hepatic and renal enzymatic activities of glutathione reeducates (GR). Data are presented as mean ± SD ($n = 6$) and statistically analyzed by one way ANOVA followed by Tukey-Kramer post hoc test. Significance was considered when * $p \leq 0.05$, ** $p \leq 0.01$, or *** $p \leq 0.001$.

2.2.4. EA9 Effects on Liver and Kidney Histology

The results of the semiquantitative histological analysis revealed that hepatic and renal slides from SHAM and SHAM + EA9 groups almost normal appearance without inflammatory cells infiltration or necrotic areas. Hepatic and renal slides from CLP group had a significantly ($p \leq 0.01$) higher number of inflammatory cells infiltration and exudates as well as multiple spots of necrosis as compared to SHAM and SHAM + EA9 groups. Although CLP animals with *Enterococcus faecium* EA9 showed reduced hepatic and renal inflammatory foci and necrotic tissues, the difference from the CLP was not statistically significant (Figure 6A–J).

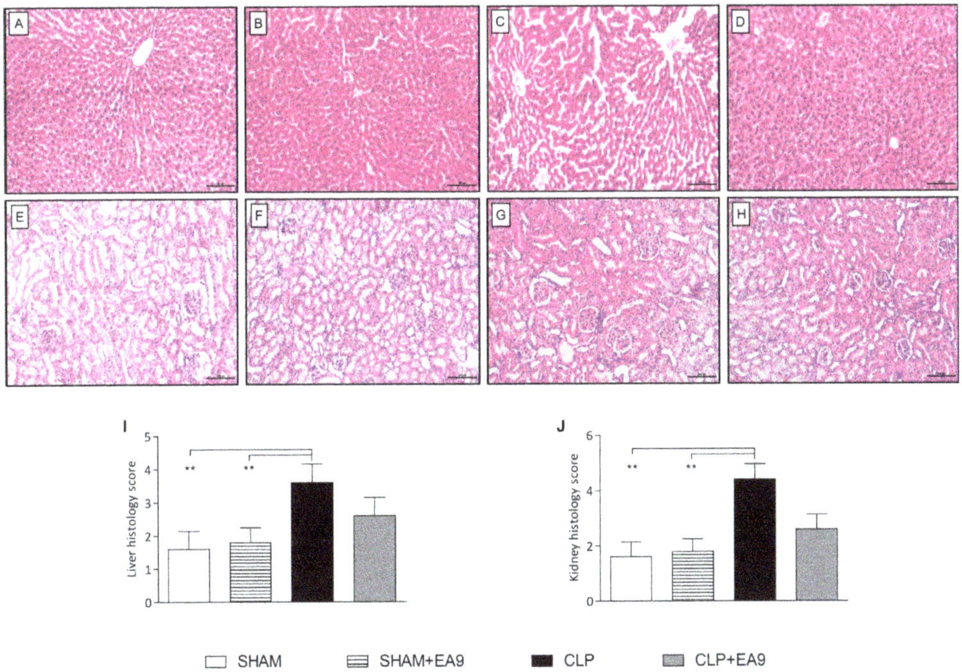

Figure 6. Histological assessment of hepatic (**A–D**) and renal (**E–H**) tissues following treatment with *Enterococcus faecium* EA9 probiotic in septic Wistar rats (scale bar = 100 μm). (**A**) Liver tissue from sham group showing regular hepatic cells, (**B**) Liver tissue from sham + *Enterococcus faecium* EA9 group showing normal hepatic cells, (**C**) Liver tissue from CLP group showing necrotic hepatic ells with multiple inflammatory infiltration, (**D**) Liver tissue from sham + *Enterococcus faecium* EA9 group showing declined hepatic necrosis and inflammatory cells, (**E**) kidney tissue from sham group showing regular renal structure, (**F**) Kidney tissue from sham + *Enterococcus faecium* EA9 group showing ordinary looking renal tubules, (**G**) Kidney tissue from CLP group showing several necrotic renal ells with diffused inflammatory cells, and (**H**) Kidney tissue from sham + *Enterococcus faecium* EA9 group showing enhanced renal structure and fewer inflammatory cells. Semi-quantitative histological scoring of the (**I**) liver and (**J**) kidney slides. Data are presented as mean ± SD (n = 5) and statistically analyzed by Kruskal-Wallis test and Dunn's multiple comparison post hoc test. Significance was considered when ** $p \leq 0.01$.

2.2.5. EA9 Effects on Liver and Kidney Gene Expressions

In the hepatic tissues, gene expressions of IL-1β ($p \leq 0.01$), INF-γ ($p \leq 0.01$), COX-2 ($p \leq 0.01$), iNOS ($p \leq 0.01$), and STAT-3 ($p \leq 0.01$) were significantly up-regulated while the gene expressions of IL-10 ($p \leq 0.05$), SOD-1 ($p \leq 0.01$), SOD-2 ($p \leq 0.01$), HO-1 ($p \leq 0.01$), AKT ($p \leq 0.05$), and mTOR ($p \leq 0.01$) were significantly down-regulated in the CLP group as compared to SHAM and SHAM + EA9 groups. The treatment of the CLP animals with the probiotic *Enterococcus faecium* EA9 significantly inhibited the hepatic gene expressions of IL-1β ($p \leq 0.05$), INF-γ ($p \leq 0.05$), iNOS ($p \leq 0.05$), and STAT-3 ($p \leq 0.05$) while it promoted the hepatic gene expressions of HO-1 ($p \leq 0.05$), AKT ($p \leq 0.01$), and mTOR ($p \leq 0.01$) as compared to CLP group (Figure 7A,B). In the renal tissues, gene expressions of IL-1β ($p \leq 0.01$), INF-γ ($p \leq 0.01$), COX-2 ($p \leq 0.01$), and STAT-3 ($p \leq 0.01$) were significantly up-regulated while the gene expressions of IL-10 ($p \leq 0.01$), SOD-1 ($p \leq 0.05$), SOD-2 ($p \leq 0.05$), HO-1 ($p \leq 0.05$), AKT ($p \leq 0.05$), and mTOR ($p \leq 0.05$) were significantly down-regulated in the CLP group as compared to SHAM and SHAM + EA9 groups. Septic animals with the *Enterococcus faecium* EA9 treatment has significantly lower renal gene expressions of

INF-γ ($p \leq 0.05$), COX-2 ($p \leq 0.01$), and STAT-3 ($p \leq 0.15$), along with enhanced renal gene expressions of IL-10 ($p \leq 0.05$), AKT ($p \leq 0.01$), and mTOR ($p \leq 0.01$) as compared to the CLP group (Figure 7C,D).

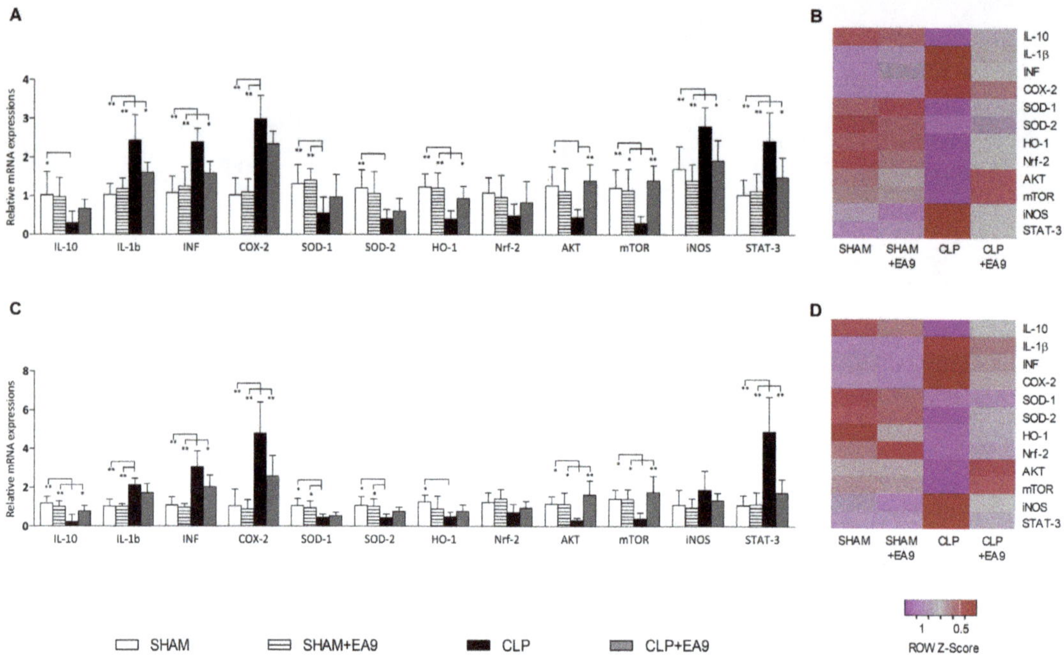

Figure 7. Assessment of mRNA expressions of multiple genes in hepatic (**A**,**B**) and renal (**C**,**D**) tissues following treatment with *Enterococcus faecium* EA9 probiotic in septic Wistar rats. Data are presented as mean ± SD (n = 6) and statistically analyzed by one way ANOVA followed by Tukey-Kramer post hoc test. Significance was considered when * $p \leq 0.05$ or ** $p \leq 0.01$.

3. Discussion

The marine environment represents an untapped niche for novel LAB isolates [24]. Marine LAB have promising possibilities as probiotics and can be considered as a source of new bioactive compounds since their characteristics and biological activity vary from their terrestrial counterparts. Their ability to live under the harsh marine circumstances makes them more competitive than their terrestrial counterparts. Therefore, the marine isolate *Enterococcus faecium* EA9 was assessed in this study for the primary safety criteria for use as a probiotic. Additionally, the antisepsis effects of *Enterococcus faecium* EA9 probiotic was evaluated as a potential therapeutic application using an in vivo model of sepsis in rats.

Enterococcus faecium EA9 demonstrated non-hemolytic activity, which confirms its safety. Previous studies have also indicated the non-hemolytic ability of *Enterococcus* sp. Pieniz et al. [25] observed that *Enterococcus durans* strain LAB18s isolated from Minas Frescal cheese did not possess any hemolytic activity. Moreover, De Vuyst et al. [26] found that several *Enterococcus* species collected from various environments were all non-hemolytic. The results showed that the marine isolate *Enterococcus faecium* EA9 is sensitive to 6 types of the tested antibiotics. On the other hand, EA9 showed resistance to 5 types of antibiotics, which is a very important criterion in developing combination therapies of antibiotic/probiotic [27]. The probiotic strain must be able to survive in presence of some types of antibiotics. These antibiotics are widely employed in medication and as preservatives in food products; as a result, a probiotic strain that is highly susceptible to most antibiotics will be quickly cleared from the gastrointestinal system, rendering it useless [28]. The antibiotic resistance genes of many probiotic strains are found in a non-transferable chromosome [29], which reduce the likelihood of transmission to other bacteria in the gut [30]. In addition, EA9 was shown to be sensitive to vancomycin, which is noteworthy given that vancomycin is often employed as a last-resort antibiotic owing to its efficiency against multiple drug-resistant bacteria. Furthermore, the resistance of probiotics to various antibiotics has previously been reported by Bs et al. [28], who reported that *Enterococcus faecium* BS5 is susceptible to tetracycline, rifamycin, and chloramphenicol but resistant to erythromycin, kanamycin, penicillin, streptomycin, and ampicillin. Notably, different probiotic strains often exhibit varying degrees of resistance to various kinds of antibiotics [31]. For instance, *Lactobacillus bulgaricus* and *Streptococcus thermophilus* have been shown to have a high level of resistance to kanamycin, ampicillin, and streptomycin [32].

Enterococcus faecium EA9 showed a good potential to resist hard surviving conditions when the pH values were reduced to 4, 3, and 2. The survival rates of *Enterococcus faecium* EA9 in this condition was 98%, 84%, and 69%. Similarly, Pimentel et al. [33] obtained 73 *enterococci* isolate from Terrincho cheese, and he reported that most of the obtained *Enterococcus durans* and *Enterococcus faecium* strains were resistant to a low pH environment. Bs et al. [28] recorded the strong ability of *Enterococcus faecium* BS5 to withstand low pH with respective survival rates of 78% (pH 4), 78% (pH 3), and 25% (pH 2). Moreover, *Enterococcus faecium* EA9 managed to endure high concentrations of bile salts, which is compatible with Bs et al. [28], who discovered that *Enterococcus faecium* BS5 managed to endure and grow at varied bile contents. Since bile salts alter the cell membrane structure, they are very hazardous to living organisms. The probiotic strains' resilience to high bile contents is mainly due to bile salts hydrolase (BSH), which hydrolyzes the toxic conjugated bile salts [34]. The capacity of the probiotic strain to tolerate the circumstances of the small intestine could be more crucial than the capacity to withstand low pH. Some investigations suggest that acid labile bacteria may be buffered throughout the stomach. The probiotic strain should, however, be able to endure destruction by bile salts and hydrolytic enzymes present in the intestine and must be able to survive and colonize there in order to be beneficial to the host. This trait is crucial for the choice of innovative probiotic strains [25].

Enterococcus faecium EA9 was reported to elicit potential antibacterial efficacy against all of the investigated pathogenic strains. This is critical when choosing probiotic strains since they must be capable of eliminating competitors. LAB generates a variety of antimicrobial agents, giving it an edge over other gut pathogens [35]. The production of the antimicrobial agents is highly affected by strain and growth condition differences [36]. Among the many antimicrobial agents produced by LAB are lactic acid, fatty acids, acetic acid, inhibitory peptides (bacteriocins), and hydrogen peroxide. *Enterococci* produce bacteriocins called enterocins, which have been isolated and described in many *enterococci* [37].

In the second part of the study, the antisepsis effects of the marine isolate *Enterococcus faecium* EA9 were assessed in a CLP model, which creates an intensive pathological environment characterized by cellular inflammation and oxidative stress in multiple organs including the liver [38] and kidneys [39]. The CLP resulted in severe alterations in the histological structure of hepatic and renal cells along with impaired hepatic and renal functions. Additionally, the reduced dry/wet ratio in the CLP group indicated the development of inflammatory edema. The marked increase in TNF-α, IL-1β, and IL-6 also indicated the inflammatory response. Studies proposed oxidative stress and CLP linkage, where markers of oxidative stress are triggered by CLP along with suppressed endogenous antioxidant protective mechanisms [40,41]. Likewise, our biochemical findings revealed the augmentation of cellular membrane lipid peroxidation markers (TBARS) as well as reduced endogenous sulfhydryl containing-antioxidant (GSH) and suppressed antioxidant enzymes in the CLP group, which indicates the extensive production of free radicals and decreased antioxidant defensive capabilities. This data implies that the CLP model was a successful model to evaluate the antisepsis potential of the isolate *Enterococcus faecium* EA9 probiotic.

The marine origin of the isolate *Enterococcus faecium* EA9 probiotic supported the fact that several marine-derived microorganisms and the associated-bioactive compounds could have possible medical applications, such as regulating blood pressure [42], hepatoprotective effects and antioxidant potentials [43], suppression of intestinal inflammation [44], antitumor activity [45,46], and antibacterial activity [47]. In the present study, the marine isolate *Enterococcus faecium* EA9 probiotic markedly reduced the pathological consequences of sepsis in both hepatic and renal tissues. This was indicated by the significant improvement of hepatic and renal function tests along with less reported necrotic and inflammatory spots in liver and kidney samples. Furthermore, the inflammatory parameters including dry/wet ratio and inflammatory cytokines were suppressed by the assessed probiotic. These cytokines are secreted by immune cells in response to sepsis to promote inflammation as a kind of defense mechanism to protect cellular integrity and maintain cellular functions [48]. This will create a harmful environment of free radicals. Therefore, the oxidative stress and lipid peroxidation biomarkers were augmented in septic animals. The free radicals alleviating effects of EA9 probiotic markedly suppressed lipid peroxidation and enhanced the antioxidant mechanisms, which prevented further cellular damage. Interestingly, antioxidant genes, such as CAT, SOD, and GR, are expressed, identified, and characterized in marine microorganisms [49–51].

Gene expression of IL-10, IL-1β, INF-γ, and COX-2 was determined as an indicator for inflammatory response. IL-10 is an anti-inflammatory cytokine while IL-1β, INF-γ, and COX-2 are expressed during pathological inflammatory conditions, such as sepsis [52–54]. The anti-inflammatory effects of *Enterococcus faecium* EA9 probiotic were obvious at the gene level, where hepatic tissues had significantly reduced IL-1β and INF-γ gene expressions while renal tissues had significantly improved IL-10 and reduced INF-γ and COX-2 gene expressions. The SOD-1 gene encodes for the cytoplasmic SOD while SOD-2 is present in the mitochondria. Both enzymes catalyze the transformation of harmful free radicals to less reactive oxygen species. Although *Enterococcus faecium* EA9 probiotic enhanced the CLP-suppressed SOD activity in the hepatic tissues, its effect was not prominent on the gene level. HO-1 is a critical cytoprotective factor that maintains antioxidant/oxidant homeostasis in case of cellular stress and inflammation [55]. HO-1 and Nrf-2 are regulator for cellular oxidant resistance [56]. *Enterococcus faecium* EA9 probiotic abolished sepsis-induced down-regulation of HO-1 in the liver. Moreover, we did not report a significant effect of CLP or *Enterococcus faecium* EA9 probiotic on Nrf-2 gene expression. AKT/mTOR signaling pathway is well-known as a regulator of cell cycle and apoptosis. The sepsis protective effects following activation of this pathway have been documented in previous studies [57,58]. Our results showed similar effects, where *Enterococcus faecium* EA9 probiotic improved gene expression of AKT in both tissues and mTOR in kidney samples. The iNOS stimulation during sepsis might result in elevated levels of NO, which is associated with the development of nitrosative stress and reactive nitrogen species [59]. *Enterococcus faecium* EA9 probiotic inhibited iNOS gene expression in the liver tissues suggesting a site-specific effect of the probiotic in relation to the iNOS pathway. STAT-3 is a transcription factor that regulates multiple inflammatory cascades in many infectious and autoimmunity conditions, such as sepsis [60]. In line with the biochemical, histological, and molecular analysis of the inflammatory response, the STAT-3 gene was markedly up-regulated in the CLP group while *Enterococcus faecium* EA9 probiotic treatment lowered its expression in hepatic and renal tissues.

Despite our efforts to reduce variabilities during the surgical procedure by controlling multiple factors, such as differences in the ligated length of the cecum, the size and number of punctures, and the amount the extruded stool, the procedure remains subjective. Furthermore, some revealing factors of the antisepsis effects of the tested probiotic were not recorded, such as body temperature and survival rate, which could have added further evidence of the protective antisepsis effects of the probiotic.

4. Materials and Methods

4.1. Assessment of Probiotic Potential of Isolate EA9

4.1.1. Phenotypic and Biochemical Characterization of Isolate EA9

For the biochemical identification and characterization of isolate EA9, overnight culture of EA9 were inoculated in Gram-positive (GP) cards and run on the VITEK 2 compact fully automated microbial identification system version 07.01 (BioMérieux, Craponne, France). The system analyzes the cells for sugar fermentation, antibiotic resistance, and enzyme hydrolysis, among other biochemical assays. Moreover, scanning electron microscope (SEM) (JSM-5300, JEOL, Tokyo, Japan) was also used to observe the typical cell shape and size of isolate EA9 [61].

4.1.2. Molecular Identification and Phylogenetic Analysis of Isolate EA9

Enterococcus faecium EA9 was grown for 24 h at 37 °C in MRS medium, then the total genomic DNA of EA9 was extracted and purified according to the manufacturer's instructions for the DNA extraction kit (QIAGEN, Hilden, Germany). The prepared DNA was loaded with the loading dye in a 1% agarose gel prepared in Tris EDTA Acetic acid (TEA) buffer with 1 μg/mL ethidium bromide (Sigma, St. Louis, MO, USA), and the voltage was applied (90 *v*/cm) for electrophoresis after soaking in TEA buffer. The UV transilluminator (Bio-Rad, Hercules, CA, USA) was used to visualize the bands that had been obtained. For molecular identification of the marine isolate EA9, the PCR product for EA9 16S rRNA gene was performed and sequenced by Applied Biotechnology Co., Ismailia, Egypt. Under the accession number MW218438, the obtained sequence was submitted to the National Center for Biotechnology Information (NCBI) GenBank. Furthermore, the sequence was examined using NCBI's BLASTn tools, which revealed that isolate EA9 has a 99.88% similarity with *Enterococcus faecium* strain DSM 20477 (accession number NR114742.1). The isolate EA9 16S rRNA gene sequence was then aligned with related species using the ClustalW tool (Version 2.1), and a phylogenetic tree was generated through www.phylogeny.fr (accessed on 12 February 2021) using the Maximum Likelihood technique [62].

4.1.3. Hemolytic Activity

Blood agar plates (supplemented with 7% human blood) were streaked with isolate EA9 and incubated at 37 °C for 24 h to test hemolytic activity. The plates were then examined for evidence of blood hemolysis, such as the creation of clear zones (β-hemolysis), development of green zones of partial hemolysis (α-hemolysis), or the absence of clear zones around the colonies (γ-hemolysis) [63].

4.1.4. Antibiotic Susceptibility

The antibiotic susceptibility of EA9 was determined using the Kirby-Bauer disc diffusion method. Antibiotic discs (Oxoid, Basingstoke, UK) of various types were placed on the surface of agar plates that had previously been inoculated with EA9. After a 24 h incubation period at 37 °C, the plates were checked for the development of clear zones around discs, indicating EA9 sensitivity to the tested antibiotic [64].

4.1.5. Resistance to Low pH

The ability of isolate EA9 to withstand low pH levels was evaluated according to Nawaz et al. study [65]. The isolate was inoculated (1% *v*/*v*) into sterile MRS broth adjusted to several pH values (2, 3, 4, and 6.5) with 0.1 N HCl and incubated at 37 °C. The absorbance was measured at 620 nm using a spectrophotometer (Unico, Franksville, WI, USA) at hourly intervals for 6 h. The isolate EA9 was cultured at various pH levels to assess its capacity to resist low pH conditions, mimicking the pH of the stomach, which is estimated to be around 3 with a stay period of about 4 h.

4.1.6. Bile Salts Tolerance

The ability of isolate EA9 to withstand high bile salts concentrations was investigated according to the method described by Harsa and Yavuzdurmaz [66]. An overnight culture of EA9 (1% v/v) was inoculated in 15 mL sterile MRS broth with various bile salts concentrations (0, 0.1, and 0.3% (w/v)) and incubated for 4 h at 37 °C. The absorbance of the culture was measured at hourly intervals with a spectrophotometer (Unico, Franksville, WI, USA) at 620 nm. The average bile salts content in the intestine is variable, although it is approximated to be around 0.3% w/v, with a stay time of 4 h.

4.1.7. Antimicrobial Activity

Different Gram-negative bacterial pathogens (*Klebsiella pneumonia* ATCC 13883, *Escherichia coli* ATCC 8739, *Aeromonas hydrophila* ATCC 13037, *Pseudomonas fleurescence* ATCC 13525) as well as Gram-positive bacterial pathogens (*Staphylococcus aureus* ATCC 25923, *Enterococcus faecalis* ATCC 29212, *Streptococcus agalactiae* ATCC 13813) were used to evaluate the antimicrobial activity of isolate EA9 using the agar well cut diffusion technique. Each pathogenic strain was inoculated (1% v/v) in nutrient agar plates. Using a sterile cork borer, wells of 8 mm diameter were cut in solidified agar and filled with 100 μL of filter sterilized cell free supernatant (CFS) of EA9. The antibacterial activity was assessed by measuring the clear zone diameter around each well after 24 h of incubation at 37 °C.

4.2. In Vivo Testing of the Antisepsis Action of the Isolated EA9 Probiotic

4.2.1. Ethical Approval and Experimental Model

The experimental animal part of this study was conducted in the animal house at Institute of Graduate Studies and Research, Alexandria University under an ethical approval number AU14-210126-3-2. The experimental procedures were in accordance with the 8th edition of the National Institute of Health (NIH) guidelines for the care and use of laboratory animals and ARRIVE guidelines. A number of twenty-four healthy male Wistar albino rats weighting approximately 200 g were included in this study, whereas animals with abnormalities or low weight were excluded. Animals were simply randomized into four groups (n = 6) in labeled polystyrene cages. The first group (SHAM) included animals, who were subjected to the surgical procedure without the ligation and puncture steps. The second group (SHAM + EA9) included animals, who were subjected to the surgical procedure without the ligation and puncture steps as well as probiotic pretreatment for 10 days before the surgical procedure. The third group (CLP) included animals, who were subjected to the CLP procedure. The fourth group (CLP + EA9) included animals, who were subjected to the CLP procedure as well as probiotic pretreatment for 10 days before the surgical procedure. The sepsis model (CLP surgical procedure) was preformed according to the method described by Zubrow et al. [67]. The cecum of the generally anesthetized animals by 60 mg/kg ketamine and 5 mg/kg xylazine was visualized and exposed after a median laparotomy incision of 15 mm long. Then, the distal portion of the cecum was ligated with a sterile 4-0 silk thread. After this, the ligated sections were exposed to 2 or 3 punctures using an 18-gauge syringe needle followed by squeezing the small volume of the feces. A sterile saline (20 mL/kg body weight) were provided subcutaneously, and the ligated tissues were replaced in the abdominal cavity. Pain following the surgical procedures was controlled by ketorolac (30 mg/kg) every 12 h, intraperitoneally. Ten days before the sepsis model, the tested EA9 probiotic in a freeze-dried format was diluted in 1 M of NaCl solution (1×10^7 CFU/mL) and a 250 μL (per animal) were provided consecutively and blindly by oral gavage to animals in the third and fourth groups. Three days after the CLP, all rats were euthanized with a lethal inhaled dose of general anesthesia, and the biological samples including blood, liver, and kidneys were collected for histological, biochemical, and molecular analysis.

4.2.2. Serum Biochemistry

Blood samples were centrifugated at 4000 RPM for 15 min to obtain serum samples, which were used to determine the concentrations of serum aspartate aminotransferase (AST) and serum alanine transaminase (ALT) as markers for liver disfunction, as well as blood urea nitrogen (BUN) and creatinine as markers for renal disfunction using a fully-automated Cobas C311 chemistry analyzer (Roche diagnostics Ltd., Risch-Rotkreuz, Switzerland).

4.2.3. Evaluation of Inflammation

The concentrations of inflammatory cytokines, such as tumor necrosis factor alpha (TNF-α), interleukin-6 (IL-6), and interleukin-1 beta (IL-1β) in the hepatic and renal homogenates, were measured with an enzyme-linked immunosorbent assay (ELISA) technique following instructions provided by the kits (R & D systems Inc., Minneapolis, MN, USA). Tissue edemas were determined by measuring wet/dry. In brief, a small portion from liver and kidney samples were isolated and weighed (wet weight). These portions were re-weighed after drying (dry weight) overnight at 60 °C.

4.2.4. Evaluation of Oxidative Damage

After homogenization in a phosphate buffer (1:10, *w*/*v*), the liver and kidney homogenates were used to determine the levels of thiobarbituric acid reactive substances (TBARS), as a lipid peroxidation biomarker and reduced glutathione (GSH), and as an endogenous antioxidant, using commercial diagnostic kits (Cayman Chemical Co., Ann Arbor, MI, USA). Though, the post-mitochondria supernatants obtained after centrifugation of the homogenates were used to determine the enzymatic activities of catalase (CAT), superoxide dismutase (SOD), glutathione reeducates (GR), and glutathione peroxidase (GPx) using their commercially assay kits (R & D systems Inc., Minneapolis, MN, USA).

4.2.5. Histological Evaluation

Liver and kidney samples were collected in 10% formalin solution, fixed in plastic cassettes, and dehydrated by graded concentration of ethanol. Afterward, each sample was embedded in a single paraffin block. All blocks were sectioned using a microtome into a 4 µm thickness sections. After picking up on glass slides, these sections were stained with hematoxylin and eosin (H & E) for histological evaluation. The stained slides were assessed microscopically and scored in a blind manner. Tissue necrosis was evaluated based on the following criteria: (score-0) normal looking cells with no necrosis, (score-1) 1 or 2 spots of necrotic tissues, (score-2) 3 or 5 spots necrotic tissues, and (score-3) more than 5 spots necrotic tissues. Tissue inflammation was assessed according to the following measures: (score-0) normal looking cells lacking inflammatory cells, (score-1) scattered inflammatory cells, (score-2) inflammatory foci, and (score-3) multiple diffused inflammatory cells. The final necrotic and inflammatory scores were summed and presented per each group.

4.2.6. Polymerase Chain Reaction (PCR) Assessment of Genes Expressions

Total RNA was extracted from the liver and kidney samples using the TRIzol method (easy-RED, iNtRON Biotechnology). At the end of the experiment, RNA purity was determined at 260/280 nm by a NanoDrop system (BioDrop, Biochrom Ltd., Cambridge, UK). The cDNA (1 ng/uL) was synthesized from the RNA samples that showed the highest A260/A280 ratio by adding DNase I (New England Biolabs, Ipswich, MA, USA) as the template. The reverse transcriptase (RT-PCR beads, Enzynomics, Daejeon, Korea) was employed, and the PCR amplification was performed by (Applied Biosystems, Veriti 96-Well Thermal Cycler, Waltham, MA, USA). The synthetized cDNA was incorporated in the Real-Time PCR reaction (Bico, Thermo-Fisher, Waltham, MA, USA) under the following procedure: 15 min of initial denaturation (95 °C) followed by 40 cycles of 95 °C for 10 s, 55–60 °C for 20 s, and 72 °C for 30 s. Unique and specific products were noticed when a melting curve at the end of the last cycle where the temperature increased from (55–60 to 95 °C) in increments of 0.5 °C. Primers list of the targeted genes in this study are present in

Table 5. β-actin served as a housekeeping gene. The values given out in an n-fold difference relative to the calibrator (control) when the $2^{\Delta\Delta Ct}$ method was applied to normalize the critical threshold (Ct) quantities of target genes relative to those of β-actin [68].

Table 5. Primers sequences according to NCBI gene database.

Gene Name	Accession Number	Sequences
Interleukin-10 (IL-10)	XM_006249712.4	F: 5′-TGCCTTCAGTCAAGTGAAGAC-3′
		R: 5′-AAACTCATTCATGGCCTTGTA-3′
Interleukin-1β (IL-1β)	NM_031512.2	F: 5′-CACCTCTCAAGCAGAGCACAG-3′
		R: 5′-GGGTTCCATGGTGAAGTCAAC-3′
Interferon-γ (INF-γ)	NM_138880.3	F: 5′-GTGTCATCGAATCGCACCTG-3′
		R: 5′-GTTCACCTCGAACTTGGCGA-3′
Cyclooxygenase-2 (COX-2)	S67722.1	F: 5′-TGAGTACCGCAAACGCTTCT-3′
		R: 5′-ACACAGGAATCTTCACAAATGGA-3′
Superoxide dismutase-1 (SOD-1)	NM_017050.1.1	F: 5′-TAACTGAAGGCGAGCATGGG-3′
		R: 5′-CCTCTCTTCATCCGCTGGAC-3′
Superoxide dismutase-2 (SOD-2)	NM_017051.2	F: 5′-AATCAACAGACCCAAGCTAGGC-3′
		R: 5′-CACAATGTCACTCCTCTCCGAA-3′
Heme oxygenase-1 (HO-1)	XM_039097470.1	F: 5′-GTAAATGCAGTGTTGGCCCC-3′
		R: 5′-ATGTGCCAGGCATCTCCTTC-3′
Nuclear factor erythroid 2-related factor-2 (Nrf-2)	NM_031789.3	F: 5′-TTGTAGATGACCATGAGTCGC-3′
		R: 5′-TGTCCTGCTGTATGCTGCTT-3′
Protein kinase B (AKT)	XM_006240631.3	F: 5′-ACCTCTGAGACCGACACCAG-3′
		R: 5′-AGGAGAACTGGGGAAAGTGC-3′
Mammalian target of rapamycin (mTOR)	NM_019906.2	F: 5′-GACAACAGCCAGGGCCGCAT-3′
		R: 5′-ACGCTGCCTTTCTCGACGGC-3′
Inducible nitric oxide synthase (iNOS)	XM_039085203.1	F: 5′-ATGGAACAGTATAAGGCAAACACC-3′
		R: 5′-GTTTCTGGTCGATGTCATGAGCAAAGG-3′
Signal transducer and activator of transcription-3 (STAT-3)	XM_006247257.4	F: 5′-GGGCCTGGTGTGAACTACTC-3′
		R: 5′-ATGGTATTGCTGCAGGTCGT-3′
Beta actin (β-actin)	NM_031144.3	F: 5′-CCGCGAGTACAACCTTCTTG-3′
		R: 5′-CAGTTGGTGACAATGCCGTG-3′

4.2.7. Statistical Analysis

The normality of the data was assessed with the Shapiro-Wilk test. Results of the biochemical and molecular analysis were expressed as mean ± standard deviation (SD) and statistically analyzed using one way analysis of variance (ANOVA) followed by of Tukey-Kramer post hoc test. Scores of the histological evaluation of necrosis and inflammation were also expressed as mean ± SD and analyzed by Kruskal-Wallis test and Dunn's multiple comparison post hoc test. Means differences were judged as significant when p values were ≤ 0.05. GraphPad Prism 5 (GraphPad Software, Inc., La Jolla, CA, USA) was utilized as the analysis software. Heatmaps were executed using Heatmapper (Wishart Research Group, University of Alberta, Ottawa, Canada) [69]. Graphical abstract was prepared using Biorender.com (Agreement number: FV24JOFCRX, 20 October 2022).

5. Conclusions

In the present work, the marine isolate EA9 was tested for its safety and probiotic potential in addition to its possible health benefits. According to the acquired results, this isolate was identified as *Enterococcus faecium,* and it possesses potential probiotic and technical qualities, include resistance to low pH, high bile salts concentrations, and broad antibacterial activity. In addition, the results of the in vivo investigation show that *Enterococcus faecium* EA9 isolate may have possible therapeutic use in pathological conditions, where inflammation and oxidative stress are deemed as contributing factors. The efficacy against experimental sepsis was proven in this study while future research might provide further evidence for other medical applications of the probiotic.

Author Contributions: E.H.Z. Conceptualization; formal analysis; methodology; writing—original draft preparation. H.M.A. Conceptualization; formal analysis; methodology; writing—original draft preparation. A.S.E.S. Investigation; methodology; visualization. E.M.A. Investigation; methodology; resources. M.M.A. Investigation; methodology; writing—Review & Editing; S.S.A.-R. Funding acquisition; formal analysis; writing—Review & Editing. All authors have read and agreed to the published version of the manuscript.

Funding: This research was funded by the Deanship of Scientific Research at King Saud University for funding the work through Research Group Project No. RGP-179.

Institutional Review Board Statement: The experimental animal part of this study was conducted in the animal house at Institute of Graduate Studies and Research, Alexandria University under an ethical approval number AU14-210126-3-2. The experimental procedures were in accordance with the 8th edition of the national institute of health (NIH) guidelines for the care and use of laboratory animals and ARRIVE guidelines.

Informed Consent Statement: Not applicable.

Data Availability Statement: The datasets generated and/or analyzed in the present study are included in the manuscript.

Acknowledgments: The authors express their appreciation to the Deanship of Scientific Research at King Saud University for funding the work through Research Group Project No. RGP-179.

Conflicts of Interest: The authors declare no conflict of interests.

References

1. Florou-Paneri, P.; Christaki, E.; Bonos, E. Lactic Acid Bacteria as Source of Functional Ingredients. In *Lactic Acid Bacteria-R & D for Food, Health and Livestock Purposes*; Kongo, M., Ed.; IntechOpen: London, UK, 2013.
2. Ou, Y.; Chen, S.; Ren, F.; Zhang, M.; Ge, S.; Guo, H.; Zhang, H.; Zhao, L. Lactobacillus casei Strain Shirota Alleviates Constipation in Adults by Increasing the Pipecolinic Acid Level in the Gut. *Front. Microbiol.* **2019**, *10*, 324. [CrossRef]
3. Wang, J.-J.; Li, S.-H.; Li, A.-L.; Zhang, Q.-M.; Ni, W.-W.; Li, M.-N.; Meng, X.-C.; Li, C.; Jiang, S.-L.; Pan, J.-C.; et al. Effect of *Lactobacillus acidophilus* KLDS 1.0738 on miRNA expression in in vitro and in vivo models of β-lactoglobulin allergy. *Biosci. Biotechnol. Biochem.* **2018**, *82*, 1955–1963. [CrossRef] [PubMed]
4. Fooladi, A.A.I.; Hosseini, H.M.; Nourani, M.R.; Khani, S.; Alavian, S.M. Probiotic as a Novel Treatment Strategy Against Liver Disease. *Hepat. Mon.* **2013**, *13*, e7521. [CrossRef]
5. Mokoena, M.P.; Mutanda, T.; Olaniran, A.O. Perspectives on the probiotic potential of lactic acid bacteria from African traditional fermented foods and beverages. *Food Nutr. Res.* **2016**, *60*, 29630. [CrossRef] [PubMed]
6. Kumar, H.; Salminen, S.; Verhagen, H.; Rowland, I.; Heimbach, J.; Bañares, S.; Young, T.; Nomoto, K.; Lalonde, M. Novel probiotics and prebiotics: Road to the market. *Curr. Opin. Biotechnol.* **2015**, *32*, 99–103. [CrossRef]
7. Román, L.; Real, F.; Sorroza, L.; Padilla, D.; Acosta, B.; Grasso, V.; Bravo, J.; Acosta, F. The in vitro effect of probiotic Vagococcus fluvialis on the innate immune parameters of Sparus aurata and Dicentrarchus labrax. *Fish Shellfish. Immunol.* **2012**, *33*, 1071–1075. [CrossRef]
8. Sorroza, L.; Padilla, D.; Acosta, F.; Román, L.; Grasso, V.; Vega, J.; Real, F. Characterization of the probiotic strain Vagococcus fluvialis in the protection of European sea bass (Dicentrarchus labrax) against vibriosis by Vibrio anguillarum. *Vet. Microbiol.* **2012**, *155*, 369–373. [CrossRef]
9. Krawczyk, B.; Wityk, P.; Gałęcka, M.; Michalik, M. The Many Faces of *Enterococcus* spp.—Commensal, Probiotic and Opportunistic Pathogen. *Microorganisms* **2021**, *9*, 1900. [CrossRef] [PubMed]

10. Yakovenko, E.P.; Strokova, T.V.; Ivanov, A.N.; Iakovenko, A.V.; Gioeva, I.Z.; Aldiyarova, M.A. The effectiveness of a probiotic containing Bifidobacterium longum BB-46 and Enterococcus faecium ENCfa-68 in the treatment of post-infectious irritable bowel syndrome. Prospective randomized comparative study. *Ter. arkhiv* **2022**, *94*, 180–187. [CrossRef]
11. Ghazisaeedi, F.; Meens, J.; Hansche, B.; Maurischat, S.; Schwerk, P.; Goethe, R.; Wieler, L.H.; Fulde, M.; Tedin, K. A virulence factor as a therapeutic: The probiotic *Enterococcus faecium* SF68 arginine deiminase inhibits innate immune signaling pathways. *Gut Microbes* **2022**, *14*, 2106105. [CrossRef]
12. Wu, Y.; Zhen, W.; Geng, Y.; Wang, Z.; Guo, Y. Pretreatment with probiotic Enterococcus faecium NCIMB 11181 ameliorates necrotic enteritis-induced intestinal barrier injury in broiler chickens. *Sci. Rep.* **2019**, *9*, 10256. [CrossRef] [PubMed]
13. Rovenský, J.; Švík, K.; Mat'ha, V.; Ištok, R.; Kamarád, V.; Ebringer, L.; Ferenčík, M.; Stančíková, M. Combination Treatment of Rat Adjuvant-Induced Arthritis with Methotrexate, Probiotic Bacteria *Enterococcus faecium*, and Selenium. *Ann. N. Y. Acad. Sci.* **2005**, *1051*, 570–581. [CrossRef] [PubMed]
14. Divyashri, G.; Krishna, G.; Muralidhara; Prapulla, S.G. Probiotic attributes, antioxidant, anti-inflammatory and neuromodulatory effects of Enterococcus faecium CFR 3003: In vitro and in vivo evidence. *J. Med. Microbiol.* **2015**, *64*, 1527–1540. [CrossRef] [PubMed]
15. Pitsouni, E.; Alexiou, V.; Saridakis, V.; Peppas, G.; Falagas, M.E. Does the use of probiotics/synbiotics prevent postoperative infections in patients undergoing abdominal surgery? A meta-analysis of randomized controlled trials. *Eur. J. Clin. Pharmacol.* **2009**, *65*, 561–570. [CrossRef] [PubMed]
16. Lukacs, S.L.; Schoendorf, K.C.; Schuchat, A. Trends in Sepsis-Related Neonatal Mortality in the United States, 1985–1998. *Pediatr. Infect. Dis. J.* **2004**, *23*, 599–603. [CrossRef] [PubMed]
17. Wynn, J.L.; Scumpia, P.O.; Delano, M.J.; O'Malley, K.A.; Ungaro, R.; Abouhamze, A.; Moldawer, L.L. Increased mortality and altered immunity in neonatal sepsis produced by generalized peritonitis. *Shock* **2007**, *28*, 675–683. [CrossRef]
18. Angurana, S.K.; Bansal, A.; Singhi, S.; Aggarwal, R.; Jayashree, M.; Salaria, M.; Mangat, N.K. Evaluation of Effect of Probiotics on Cytokine Levels in Critically Ill Children With Severe Sepsis: A Double-Blind, Placebo-Controlled Trial. *Crit. Care Med.* **2018**, *46*, 1656–1664. [CrossRef]
19. Yılmaz, M.; Erdem, A.O. The Protective Role of Probiotics in Sepsis Induced Rats. *Turk. J. Trauma Emerg. Surg.* **2020**, *26*, 843–846. [CrossRef]
20. Feng, G.; Dong, S.; Huang, M.; Zeng, M.; Liu, Z.; Zhao, Y.; Wu, H. Biogenic Polyphosphate Nanoparticles from a Marine Cyanobacterium *Synechococcus* sp. PCC 7002: Production, Characterization, and Anti-Inflammatory Properties In Vitro. *Mar. Drugs* **2018**, *16*, 322. [CrossRef] [PubMed]
21. Wasana, W.P.; Senevirathne, A.; Nikapitiya, C.; Eom, T.-Y.; Lee, Y.; Lee, J.-S.; Kang, D.-H.; Oh, C.; De Zoysa, M. A Novel *Pseudoalteromonas xiamenensis* Marine Isolate as a Potential Probiotic: Anti-Inflammatory and Innate Immune Modulatory Effects against Thermal and Pathogenic Stresses. *Mar. Drugs* **2021**, *19*, 707. [CrossRef]
22. Hongpattarakere, T.; Cherntong, N.; Wichienchot, S.; Kolida, S.; Rastall, R.A. In vitro prebiotic evaluation of exopolysaccharides produced by marine isolated lactic acid bacteria. *Carbohydr. Polym.* **2012**, *87*, 846–852. [CrossRef] [PubMed]
23. Honkanen, P. *Consumer Acceptance of (Marine) Functional Food*; Academic Publishers: Wageningen, The Netherlands, 2009; pp. 141–154.
24. Otaru, N.; Ye, K.; Mujezinovic, D.; Berchtold, L.; Constancias, F.; Cornejo, F.A.; Krzystek, A.; de Wouters, T.; Braegger, C.; Lacroix, C.; et al. GABA Production by Human Intestinal *Bacteroides* spp.: Prevalence, Regulation, and Role in Acid Stress Tolerance. *Front. Microbiol.* **2021**, *12*, 656895. [CrossRef] [PubMed]
25. Pieniz, S.; Andreazza, R.; Anghinoni, T.; Camargo, F.; Brandelli, A. Probiotic potential, antimicrobial and antioxidant activities of Enterococcus durans strain LAB18s. *Food Control* **2014**, *37*, 251–256. [CrossRef]
26. De Vuyst, L.; Moreno, M.F.; Revets, H. Screening for enterocins and detection of hemolysin and vancomycin resistance in enterococci of different origins. *Int. J. Food Microbiol.* **2003**, *84*, 299–318. [CrossRef]
27. Allen, H.K.; Trachsel, J.; Looft, T.; Casey, T.A. Finding alternatives to antibiotics. *Ann. N. Y. Acad. Sci.* **2014**, *1323*, 91–100. [CrossRef]
28. Bs, S.; Thankappan, B.; Mahendran, R.; Muthusamy, G.; Selta, D.R.F.; Angayarkanni, J. Evaluation of GABA Production and Probiotic Activities of Enterococcus faecium BS5. *Probiotics Antimicrob. Proteins* **2021**, *13*, 993–1004. [CrossRef]
29. Khanlari, Z.; Moayedi, A.; Ebrahimi, P.; Khomeiri, M.; Sadeghi, A. Enhancement of γ-aminobutyric acid (GABA) content in fermented milk by using *Enterococcus faecium* and *Weissella confusa* isolated from sourdough. *J. Food Process. Preserv.* **2021**, *45*, e15869. [CrossRef]
30. Abriouel, H.; Ben Omar, N.; Molinos, A.C.; López, R.L.; Grande, M.J.; Martínez-Viedma, P.; Ortega, E.; Cañamero, M.M.; Galvez, A. Comparative analysis of genetic diversity and incidence of virulence factors and antibiotic resistance among enterococcal populations from raw fruit and vegetable foods, water and soil, and clinical samples. *Int. J. Food Microbiol.* **2008**, *123*, 38–49. [CrossRef]
31. Mathur, S.; Singh, R. Antibiotic resistance in food lactic acid bacteria—A review. *Int. J. Food Microbiol.* **2005**, *105*, 281–295. [CrossRef]
32. Zhou, N.; Zhang, J.; Fan, M.; Wang, J.; Guo, G.; Wei, X. Antibiotic resistance of lactic acid bacteria isolated from Chinese yogurts. *J. Dairy Sci.* **2012**, *95*, 4775–4783. [CrossRef]

33. Pimentel, L.L.; Semedo, T.; Tenreiro, R.; Crespo, M.T.B.; Pintado, M.M.E.; Malcata, F.X. Assessment of Safety of Enterococci Isolated throughout Traditional Terrincho Cheesemaking: Virulence Factors and Antibiotic Susceptibility. *J. Food Prot.* **2007**, *70*, 2161–2167. [CrossRef] [PubMed]
34. du Toit, M.; Franz, C.; Dicks, L.; Schillinger, U.; Haberer, P.; Warlies, B.; Ahrens, F.; Holzapfel, W. Characterisation and selection of probiotic lactobacilli for a preliminary minipig feeding trial and their effect on serum cholesterol levels, faeces pH and faeces moisture content. *Int. J. Food Microbiol.* **1998**, *40*, 93–104. [CrossRef] [PubMed]
35. Radulović, Z.; Petrović, T.; Nedović, V.; Dimitrijević, S.; Mirković, N.; Petrušić, M.; Paunović, D. Characterization of autochthonous Lactobacillus paracasei strains on potential probiotic ability. *J. Dairy Prod. Process. Improv.* **2010**, *60*, 86–93.
36. Tannock, G.W. A Special Fondness for Lactobacilli. *Appl. Environ. Microbiol.* **2004**, *70*, 3189–3194. [CrossRef] [PubMed]
37. Todorov, S.D.; Prévost, H.; Lebois, M.; Dousset, X.; LeBlanc, J.G.; Franco, B.D. Bacteriocinogenic Lactobacillus plantarum ST16Pa isolated from papaya (Carica papaya)—From isolation to application: Characterization of a bacteriocin. *Food Res. Int.* **2011**, *44*, 1351–1363. [CrossRef]
38. Ding, L.; Gong, Y.; Yang, Z.; Zou, B.; Liu, X.; Zhang, B.; Li, J. Lactobacillus rhamnosus GG Ameliorates Liver Injury and Hypoxic Hepatitis in Rat Model of CLP-Induced Sepsis. *Dig. Dis. Sci.* **2019**, *64*, 2867–2877. [CrossRef]
39. Panpetch, W.; Sawaswong, V.; Chanchaem, P.; Ondee, T.; Dang, C.P.; Payungporn, S.; Tumwasorn, S.; Leelahavanichkul, A. Candida Administration Worsens Cecal Ligation and Puncture-Induced Sepsis in Obese Mice Through Gut Dysbiosis Enhanced Systemic Inflammation, Impact of Pathogen-Associated Molecules From Gut Translocation and Saturated Fatty Acid. *Front. Immunol.* **2020**, *11*, 561652. [CrossRef]
40. Ahmad, A.; Druzhyna, N.; Szabo, C. Delayed Treatment with Sodium Hydrosulfide Improves Regional Blood Flow and Alleviates Cecal Ligation and Puncture (CLP)-Induced Septic Shock. *Shock* **2016**, *46*, 183–193. [CrossRef]
41. Talarmin, H.; Derbré, F.; Lefeuvre-Orfila, L.; Léon, K.; Droguet, M.; Pennec, J.-P.; Giroux-Metgès, M.-A. The diaphragm is better protected from oxidative stress than hindlimb skeletal muscle during CLP-induced sepsis. *Redox Rep.* **2016**, *22*, 218–226. [CrossRef] [PubMed]
42. He, H.-L.; Liu, D.; Ma, C.-B. Review on the Angiotensin-I-Converting Enzyme (ACE) Inhibitor Peptides from Marine Proteins. *Appl. Biochem. Biotechnol.* **2012**, *169*, 738–749. [CrossRef]
43. Chen, H.; Yan, X.; Zhu, P.; Lin, J. Antioxidant activity and hepatoprotective potential of agaro-oligosaccharides in vitro and in vivo. *Nutr. J.* **2006**, *5*, 31. [CrossRef]
44. Higashimura, Y.; Naito, Y.; Takagi, T.; Mizushima, K.; Hirai, Y.; Harusato, A.; Ohnogi, H.; Yamaji, R.; Inui, H.; Nakano, Y.; et al. Oligosaccharides from agar inhibit murine intestinal inflammation through the induction of heme oxygenase-1 expression. *J. Gastroenterol.* **2012**, *48*, 897–909. [CrossRef] [PubMed]
45. Farag, A.M.; Hassan, S.W.; Beltagy, E.A.; El-Shenawy, M.A. Optimization of production of anti-tumor l-asparaginase by free and immobilized marine Aspergillus terreus. *Egypt. J. Aquat. Res.* **2015**, *41*, 295–302. [CrossRef]
46. Pandian, S.R.K.; Deepak, V.; Sivasubramaniam, S.D.; Nellaiah, H.; Sundar, K. Optimization and purification of anticancer enzyme L-glutaminase from Alcaligenes faecalis KLU102. *Biologia* **2014**, *69*, 1644–1651. [CrossRef]
47. Indarmawan, T.; Mustopa, A.Z.; Budiarto, B.R.; Tarman, K. Antibacterial Activity of Extracellular Protease Isolated From an Algicolous Fungus Xylaria psidii KT30 Against Gram-Positive Bacteria. *HAYATI J. Biosci.* **2016**, *23*, 73–78. [CrossRef]
48. Chen, H.-Y.; Zhang, S.; Li, J.; Huang, N.; Sun, J.; Li, B.-H.; Yang, J.; Li, Z.-F. Baicalein Attenuates Severe Polymicrobial Sepsis via lleviating Immune Dysfunction of T Lymphocytes and Inflammation. *Chin. J. Integr. Med.* **2022**, *28*, 711–718. [CrossRef]
49. Wang, W.; Ji, X.; Yuan, C.; Dai, F.; Zhu, J.; Sun, M. A Method for Molecular Analysis of Catalase Gene Diversity in Seawater. *Indian J. Microbiol.* **2013**, *53*, 477–481. [CrossRef] [PubMed]
50. Wang, Q.-F.; Wang, Y.-F.; Hou, Y.-H.; Shi, Y.-L.; Han, H.; Miao, M.; Wu, Y.-Y.; Liu, Y.-P.; Yue, X.-N.; Li, Y.-J. Cloning, expression and biochemical characterization of recombinant superoxide dismutase from Antarctic psychrophilic bacterium *Pseudoalteromonas* sp. ANT506. *J. Basic Microbiol.* **2015**, *56*, 753–761. [CrossRef]
51. Ji, M.; Barnwell, C.V.; Grunden, A.M. Characterization of recombinant glutathione reductase from the psychrophilic Antarctic bacterium Colwellia psychrerythraea. *Extremophiles* **2015**, *19*, 863–874. [CrossRef]
52. Reboldi, A.; Dang, E.V.; Mcdonald, J.G.; Liang, G.; Russell, D.W.; Cyster, J.G. 25-Hydroxycholesterol suppresses interleukin-1–driven inflammation downstream of type I interferon. *Science* **2014**, *345*, 679–684. [CrossRef]
53. Yin, K.; Hock, C.E.; Lai, P.-S.; Ross, J.T.; Yue, G. Role Of Interferon-Gamma In Lung Inflammation Following Cecal Ligation And Puncture In Rats. *Shock* **1999**, *12*, 215–221. [CrossRef] [PubMed]
54. Zhang, Y.-F.; Sun, C.-C.; Duan, J.-X.; Yang, H.-H.; Zhang, C.-Y.; Xiong, J.-B.; Zhong, W.-J.; Zu, C.; Guan, X.-X.; Jiang, H.-L.; et al. A COX-2/sEH dual inhibitor PTUPB ameliorates cecal ligation and puncture-induced sepsis in mice via anti-inflammation and anti-oxidative stress. *Biomed. Pharmacother.* **2020**, *126*, 109907. [CrossRef] [PubMed]
55. Araujo, J.A.; Zhang, M.; Yin, F. Heme Oxygenase-1, Oxidation, Inflammation, and Atherosclerosis. *Front. Pharmacol.* **2012**, *3*, 119. [CrossRef]
56. Ma, Q. Role of Nrf2 in Oxidative Stress and Toxicity. *Annu. Rev. Pharmacol. Toxicol.* **2013**, *53*, 401–426. [CrossRef]
57. Qi, Z.; Wang, R.; Liao, R.; Xue, S.; Wang, Y. Neferine Ameliorates Sepsis-Induced Myocardial Dysfunction Through Anti-Apoptotic and Antioxidative Effects by Regulating the PI3K/AKT/mTOR Signaling Pathway. *Front. Pharmacol.* **2021**, *12*, 706251. [CrossRef]
58. Sang, Z.; Dong, S.; Zhang, P.; Wei, Y. miR-214 ameliorates sepsis-induced acute kidney injury via PTEN/AKT/mTOR-regulated autophagy. *Mol. Med. Rep.* **2021**, *24*, 683. [CrossRef]

59. Heemskerk, S.; Masereeuw, R.; Russel, F.G.M.; Pickkers, P. Selective iNOS inhibition for the treatment of sepsis-induced acute kidney injury. *Nat. Rev. Nephrol.* **2009**, *5*, 629–640. [CrossRef]
60. Lei, W.; Liu, D.; Sun, M.; Lu, C.; Yang, W.; Wang, C.; Cheng, Y.; Zhang, M.; Shen, M.; Yang, Z.; et al. Targeting STAT3: A crucial modulator of sepsis. *J. Cell. Physiol.* **2021**, *236*, 7814–7831. [CrossRef]
61. Zaghloul, E.H.; Ibrahim, H.A.H.; El-Badan, D.E.S. Production of biocement with marine bacteria; Staphylococcus epidermidis EDH to enhance clay water retention capacity. *Egypt. J. Aquat. Res.* **2020**, *47*, 53–59. [CrossRef]
62. Amer, M.S.; Zaghloul, E.H.; Ibrahim, M.I.A. Characterization of exopolysaccharide produced from marine-derived Aspergillus terreus SEI with prominent biological activities. *Egypt. J. Aquat. Res.* **2020**, *46*, 363–369. [CrossRef]
63. Guttmann, D.M.; Ellar, D.J. Phenotypic and genotypic comparisons of 23 strains from the *Bacillus cereus* complex for a selection of known and putative *B. thuringiensis* virulence factors. *FEMS Microbiol. Lett.* **2000**, *188*, 7–13. [CrossRef] [PubMed]
64. Campedelli, I.; Mathur, H.; Salvetti, E.; Clarke, S.; Rea, M.C.; Torriani, S.; Ross, R.P.; Hill, C.; O'Toole, P.W. Genus-Wide Assessment of Antibiotic Resistance in *Lactobacillus* spp. *Appl. Environ. Microbiol.* **2019**, *85*, e01738-18. [CrossRef]
65. Muhammad, N.; Juan, W.; Aiping, Z.; ChaoFeng, M.; Xiaokang, W.; Jiru, X.; Nawaz, M.; Wang, J.; Zhou, A.; Wu, X.; et al. Screening and characterization of new potentially probiotic lactobacilli from breast-fed healthy babies in Pakistan. *Afr. J. Microbiol. Res.* **2011**, *5*, 1428–1436. [CrossRef]
66. Harsa, H.Ş.; Yavuzdurmaz, H. Isolation, Characterization, Determination of Probiotic Properties of Lactic Acid Bacteria from Human Milk. MSc. Thesis, Izmir Institute of Technology, Urla, Turkey, October 2007.
67. Zubrow, M.E.; Margulies, S.S.; Yehya, N. Nordihydroguaiaretic acid reduces secondary organ injury in septic rats after cecal ligation and puncture. *PLoS ONE* **2020**, *15*, e0237613. [CrossRef] [PubMed]
68. Livak, K.J.; Schmittgen, T.D. Analysis of relative gene expression data using real-time quantitative PCR and the $2^{-\Delta\Delta CT}$ Method. *Methods* **2001**, *25*, 402–408. [CrossRef]
69. Babicki, S.; Arndt, D.; Marcu, A.; Liang, Y.; Grant, J.R.; Maciejewski, A.; Wishart, D.S. Heatmapper: Web-enabled heat mapping for all. *Nucleic Acids Res.* **2016**, *44*, W147–W153. [CrossRef]

marine drugs

MDPI

Article

Eckmaxol Isolated from *Ecklonia maxima* Attenuates Particulate-Matter-Induced Inflammation in MH-S Lung Macrophage

D. P. Nagahawatta [1,†], N. M. Liyanage [1,†], H. H. A. C. K. Jayawardhana [1], Thilina U. Jayawardena [1,2], Hyo-Geun Lee [1], Moon-Soo Heo [1,*] and You-Jin Jeon [1,3,*]

1 Department of Marine Life Sciences, Jeju National University, Jeju 63243, Jeju Self-Governing Province, Republic of Korea

2 Department of Chemistry, Biochemistry and Physics, Université du Québec à Trois-Rivières, Trois-Rivières, QC G8Z 4M3, Canada

3 Marine Science Institute, Jeju National University, Jeju 63333, Jeju Self-Governing Province, Republic of Korea

* Correspondence: msheo@jejunu.ac.kr (M.-S.H.); youjin2014@gmail.com (Y.-J.J.); Tel.: +82-64-754-3475 (M.-S.H.)

† These authors contributed equally to this work.

Abstract: Airborne particulate matter (PM) originating from industrial processes is a major threat to the environment and health in East Asia. PM can cause asthma, collateral lung tissue damage, oxidative stress, allergic reactions, and inflammation. The present study was conducted to evaluate the protective effect of eckmaxol, a phlorotannin isolated from *Ecklonia maxima*, against PM-induced inflammation in MH-S macrophage cells. It was found that PM induced inflammation in MH-S lung macrophages, which was inhibited by eckmaxol treatment in a dose-dependent manner (21.0–84.12 μM). Eckmaxol attenuated the expression of cyclooxygenase-2 (COX-2) and inducible nitric oxide synthase (iNOS) in PM-induced lung macrophages. Subsequently, nitric oxide (NO), prostaglandin E-2 (PGE-2), and pro-inflammatory cytokines (IL-1β, IL-6, and TNF-α) were downregulated. PM stimulated inflammation in MH-S lung macrophages by activating Toll-like receptors (TLRs), nuclear factor-kappa B (NF-κB), and mitogen-activated protein kinase (MAPK) pathways. Eckmaxol exhibited anti-inflammatory properties by suppressing the activation of TLRs, downstream signaling of NF-κB (p50 and p65), and MAPK pathways, including c-Jun N-terminal kinase (JNK) and p38. These findings suggest that eckmaxol may offer substantial therapeutic potential in the treatment of inflammatory diseases.

Keywords: eckmaxol; *Ecklonia maxima*; anti-inflammation; particulate matter; lung macrophage; chronic diseases; bioactive compound

Citation: Nagahawatta, D.P.; Liyanage, N.M.; Jayawardhana, H.H.A.C.K.; Jayawardena, T.U.; Lee, H.-G.; Heo, M.-S.; Jeon, Y.-J. Eckmaxol Isolated from *Ecklonia maxima* Attenuates Particulate-Matter-Induced Inflammation in MH-S Lung Macrophage. *Mar. Drugs* **2022**, *20*, 766. https://doi.org/10.3390/md20120766

Academic Editors: Elena Talero and Javier Ávila-Román

Received: 4 November 2022
Accepted: 4 December 2022
Published: 7 December 2022

1. Introduction

Particulate matter (PM) is a known threat under air pollution in urban areas. PM includes both organic materials, such as biological materials (endotoxins, fungal spores, and pollen), and inorganic elements, such as metals, salts, and carbonaceous materials [1]. Ultrafine PM (≥100 nm) does not sediment or flocculate easily and is retained in the atmosphere for a longer period as compared to other particles (2.5–10 μM). This results in the transport of PM with the wind [2,3]. Prolonged exposure to PM results in adverse effects on human health, particularly on normal lung function, by inducing inflammation and oxidative stress in lung cells [4,5]. The inhalation of PM depends on its penetration depth, deposition, particle size, shape, and density [6]. The World Health Organization (WHO) reported that PM exposure was responsible for more than seven billion deaths in 2012, whereas the American Cancer Society indicated that the rise in PM increased the total death rate by 7% [5,7,8].

Lungs are vital organs in the human body as they supply oxygen. Continuous PM inhalation has severe negative effects on the lungs. PM is known to be important in the development of chronic inflammatory lung diseases, such as asthma, chronic obstructive pulmonary disorder, and lung cancer [9]. PM mainly mediates oxidative stress in cells via reactive oxygen species (ROS), which are generated by free radicals contained in particle surfaces [10]. Both oxidative stress and PM itself activate redox-sensitive signaling pathways, which result in inflammatory responses [11]. Inflammatory responses are considered as the primary protective actions and involve the upregulation and activation of several important genes for signaling molecules such as cytokines (TNF-α, IL-6), chemokines (IL-8), and adhesion molecules [12,13].

Inflammatory cytokines are key signaling proteins synthesized and released by macrophages under stress conditions, and their quantity and interaction with the receptors control the activation of immune cells and subsequent signaling cascades [14,15]. This proves that macrophages and cytokine signaling play major regulatory roles in inflammatory processes in the lungs.

Ecklonia maxima is a brown seaweed commonly found in the South African coastal region that has high bioactivity. The leaves of *E. maxima* are commonly used as a source of alginates, animal feed, fertilizers, nutrient supplements, and medication preparations. Polysaccharides from this seaweed species and antioxidant have been found to have anti-diabetes and anti-cancer activities [16,17]. Eckmaxol is a phlorotannin isolated from *E. maxima* leaves [18]. Phlorotanins are polyphenolic compounds with a wide range of molecular weights that possess many potential health benefits [19]. They are formed by the polymerization of aromatic precursors through the acetate–malonate pathway [20]. Phlorotanins are considered to possess numerous bioactivities, such as anti-diabetic, anti-inflammatory, antioxidant, anti-bacterial, and anti-cancer activities [21–25]. The activity of eckmaxol against neurotoxicity, melanogenesis, and LPS-induced inflammation was previously evaluated [18,26,27]. However, to the best of our knowledge, the protective effect of eckmaxol against PM-induced lung inflammation has not been systematically investigated. Thus, the identification of Toll-like receptors (TLRs) responsible for the induction of inflammation in PM-exposed lung macrophages and their downstream signal transduction may be a possible target for the development of a therapeutic agent for the treatment of PM-induced lung inflammation. The purpose of the present study was to investigate the influence of PM on the inflammatory responses of lung macrophages and the protective effect of eckmaxol isolated from *E. maxima* against PM-induced inflammatory responses in the lungs.

2. Results

2.1. Characterization of PM and Identification Eckmaxol

Certified reference material No.28 (Chinese PM) was used for this experiment. The detailed procedure for collecting PM via mechanical vibration and chemical characterization was described previously [28]. The particle size and distribution were evaluated by scanning electron microscopy (SEM), as shown in Figure 1a,b. This provides evidence that the majority of the particles had an average diameter of less than 5 μm. In addition, the data were provided by the National Institute for Environmental Studies (NIES), Ibaraki, Japan. The isolated compound was characterized using high-performance liquid chromatography (HPLC) and electrospray ionization (ESI). The HPLC analysis solidified that eckmaxol with high purity via analysis peak characterization (Figure S1). The recorded purity was more than 90%. ESI-MS (positive) evaluation based on the HPLC analysis confirmed the molecular weight of the eckmaxol that aligned with the previously published results [27]. The chemical structure of the eckmaxol ($C_{36}H_{24}O_{18}$) was demonstrated in Figure 1c.

2.2. Effect of Eckmaxol on MH-S Lung Macrophages and PM-Stimulated Cell Viability and NO Production

According to the results, eckmaxol concentrations higher than 84.12 μM showed a cytotoxic effect on MH-S lung macrophages (Figure 2a). Therefore, eckmaxol concentrations between 21.00–84.12 μM were used for further experiments. As shown in Figure 2b, the protective effect of eckmaxol was examined against PM-stimulated MH-S lung macrophages.

Cell viability was significantly affected by PM, whereas eckmaxol exhibited a significant recovery effect in a dose-dependent manner. Owing to PM, NO production was significantly upregulated, while treatment with eckmaxol significantly downregulated NO production in a dose-dependent manner (Figure 2c).

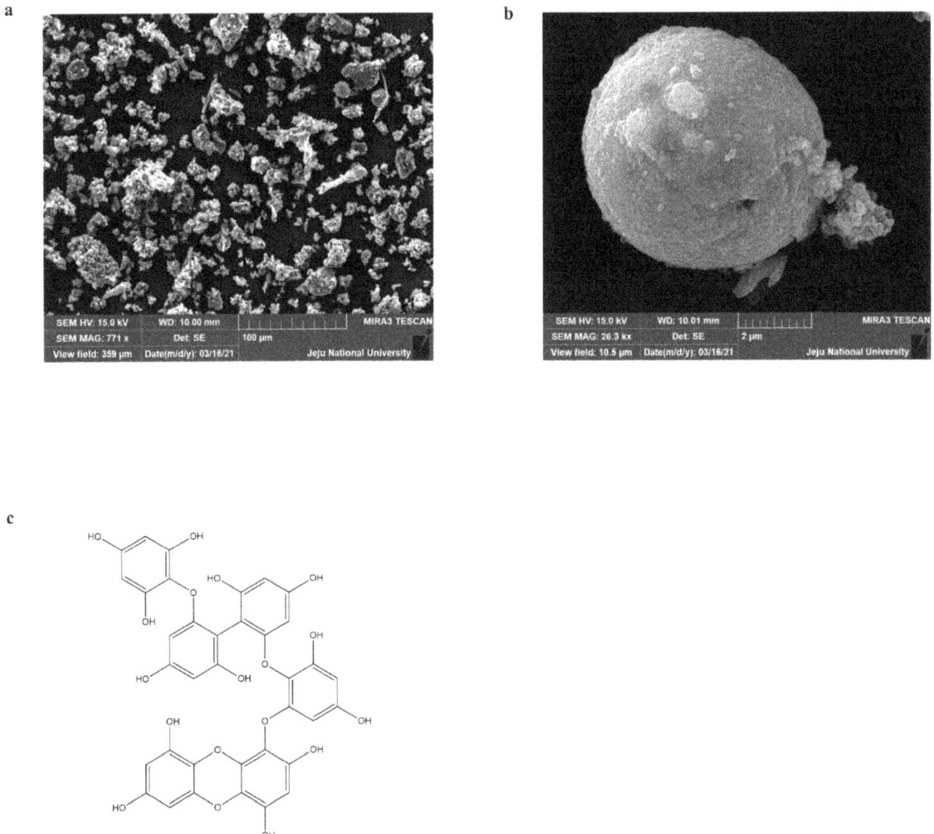

Figure 1. Physical parameters of particulate matter (PM) and chemical structure of eckmaxol. (**a**) Scanning electron microscopic (SEM) image, (**b**) magnified SEM image of PM particle of certified CRM No.28, National Institute for Environmental Studies (NIES), Ibaraki, Japan, and (**c**) the chemical structure of eckmaxol.

2.3. Preventive Effect of Eckmaxol on Prostaglandin E2 (PGE-2) and Pro-Inflammatory Cytokine Production in PM-Induced MH-S Cells

Further confirmation of in vitro anti-inflammatory properties was investigated by examining the secretion levels of PGE-2 and pro-inflammatory cytokines (TNF-α, IL-6, and IL-1β) in PM-induced MH-S cells using enzyme-linked immunosorbent assay (ELISA). As shown in Figure 3, the production levels of PGE-2 and pro-inflammatory cytokines were strongly stimulated by PM, whereas the eckmaxol-treated groups showed significant and dose-dependent suppression of their production.

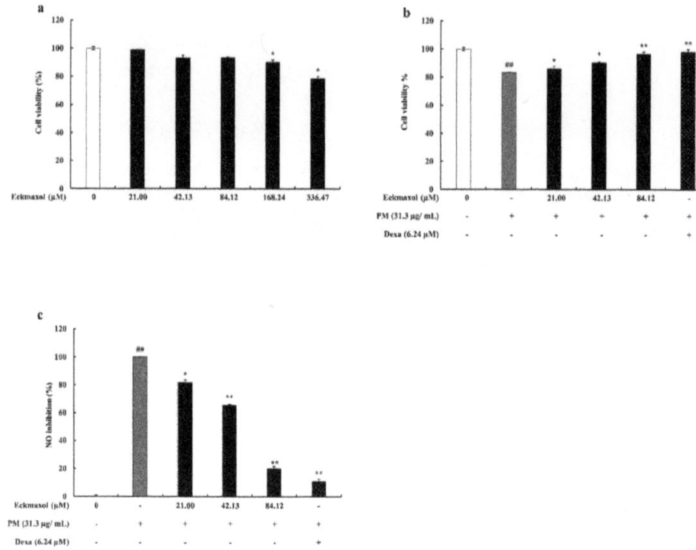

Figure 2. Dose-range determination and cytoprotective activity evaluation of eckmaxol on particulate matter (PM)-induced MH-S lung macrophages. (**a**) Cytotoxicity of eckmaxol, (**b**) cytoprotective effect of eckmaxol against PM, (**c**) NO production inhibition effect of eckmaxol against PM. Triplicate experiments were used to evaluate the data and the mean value is expressed with ± SD. * $p < 0.05$, ** $p < 0.01$, against PM-treated group or ## $p < 0.01$, against control (ANOVA, Duncan's multiple range test).

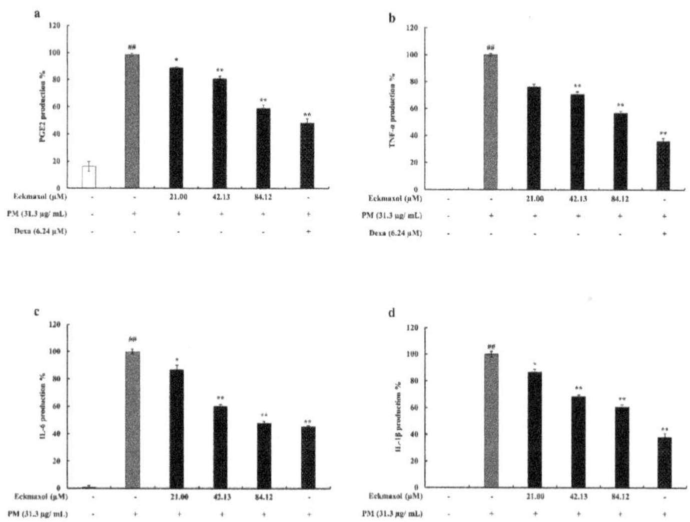

Figure 3. Inhibitory effect of eckmaxol against particulate matter (PM)-induced PGE-2 and pro-inflammatory cytokine produciton in MH-S lung macrophages (TNF-α, IL-6, and IL-1β) production. Inhibitory effect on (**a**) PGE-2 production, (**b**) TNF-α production, (**c**) IL-6 production, and (**d**) IL-1β production. Triplicate experiments were used to evaluate the data and the mean value is expressed with ± SD. * $p < 0.05$, ** $p < 0.01$, against PM-treated group or ## $p < 0.01$, against control (ANOVA, Duncan's multiple range test).

2.4. Potential of Eckmaxol to Inhibit Inducible Nitric Oxide Synthase (iNOS) and Cyclooxygenase-2 (COX-2) Gene and Protein Expression in PM Stimulated MH-S Lung Macrophages

Protein expression of iNOS and COX-2 revealed the anti-inflammatory activity of eckmaxol via Western blotting. The results of the gene expression analysis confirmed this. According to the Western blots, the upregulated gene expression of iNOS and COX-2 was significantly decreased by eckmaxol treatment. The gene expression results for iNOS and COX-2 exhibited a similar trend (Figure 4).

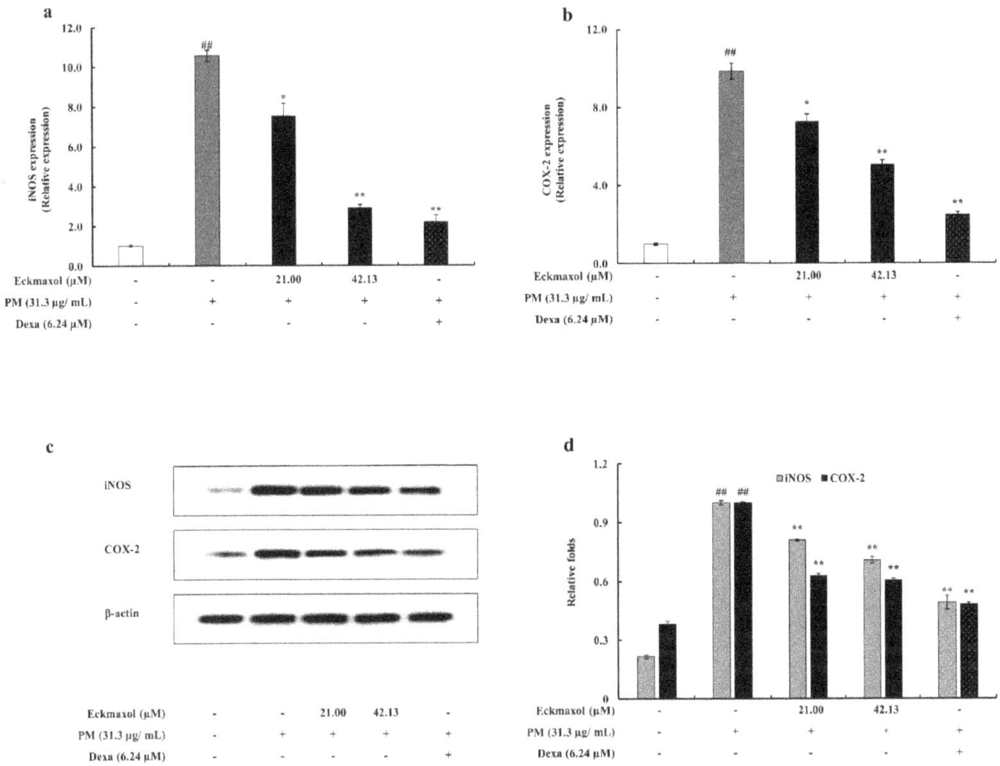

Figure 4. Inflammation-associated protein and gene expression of iNOS and COX-2 in particulate matter (PM)-induced MH-S lung macrophage attenuated by eckmaxol. (**a**) iNOS gene expression, (**b**) COX-2 gene expression, (**c**) iNOS and COX-2 protein expression, and (**d**) quantification of iNOS and COX-2 protein expression. Triplicate experiments were used to evaluate the data and the mean value is expressed with ± SD. * $p < 0.05$, ** $p < 0.01$, against PM-treated group or ## $p < 0.01$, against control (ANOVA, Duncan's multiple range test). β-actin was used as the house-keeping gene. Quantitative data were analyzed using Image J software.

2.5. Eckmaxol Suppresses the Pro-Inflammatory Cytokine Gene Expressions

The mRNA expression levels of selected proinflammatory cytokines (TNF-α, IL-6, and IL-1β) were evaluated to measure the inhibitory effect of eckmaxol on mRNA expression. According to the qPCR results, the gene expression levels of pro-inflammatory cytokines were significantly increased by PM exposure and the upregulated gene expression was significantly and dose-dependently decreased by eckmaxol treatment (Figure 5a–c).

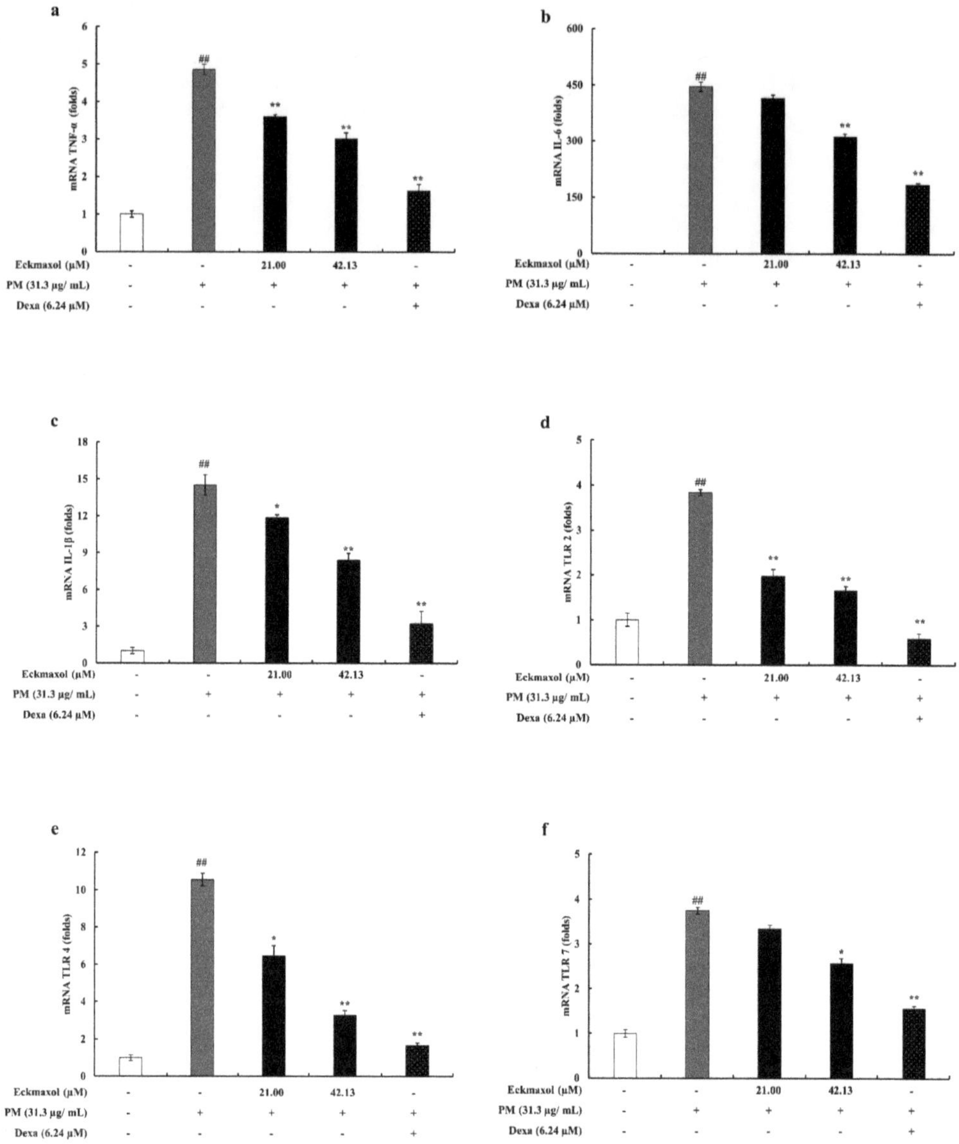

Figure 5. Gene expression levels evaluation in particulate matter (PM)-stimulated MH-S lung macrophages attenuated by eckmaxol. (**a**) TNF-α, (**b**) IL-6, (**c**) IL-1β, (**d**) TLR-2, (**e**) TLR-4, and (**f**) TLR-7. The mRNA expression levels were measured via RT-qPCR techniques. Triplicate experiments were used to evaluate the data and the mean value is expressed with ± SD. * $p < 0.05$, ** $p < 0.01$, against PM-treated group or ## $p < 0.01$, against control (ANOVA, Duncan's multiple-range test).

2.6. Inhibitory Activity of Eckmaxol on the Expression of TLRs

The gene expression levels of TLRs in MH-S lung macrophages were measured using qPCR. The results revealed elevated expression of TLR-2, TLR-4, and TLR-7 in PM-stimulated MH-S lung macrophages. However, these inclined expressions were significantly downregulated by eckmaxol treatment (Figure 5d–f).

2.7. Eckmaxol Inhibited the Nuclear Factor-κB (NF-κB) Nuclear Translocation and Mitogen-activated Protein Kinase (MAPK) Phosphorylation Induced via PM

Nuclear translocation of NF-κB was evaluated by measuring the phosphorylation levels of NF-κB subunits p65 and p50 in the cytoplasm and their expression levels in the nucleus. As shown in Figure 6, phosphorylation of p65 and p50 in the cytosol was significantly increased by PM treatment. This upregulation was significantly decreased by eckmaxol treatment. The expression levels of these subunits in the nucleus were also significantly increased by PM exposure and downregulated by eckmaxol in a significant and dose-dependent manner. This was further analyzed using an immunofluorescence assay. Phosphorylation levels of p50 and p65 were detected using green and red fluorescence-conjugated secondary antibodies, respectively. According to the results, PM significantly stimulated the phosphorylation of p50 and p65 in the macrophages. However, eckmaxol significantly reduced the phosphorylation levels of p50 and p65. These results further confirmed the results of the Western blot analysis (Figure 6e).

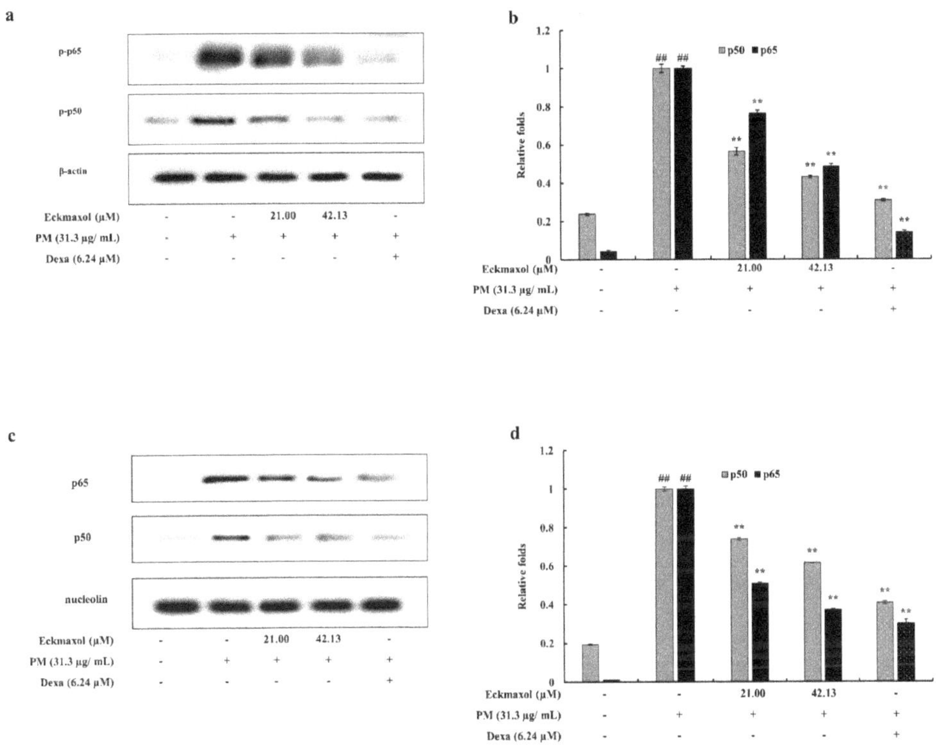

Figure 6. *Cont.*

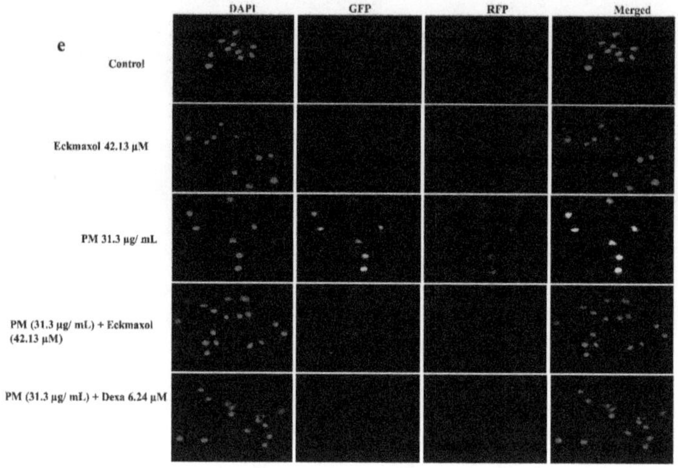

Figure 6. Influence of eckmaxol on the nuclear translocation of NF-κB in particulate matter (PM)-indued MH-S lung macrophages. (**a**) Phosphorylation of p65 and p50 in cytoplasm, (**b**) quantification of p50 and p65 in cytoplasm, (**c**) protein expression of p65 and p50 in the nucleus, and (**d**) quantification of protein expression of p65 and p50. (**e**) The cells were stained using 4′,6-diamidino-2-phenylindole (DAPI), green flourecesence protein (GFP), and red fluorescensce protein (RFP) stainings. Triplicate experiments were used to evaluate the data and the mean value is expressed with ± SD. ** $p < 0.01$, against PM-treated group or ## $p < 0.01$, against control (ANOVA, Duncan's multiple range test). β-actin (cytoplasm) and nucleolin (nucleus) were used as an internal control. Quantitative data were analyzed using Image J software.

Phosphorylation of transcription factors, such as c-Jun N-terminal Kinase (JNK) and p38, leads to gene expression and cytokine production. PM significantly upregulated the phosphorylation of JNK and p38. However, this was significantly reduced by eckmaxol treatment. Thus, eckmaxol significantly downregulated PM-stimulated MAPK phosphorylation (Figure 7).

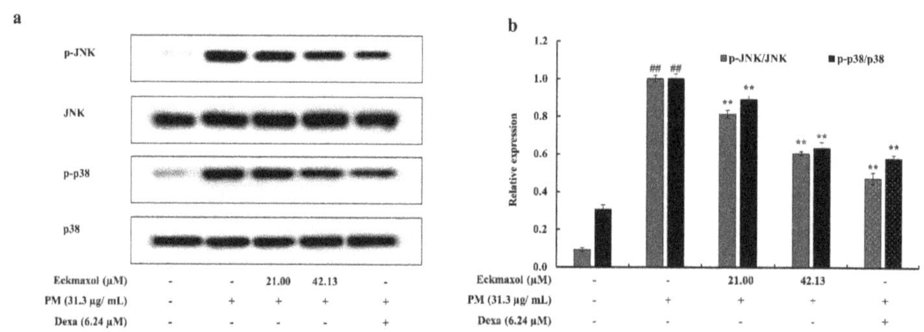

Figure 7. Evaluation of the effect of eckmaxol treatments on mitogen-activated protein kinase (MAPK) pathway proteins in pariculate matter (PM)-stimulated MH-S lung macrophages. (**a**) Western blot results of JNK, p38, and their phosphorylated forms and (**b**) quantitative data. Triplicated experiments were used to evaluate the data and the mean value is expressed with ± SD. ** $p < 0.01$, against PM-treated group or ## $p < 0.01$, against control (ANOVA, Duncan's multiple range test). β-actin was used as a internal control. Quantitative data were analyzed using Image J software.

3. Discussion

The investigation of marine resources has become a major topic in the current scientific world. Marine algae are considered a major resource that consists of various secondary metabolites, such as polyphenols, polysaccharides, proteins, and peptides, which have many valuable bioactivities [29–31]. Among these marine algae, *E. maxima* contain various phloroglucinol-derived polyphenols such as eckmaxol, which exhibits valuable bioactivities, such as neuroprotective effects [18]. The results of the present study reveal the potential of eckmaxol as an anti-inflammatory agent against PM-induced inflammation in lung macrophages.

Dose-range determination analysis revealed a safe range of eckmaxol in MH-S lung macrophages. Furthermore, these optimized doses of eckmaxol exhibited cytoprotectivity by downregulating cell death and NO production in PM-induced MH-S lung macrophages. NO and PGE-2 act as inflammatory mediators and play a crucial role in chronic inflammation and host defenses. NO is derived from L-arginine through the enzymatic activity of iNOS and PGE-2 is generated by the enzymatic activity of COX-2 that converts arachidonic acid into PGE-2. Certain types of inflammation, such as asthma, generate NO at high levels, which acts as a pro-inflammatory agent, and COX-2 contributes to the production of autoregulatory, homeostatic prostanoid, and prostanoid release during inflammation. Therefore, regulation of NO and PGE-2 production is a versatile way to regulate inflammatory responses in macrophages [32,33]. The Western blot and qPCR results of iNOS and COX-2 solidified the anti-inflammatory potential of eckmaxol. Further, this solidified the potential of eckmaxol to down-regulate inflammatory mediators such as NO and PGE-2.

Small secreted proteins, called cytokines, are considered key regulators of inflammation. Cytokines are generated in response to invading pathogens and stimulate, proliferate, and recruit immune cells. Therefore, regulating cytokine levels could lead to manipulating the ultimate PGE-2 and NO production. The present study evaluated the regulation of PM-stimulated TNF-α, IL-6, and IL-1 production by eckmaxol based on their crucial role in inflammation [34,35]. ELISA results showed a significant up-regulation of these pro-inflammatory cytokine productions, significantly and dose-dependently declined by eckmaxol treatments. These results were consistent with the gene expression results of IL-1β, IL-6, and TNF-α. This further confirmed the anti-inflammatory activity of eckmaxol in the MH-S lung macrophages. These results emphasize the studying of the regulation of inflammatory signaling pathways by eckmaxol.

Therefore, the present study evaluated the activation of TLRs on the cell surface and endosomes to measure the effect of eckmaxol on inflammatory responses. The downstream signaling pathways initiated by TLRs activate the NF-κB and MAPK signaling pathways through myeloid differentiation primary response 88 (MYD88) and TNF receptor-associated factor 6 (TRAF6) [36]. Thus, identification of TLR activation and expression levels provides insight into PM-stimulated inflammation and the anti-inflammatory potential of eckmaxol. According to a previous study, TLR-2 and TLR-4 null mice stimulated by PM expressed lower levels of inflammatory responses than normal mice [37]. Furthermore, upregulation of TLR-7 by PM in MH-S lung macrophages has been previously reported [7]. This indicates the importance of TLRs in PM-stimulated inflammation. Gene expression evaluation in the present study highlighted that the upregulation of TLR-2, TLR-4, and TLR-7 by PM was significantly and dose-dependently decreased by eckmaxol treatment.

The uncontrolled production of pro-inflammatory cytokines and activation of cell signaling pathways, such as NF-κB and MAPK, play a pivotal role in pro-inflammatory responses [4]. NF-κB comprises a family of transcription factors that initiate gene expression to produce pro-inflammatory cytokines. Under non-stimulated conditions, NF-κB proteins such as p50 and p65 are bound to an inhibitor called nuclear factor of kappa light polypeptide gene enhancer in B-cells inhibitor, alpha (IκBα). This maintains p50 and p65 in their inactive forms in the cytoplasm. However, when a cell is exposed to the stimulators of pro-inflammatory mediators such as iNOS, COX-2, and pro-inflammatory cytokines, it initiates the translocation of these transcription factors from the cytoplasm to the nucleus and the transcription of genes responsible for pro-inflammatory cytokine production [38]. The current study evaluated the nuclear translocation of p50 and p65 following PM stimulation and the potential of eckmaxol for its regulation. Western blotting and immunostaining revealed that eckmaxol successfully downregulated the phosphorylation and translocation of p50 and p65. This finding strengthens the potential of eckmaxol as an anti-inflammatory agent.

The MAPK signaling pathway consists of numerous serine–threonine protein kinases that transfer signals from the cell surface to the nucleus to initiate gene expression, differentiation, mitosis, apoptosis, and survival [39]. Many studies have confirmed the important role of MAPKs, such as p38 and JNK, in inflammation-related gene expression [40,41]. Overall, regulating MAPK phosphorylation is a feasible approach to manipulate pro-inflammation and treat pro-inflammatory diseases. The phosphorylation of p38 and JNK was evaluated in the present study, which revealed that eckmaxol significantly decreased the PM-induced phosphorylation of these MAPKs. This suggests the potential of eckmaxol to regulate pro-inflammatory gene expression and cytokine production.

The experimental evidence provides mechanistic insight into the effect of eckmaxol against PM-induced pro-inflammation through evading extensive NO and PGE-2 production. Moreover, the involvement of inflammatory signaling pathways and cytokine production with PM exposure, and the effect of eckmaxol to attenuate them in MH-S lung macrophages, were confirmed. The findings further emphasize the potential of eckmaxol to develop an anti-inflammatory agent.

4. Materials and Methods

4.1. Chemicals and Regents

CRM-certified Chinese fine dust PM (CRM No.28 Urban Aerosols) was purchased from the Center for Environmental Measurement and Analysis, National Institute for Environmental Studies (Ibaraki, Japan). The murine MH-S lung macrophage cell line was purchased from American Type Culture Collection (Rockville, MD, USA). Roswell Park Memorial Institute medium (RPMI) supplemented with fetal bovine serum (FBS) and antibiotics (penicillin and streptomycin) were purchased from Gibco (Life Technologies, Grand Island, NY, USA). The antibodies used for Western blotting were purchased from Santa Cruz Biotechnology (Santa Cruz, CA, USA). The cytokine assay kits used for the experiment were purchased from eBioscience (San Diego, CA, USA), R&D Systems (Minneapolis, MN, USA), BD Optics (San Diego, CA, USA), and Invitrogen (Carlsbad, CA, USA). Unless otherwise noted, all chemicals were purchased from Sigma–Aldrich (St. Louis, MO, USA). HPLC-grade methanol and acetonitrile were purchased from Honeywell Burdick and Jackson, respectively (Muskegon, MI, USA). Analytical grade formic acid was obtained from Fluka Chemical (Buchs, Switzerland), and distilled water was purified from the Milli Q system (Millipore, Milford, MA, USA) used in this study.

4.2. Isolation and Characterization of Eckmaxol

Eckmaxol was isolated and purified as previously described with slight modifications. Briefly, *E. maxima* ethyl acetate fraction was used to isolate eckmaxol using centrifugal partition chromatography (CPC 240, Tokyo, Japan) with a ratio of n-hexane: ethyl acetate: methanol: water (3:7:4:6 v/v). The two phases were separated after the mixture had been thoroughly equilibrated in a separating funnel at room temperature (25 °C). The upper organic phase was used as the stationary phase and the lower aqueous phase was used as the mobile phase. The organic stationary phase was filled with the CPC phase and rotated at a speed of 1000 rpm. Subsequently, the aqueous mobile phase was pumped into the column in descending mode at a flow rate of 2 mL/min. Hydrodynamic equilibrium was maintained before sample injection, and 500 mg of *E. maxima* ethyl acetate fraction was dissolved in 6 mL 1:1 v/v. water: methanol and injected through the injection valve. An automatic fraction collector was used to collect the fractions (6 mL per tube) in the UV detection range of 230 nm. An HPLC system equipped with a PDA detector was used for further purification. A YMC-Pack ODS-A 10 × 250 mm, 5 μm column with acetonitrile + 0.1% formic acid and deionized water + 0.1% formic acid was used as the mobile phase at a flow rate of 2 mL/min [18,27,42].

4.3. Morphological Analysis of PM

First, the sample was coated with a platinum sputter (Quorum Technologies, Lewes, UK), and the surface morphology of the CRM No. 28 particles was observed using a JSM-6700F field-emission scanning electron microscope (JEOL, Tokyo, Japan). The device was operated at 10.0 kV [43].

4.4. Cell Culture

4.4.1. MH-S Lung Macrophage Cell Culture

Murine MH-S lung macrophages were maintained in RPMI growth medium supplemented with 10% FBS and 1% antibiotics. Cells were maintained under controlled conditions of 5% CO_2 at 37 °C. Cells were periodically sub-cultured and used in the exponential growth phase for the experiments [7].

4.4.2. Cell Viability Assay and Dose-Range Determination for PM

The cytotoxic effects of PM, eckmaxol, and their combinations on MH-S lung macrophages were assessed using a colorimetric MTT assay. The experiment was performed according to the procedure described by Sanjeewa et al. (2020). MH-S lung macrophages were seeded at a concentration of 1 × 105 cells/mL in a 96 well plate. Eckmaxol (15.6–250 μg/mL) was treated after a 24 h incubation period. The cells were then treated for 1 h with PM (31.3 μg/mL) and incubated for 24 h again. The MTT assay was conducted to assess cell viability. Absorbance was measured at 540 nm using a Model 680 plate reader (Biotek Instruments, Inc., Winooski, VT, USA) [7,44].

4.4.3. Determination of Nitric Oxide (NO) Production

A Griess assay was performed to evaluate the ability of eckmaxol to inhibit NO production in PM-induced MH-S lung macrophages. In brief, MS-H cells were seeded at a concentration of 1×10^5 cells/mL in a 96-well plate, and eckmaxol was added after a 24 h period incubation period. After 1 h, PM was added, and incubation was continued for another 24 h under controlled conditions of 5% CO_2 at 37 °C. An equal amount of Griess reagent was added to the culture supernatant and mixed in a 96 well plate. After 10 min of incubation, absorbance was measured at 540 nm [7].

4.4.4. Evaluation of Pro-Inflammatory Cytokines and Prostaglandin E-2 (PGE-2) Production

MH-S lung macrophages were seeded and treated with different eckmaxol concentrations. After 1 h of incubation, cells were stimulated with PM for 24 h. The supernatant was collected to analyze pro-inflammatory cytokine levels, including cytokines (IL-1β, IL-6, and TNF-α) and PGE-2 production using ELISA kits [45].

4.5. Western Blotting

MH-S lung macrophages were seeded in a six-well plate and treated with different concentrations of eckmaxol after 24 h of seeding. The cells were then stimulated with PM for 1 h and incubated for another 24 h to collect the cells. The harvested MH-S lung macrophages were washed with ice-cold PBS, and cytosolic proteins were collected using a cytoplasmic and nuclear protein extraction kit (Thermo Scientific, Rockford, IL, USA) according to a previously described method [46]. After extraction, the protein content of each supernatant was determined using the BCA protein assay kit. Cellular proteins were separated by electrophoresis on 12% SDS-polyacrylamide gels and transferred to polyvinylidene fluoride (PVDF) membranes (GE Healthcare, Uppsala, Sweden). Membranes were blocked with 5% skim milk in TBST at room temperature for 2 h and incubated with primary antibodies in a cold room for approximately 8 h. Anti-inducible nitric oxide synthase (iNOS) and COX-2 were used for this experiment. p38, p-p38, p50, p-p50, p65, p-p65, JNK, p-JNK, nucleolin, and β-actin (1:1000) were used. After 8 h of incubation, the blots were washed twice with Ttween 20/Tris-buffered saline and incubated with the secondary antibodies for 45 min (1:3000). The bands were visualized using the FUSION Solo Vilber Lourmat system and band intensity was quantified using the ImageJ program.

4.6. Evaluation of the NF-κB Nuclear Localization

MH-S lung macrophages were seeded in Nunc® Lab-Tek® 8-well Chamber Slide™ (Nunc, NY, USA), treated with eckmaxol at concentrations of 15.6, 31.3, 62.5 μg/mL and stimulated with PM. MH-S lung macrophages were fixed with 4% paraformaldehyde for 5 min at room temperature and washed thrice with ice-cold PBST for 5 min each. The cells were then permeabilized with 0.1% Triton-X-100 and rinsed thrice with ice-cold water with PBST for 5 min. The cells were blocked with 10% donkey serum (Abcam, Cambridge, MA, USA) and incubated overnight with NF-κB p50 and p65 at 4 °C (1:200 in donkey serum). Next, fluorescent dye-conjugated secondary antibodies (Alexa Fluor® 647, Abcam, Cambridge, MA, USA) were added and incubated for 2 h at room temperature, followed by three washes with ice-cold PBST for 5 min each. The prepared samples were incubated with 4′,6-diamidino-2-phenylindole (DAPI) nuclear stain (300 nM) for 10 min and washed thrice with PBST for 5 min to remove excess DAPI. Then, the coverslips were placed on chamber glass slides with Fluor Shield™ histology mounting medium (Sigma-Aldrich, St. Louis, MO, USA). Images of the stained slides were captured using a Lionheart™ FX Automated Microscope System (Bio-Tek Instruments, Inc., Winooski, VT, USA) [7].

4.7. Gene Expression Analysis

4.7.1. RNA Extraction and cDNA Synthesis

Total RNA from MH-S lung macrophages was extracted according to the manufacturer's instructions using TRIzol reagent (Life Technologies, Carlsbad, CA, USA). The total amount of RNA (1 μg) was reverse-transcribed using a first-strand cDNA synthesis kit (TaKaRa, Shiga, Japan) to obtain cDNA according to the manufacturer's instructions. The cDNA was amplified using the primers listed in Table 1 (Bioneer, Seoul, South Korea) [47].

Table 1. Sequences of primers used in the present study.

Gene	Primer	Sequence
GAPDH	Sense Antisense	5′-AAGGGTCATCATCTCTGCCC-3′ 5′-GTGATGGCATGGACTGTGGT-3′
iNOS	Sense Antisense	5′-ATGTCCGAAGCAAACATCAC-3′ 5′-TAATGTCCAGGAAGTAGGTG-3′
COX-2	Sense Antisense	5′-CAGCAAATCCTTGCTGTTCC-3′ 5′-TGGGCAAAGAATGCAAACATC-3′
IL-6	Sense Antisense	5′-GTACTCCAGAAGACCAGAGG-3′ 5′-TGCTGGTGACAACCACGGCC-3′
IL-1β	Sense Antisense	5′-CAGGATGAGGACATGAGCACC-3′ 5′-CTCTGCAGACTCAAACTCCAC-3′
TNF-α	Sense Antisense	5′-TTGACCTCAGCGCTGAGTTG-3′ 5′-CCTGTAGCCCACGTCGTAGC-3′
TLR-2	Sense Antisense	5′-CAGCTGGAGAACTCTGACCC-3′ 5′-CAAAGAGCCTGAAGTGGGAG-3′
TLR-4	Sense Antisense	5′-CAACATCATCCAGGAAGGC-3 5′-GAAGGCGATACAATTCCACC-3′
TLR-7	Sense Antisense	5′-TTCCTTCCGTAGGCTGAACC-3′ 5′-GTAAGCTGGATGGCAGATCC-3′

4.7.2. Real-Time Reverse Transcription-Polymerase Chain Reaction (RT-PCR)

The conditions for the PCR amplification were as follows: one cycle at 95 °C for 10 s, followed by 45 cycles at 95 °C for 5 s, 55 °C for 10 s, and 72 °C for 20 s; and a final single cycle at 95 °C for 15 s, 55 °C for 30 s, and 95 °C for 15 s. The relative levels of target genes were calculated and normalized to GAPDH levels. All experiments were performed in triplicate. mRNA expression levels were calculated using the Livak method ($2^{-\Delta\Delta CT}$) (Livak and Schmittgen, 2001).

5. Conclusions

The results of this study provide important evidence of the effects of PM on inflammatory responses in lung macrophages. Here, the stimulation of inflammation in lung macrophages by PM and downstream activation through TLRs, NF-κB, and MAPK was demonstrated. Furthermore, the study confirmed that eckmaxol isolated from *E. maxima* significantly and dose-dependently inhibited PM-stimulated inflammation in lung macrophages via these receptors and signaling pathways. Taken together, the results of gene expression analysis and protein production provide clear insight into the potential of eckmaxol as an anti-inflammatory agent. In addition, further studies, including in vivo experiments and human trials, are required to confirm the use of eckmaxol as an anti-inflammatory agent against PM-induced pro-inflammation in the lungs.

Supplementary Materials: The following supporting information can be downloaded at: https://www.mdpi.com/article/10.3390/md20120766/s1, Figure S1: Chromatogram of Liquid Chromatography Mass Spectrometry (LC/MS) analysis.

Author Contributions: Conceptualization, Y.-J.J. and D.P.N.; methodology, D.P.N. and N.M.L.; software, D.P.N., N.M.L. and H.H.A.C.K.J.; validation, D.P.N. and T.U.J.; formal analysis, D.P.N.; investigation, D.P.N. and Y.-J.J..; resources, Y.-J.J., M.-S.H. and H.-G.L.; data curation, D.P.N. and N.M.L.; writing—original draft preparation, D.P.N.; writing—review and editing, D.P.N. and N.M.L.; visualization, D.P.N.; supervision, Y.-J.J.; project administration Y.-J.J.; funding acquisition, Y.-J.J. All authors have read and agreed to the published version of the manuscript.

Funding: This research was supported by the 2022 scientific promotion program funded by Jeju National University.

Institutional Review Board Statement: Not applicable.

Data Availability Statement: Not applicable.

Conflicts of Interest: The authors declare no conflict of interest.

References

1. Anastasio, C.; Martin, S.T. Atmospheric nanoparticles. *Rev. Mineral. Geochem.* **2001**, *44*, 293–349. [CrossRef]
2. Yang, J.; Kim, Y.-K.; Kang, T.S.; Jee, Y.-K.; Kim, Y.-Y. Importance of indoor dust biological ultrafine particles in the pathogenesis of chronic inflammatory lung diseases. *Environ. Health Toxicol.* **2017**, *32*, e2017021. [CrossRef] [PubMed]
3. Pope, C.A., 3rd; Ezzati, M.; Dockery, D.W. Fine-particulate air pollution and life expectancy in the United States. *N. Engl. J. Med.* **2009**, *360*, 376–386. [CrossRef] [PubMed]
4. Raudoniute, J.; Stasiulaitiene, I.; Kulvinskiene, I.; Bagdonas, E.; Garbaras, A.; Krugly, E.; Martuzevicius, D.; Bironaite, D.; Aldonyte, R. Pro-inflammatory effects of extracted urban fine particulate matter on human bronchial epithelial cells BEAS-2B. *Environ. Sci. Pollut. Res.* **2018**, *25*, 32277–32291. [CrossRef]
5. Cao, X.-J.; Lei, F.-F.; Liu, H.; Luo, W.-Y.; Xiao, X.-H.; Li, Y.; Lu, J.-F.; Dong, Z.-B.; Chen, Q.-Z. Effects of Dust Storm Fine Particle-Inhalation on the Respiratory, Cardiovascular, Endocrine, Hematological, and Digestive Systems of Rats. *Chin. Med. J.* **2018**, *131*, 2482–2485. [CrossRef]
6. Kroll, A.; Gietl, J.K.; Wiesmuller, G.A.; Gunsel, A.; Wohlleben, W.; Schnekenburger, J.; Klemm, O. In vitro toxicology of ambient particulate matter: Correlation of cellular effects with particle size and components. *Environ. Toxicol.* **2013**, *28*, 76–86. [CrossRef]
7. Sanjeewa, K.K.A.; Jayawardena, T.U.; Kim, S.-Y.; Lee, H.G.; Je, J.-G.; Jee, Y.; Jeon, Y.-J. Sargassum horneri (Turner) inhibit urban particulate matter-induced inflammation in MH-S lung macrophages via blocking TLRs mediated NF-κB and MAPK activation. *J. Ethnopharmacol.* **2020**, *249*, 112363. [CrossRef]

8. Kim, J.H.; Lee, S.K.; Lim, J.H.; Kim, Y.S.; Moon, W.K.; Choi, C.H.; Yoon, K.J.; Moon, S.H.; Lee, S.H.; Yeom, M.H. Composition for Alleviating Skin Inflammation Caused by Yellow Dust and Fine Particulate, Comprising Natural Plant Extract. U.S. Patent No. US20180207222A1, 26 July 2018.
9. Jang, A.-S. Particulate Matter and Bronchial Asthma. *Korean J. Med.* **2015**, *88*, 150–155. [CrossRef]
10. Akhtar, U.S.; McWhinney, R.D.; Rastogi, N.; Abbatt, J.P.; Evans, G.J.; Scott, J.A. Cytotoxic and proinflammatory effects of ambient and source-related particulate matter (PM) in relation to the production of reactive oxygen species (ROS) and cytokine adsorption by particles. *Inhal. Toxicol.* **2010**, *22* (Suppl. S2), 37–47. [CrossRef]
11. Gerlofs-Nijland, M.E.; Rummelhard, M.; Boere, A.J.; Leseman, D.L.; Duffin, R.; Schins, R.P.; Borm, P.J.; Sillanpaa, M.; Salonen, R.O.; Cassee, F.R. Particle induced toxicity in relation to transition metal and polycyclic aromatic hydrocarbon contents. *Environ. Sci. Technol.* **2009**, *43*, 4729–4736. [CrossRef]
12. Gualtieri, M.; Ovrevik, J.; Holme, J.A.; Perrone, M.G.; Bolzacchini, E.; Schwarze, P.E.; Camatini, M. Differences in cytotoxicity versus pro-inflammatory potency of different PM fractions in human epithelial lung cells. *Toxicol. In Vitro* **2010**, *24*, 29–39. [CrossRef] [PubMed]
13. Schins, R.P.F.; Knaapen, A.M.; Weishaupt, C.; Winzer, A.; Borm, P.J.A. Cytotoxic and Inflammatory Effects of Coarse and Fine Particulate Matter in Macrophages and Epithelial Cells. *Ann. Occup. Hyg.* **2002**, *46* (Suppl. S1), 203–206.
14. Hoek, J.B.; Pastorino, J.G. Ethanol, oxidative stress, and cytokine-induced liver cell injury. *Alcohol* **2002**, *27*, 63–68. [CrossRef] [PubMed]
15. Borroni, E.M.; Mantovani, A.; Locati, M.; Bonecchi, R. Chemokine receptors intracellular trafficking. *Pharmacol. Ther.* **2010**, *127*, 1–8. [CrossRef]
16. Dorschmann, P.; Bittkau, K.S.; Neupane, S.; Roider, J.; Alban, S.; Klettner, A. Effects of Fucoidans from Five Different Brown Algae on Oxidative Stress and VEGF Interference in Ocular Cells. *Mar. Drugs* **2019**, *17*, 258. [CrossRef]
17. Daub, C.D.; Mabate, B.; Malgas, S.; Pletschke, B.I. Fucoidan from Ecklonia maxima is a powerful inhibitor of the diabetes-related enzyme, α-glucosidase. *Int. J. Biol. Macromol.* **2020**, *151*, 412–420. [CrossRef]
18. Wang, J.; Zheng, J.; Huang, C.; Zhao, J.; Lin, J.; Zhou, X.; Naman, C.B.; Wang, N.; Gerwick, W.H.; Wang, Q.; et al. Eckmaxol, a Phlorotannin Extracted from Ecklonia maxima, Produces Anti-beta-amyloid Oligomer Neuroprotective Effects Possibly via Directly Acting on Glycogen Synthase Kinase 3beta. *ACS Chem. Neurosci.* **2018**, *9*, 1349–1356. [CrossRef]
19. Lopes, G.; Andrade, P.B.; Valentao, P. Phlorotannins: Towards New Pharmacological Interventions for Diabetes Mellitus Type 2. *Molecules* **2016**, *22*, 56. [CrossRef]
20. Li, Y.-X.; Wijesekara, I.; Li, Y.; Kim, S.-K. Phlorotannins as bioactive agents from brown algae. *Process Biochem.* **2011**, *46*, 2219–2224. [CrossRef]
21. Heo, S.-J.; Park, E.-J.; Lee, K.-W.; Jeon, Y.-J. Antioxidant activities of enzymatic extracts from brown seaweeds. *Bioresour. Technol.* **2005**, *96*, 1613–1623. [CrossRef]
22. Yuan, Y.V.; Walsh, N.A. Antioxidant and antiproliferative activities of extracts from a variety of edible seaweeds. *Food Chem. Toxicol.* **2006**, *44*, 1144–1150. [CrossRef] [PubMed]
23. Braden, K.W.; Blanton, J.R., Jr.; Allen, V.G.; Pond, K.R.; Miller, M.F. Ascophyllum nodosum supplementation: A preharvest intervention for reducing Escherichia coli O157:H7 and Salmonella spp. in feedlot steers. *J. Food Prot.* **2004**, *67*, 1824–1828. [CrossRef] [PubMed]
24. Ahn, M.J.; Yoon, K.D.; Min, S.Y.; Lee, J.S.; Kim, J.H.; Kim, T.G.; Kim, S.H.; Kim, N.G.; Huh, H.; Kim, J. Inhibition of HIV-1 reverse transcriptase and protease by phlorotannins from the brown alga Ecklonia cava. *Biol. Pharm. Bull.* **2004**, *27*, 544–547. [CrossRef]
25. Le, Q.-T.; Li, Y.; Qian, Z.-J.; Kim, M.-M.; Kim, S.-K. Inhibitory effects of polyphenols isolated from marine alga Ecklonia cava on histamine release. *Process Biochem.* **2009**, *44*, 168–176. [CrossRef]
26. Wang, L.; Je, J.-G.; Kim, H.-S.; Wang, K.; Fu, X.; Xu, J.; Gao, X.; Jeon, Y.-J. Anti-Melanogenesis and Photoprotective Effects of Ecklonia maxima Extract Containing Dieckol and Eckmaxol. *Mar. Drugs* **2022**, *20*, 557. [CrossRef]
27. Kim, H.-S.; Je, J.-G.; An, H.; Baek, K.; Lee, J.M.; Yim, M.-J.; Ko, S.-C.; Kim, J.-Y.; Oh, G.-W.; Kang, M.-C.; et al. Isolation and Characterization of Efficient Active Compounds Using High-Performance Centrifugal Partition Chromatography (CPC) from Anti-Inflammatory Activity Fraction of Ecklonia maxima in South Africa. *Mar. Drugs* **2022**, *20*, 471. [CrossRef]
28. Mori, I.; Sun, Z.; Ukachi, M.; Nagano, K.; McLeod, C.W.; Cox, A.G.; Nishikawa, M. Development and certification of the new NIES CRM 28: Urban aerosols for the determination of multielements. *Anal. Bioanal. Chem.* **2008**, *391*, 1997–2003. [CrossRef] [PubMed]
29. Jiao, G.; Yu, G.; Zhang, J.; Ewart, H.S. Chemical Structures and Bioactivities of Sulfated Polysaccharides from Marine Algae. *Mar. Drugs* **2011**, *9*, 196–223. [CrossRef] [PubMed]
30. Santos, S.A.O.; Vilela, C.; Freire, C.S.R.; Abreu, M.H.; Rocha, S.M.; Silvestre, A.J.D. Chlorophyta and Rhodophyta macroalgae: A source of health promoting phytochemicals. *Food Chem.* **2015**, *183*, 122–128. [CrossRef]
31. Harnedy, P.A.; FitzGerald, R.J. Bioactive proteins, peptides, and amino acids from macroalgae1. *J. Phycol.* **2011**, *47*, 218–232. [CrossRef]
32. Tripathi, P.; Tripathi, P.; Kashyap, L.; Singh, V. The role of nitric oxide in inflammatory reactions. *FEMS Immunol. Med. Microbiol.* **2007**, *51*, 443–452. [CrossRef] [PubMed]
33. Ricciotti, E.; FitzGerald, G.A. Prostaglandins and inflammation. *Arterioscler. Thromb. Vasc. Biol.* **2011**, *31*, 986–1000. [CrossRef]

34. Turner, M.D.; Nedjai, B.; Hurst, T.; Pennington, D.J. Cytokines and chemokines: At the crossroads of cell signalling and inflammatory disease. *Biochim. Biophys. Acta (BBA)-Mol. Cell Res.* **2014**, *1843*, 2563–2582. [CrossRef] [PubMed]
35. Tay, M.Z.; Poh, C.M.; Rénia, L.; MacAry, P.A.; Ng, L.F. The trinity of COVID-19: Immunity, inflammation and intervention. *Nat. Rev. Immunol.* **2020**, *20*, 363–374. [CrossRef] [PubMed]
36. Cui, M.; Wu, J.; Wang, S.; Shu, H.; Zhang, M.; Liu, K.; Liu, K. Characterization and anti-inflammatory effects of sulfated polysaccharide from the red seaweed Gelidium pacificum Okamura. *Int. J. Biol. Macromol.* **2019**, *129*, 377–385. [CrossRef]
37. He, M.; Ichinose, T.; Ren, Y.; Song, Y.; Yoshida, Y.; Arashidani, K.; Yoshida, S.; Nishikawa, M.; Takano, H.; Sun, G. PM2.5-rich dust collected from the air in Fukuoka, Kyushu, Japan, can exacerbate murine lung eosinophilia. *Inhal. Toxicol.* **2015**, *27*, 287–299. [CrossRef]
38. Kiemer, A.K.; Hartung, T.; Huber, C.; Vollmar, A.M. Phyllanthus amarus has anti-inflammatory potential by inhibition of iNOS, COX-2, and cytokines via the NF-κB pathway. *J. Hepatol.* **2003**, *38*, 289–297. [CrossRef]
39. Salter, D. Connective Tissue Responses to Mechanical Stresses. In *Rheumatology*; Elsevier-Health Sciences Division: Amsterdam, The Netherlands, 2010.
40. Akira, S. Toll-like Receptors and Innate Immunity. In *Advances in Immunology*; Dixon, F.J., Ed.; Academic Press: Cambridge, MA, USA, 2001; Volume 78, pp. 1–56.
41. Sanjeewa, K.K.A.; Jayawardena, T.U.; Kim, H.-S.; Kim, S.-Y.; Ahn, G.; Kim, H.-J.; Fu, X.; Jee, Y.; Jeon, Y.-J. Ethanol extract separated from Sargassum horneri (Turner) abate LPS-induced inflammation in RAW 264.7 macrophages. *Fish. Aquat. Sci.* **2019**, *22*, 6. [CrossRef]
42. Zhou, X.; Yi, M.; Ding, L.; He, S.; Yan, X. Isolation and Purification of a Neuroprotective Phlorotannin from the Marine Algae Ecklonia maxima by Size Exclusion and High-Speed Counter-Current Chromatography. *Mar. Drugs* **2019**, *17*, 212. [CrossRef]
43. Jayawardena, T.U.; Sanjeewa, K.; Lee, H.-G.; Nagahawatta, D.; Yang, H.-W.; Kang, M.-C.; Jeon, Y.-J. Particulate Matter-Induced Inflammation/Oxidative Stress in Macrophages: Fucosterol from Padina boryana as a Potent Protector, Activated via NF-κB/MAPK Pathways and Nrf2/HO-1 Involvement. *Mar. Drugs* **2020**, *18*, 628. [CrossRef]
44. Mosmann, T. Rapid colorimetric assay for cellular growth and survival: Application to proliferation and cytotoxicity assays. *J. Immunol. Methods* **1983**, *65*, 55–63. [CrossRef] [PubMed]
45. Fernando, I.S.; Jayawardena, T.U.; Kim, H.-S.; Lee, W.W.; Vaas, A.; De Silva, H.; Abayaweera, G.; Nanayakkara, C.; Abeytunga, D.; Lee, D.-S. Beijing urban particulate matter-induced injury and inflammation in human lung epithelial cells and the protective effects of fucosterol from Sargassum binderi (Sonder ex J. Agardh). *Environ. Res.* **2019**, *172*, 150–158. [CrossRef] [PubMed]
46. Sanjeewa, K.K.A.; Fernando, I.P.S.; Kim, E.-A.; Ahn, G.; Jee, Y.; Jeon, Y.-J. Anti-inflammatory activity of a sulfated polysaccharide isolated from an enzymatic digest of brown seaweed Sargassum horneri in RAW 264.7 cells. *Nutr. Res. Pract.* **2017**, *11*, 3. [CrossRef] [PubMed]
47. Nagahawatta, D.P.; Kim, H.-S.; Jee, Y.-H.; Jayawardena, T.U.; Ahn, G.; Namgung, J.; Yeo, I.-K.; Sanjeewa, K.K.A.; Jeon, Y.-J. Sargachromenol Isolated from Sargassum horneri Inhibits Particulate Matter-Induced Inflammation in Macrophages through Toll-like Receptor-Mediated Cell Signaling Pathways. *Mar. Drugs* **2022**, *20*, 28. [CrossRef] [PubMed]

marine drugs

Article

The Inhibitory Effect of Phycocyanin Peptide on Pulmonary Fibrosis *In Vitro*

Run-Ze Liu [1,2,†], Wen-Jun Li [1,2,†], Juan-Juan Zhang [3], Zheng-Yi Liu [1,2], Ya Li [3], Chao Liu [3,*] and Song Qin [1,2,*]

1 Yantai Institute of Coastal Zone Research, Chinese Academy of Sciences, Yantai 264003, China
2 Center for Ocean Mega-Science, Chinese Academy of Sciences, Qingdao 266071, China
3 Yantai Jiahui Biotech Co., Ltd., Yantai 264003, China
* Correspondence: sanmu_363@163.com (C.L.); sqin@yic.ac.cn (S.Q.);
Tel.: +86-13455568130 (C.L.); +86-15965197088 (S.Q.)
† These authors contributed equally to this work.

Abstract: Phycocyanin is an excellent antioxidant with anti-inflammatory effects on which recent studies are growing; however, its specific target remains unclear. Linear tetrapyrrole compounds such as bilirubin have been shown to lead to the induction of heme oxygenase 1 expression *in vivo*, thus achieving antioxidant and anti-inflammatory effects. Phycocyanin is bound internally with linear tetrapyrrole phycocyanobilin in a similar structure to bilirubin. We speculate that there is probably a way of inducing the expression of heme oxygenase 1, with which tissue oxidative stress and inflammation can be inhibited, thus inhibiting pulmonary fibrosis caused by oxidative damage and inflammation of lung. By optimizing the enzymatic hydrolysis process, phycocyanobilin-bound phycocyanin peptide were obtained, and its *in vitro* antioxidant, anti-inflammatory, and anti-pulmonary fibrosis activities were investigated. The results show that the phycocyanobilin peptide was able to alleviate oxidative and inflammatory damage in cells through the Keap1-Nrf2-HO-1 pathway, which in turn relieved pulmonary fibrosis symptoms.

Keywords: phycocyanin; phycocyanin peptide; pulmonary fibrosis; inflammatory; oxidative stress; heme oxygenases-1

Citation: Liu, R.-Z.; Li, W.-J.; Zhang, J.-J.; Liu, Z.-Y.; Li, Y.; Liu, C.; Qin, S. The Inhibitory Effect of Phycocyanin Peptide on Pulmonary Fibrosis *In Vitro*. *Mar. Drugs* **2022**, *20*, 696. https://doi.org/10.3390/md20110696

Academic Editor: Fernando Albericio

Received: 27 September 2022
Accepted: 2 November 2022
Published: 6 November 2022

1. Introduction

Idiopathic pulmonary fibrosis (IPF) is a progressive form of pulmonary fibrosis in which the structure of the alveoli is damaged and gas exchange in the lungs is impaired, eventually leading to respiratory failure and death [1]. At present, the etiology and specific pathogenesis of IPF remains poorly understood. The current mainstream view is that IPF results from alveolar epithelial cell damage and consists of oxidative stress damage [2] and inflammatory damage by various interactive genetic and environmental factors. An epithelial cell injury would trigger abnormal communication between epithelial cells and fibroblasts, induce activation of myofibroblasts, secrete a large amount of extracellular matrix (ECM), deposit in the lungs, and consequently bring about non-structural remodeling of the lungs and fibrosis [3]. Unfortunately, there is no practically effective treatment for IPF. Conventional anti-pneumonia drugs, such as prednisone, dexamethasone, and other corticosteroids have no obvious effect on pulmonary fibrosis but serious side effects [4]. In recent years, new drugs such as nintedanib and pirfenidone can alleviate the symptoms of pulmonary fibrosis to a certain extent, but serious side effects (nausea, abdominal pain, diarrhea, photosensitivity, and nervous system abnormalities) are unavoidable, thus hampering the clinical application [5,6]. Therefore, looking for a safe, less toxic, and side-effect-free drug for the treatment of IPF has become an urgent task for drug research institutions.

Inflammation is widespread in the organism of IPF patients, which is the host's defense action against injury, oxidative damage, or infectious pathogens, as well as adaptive immune response [7]. In the innate immune response, macrophages play a key role as an immune cell. By secreting a variety of inflammatory factors, including tumour necrosis factor (TNF) and interleukins (IL), macrophages are able to clear pathogens from the body [8,9]. Moderate amounts of inflammatory factors can participate in the body's immune response, while excessive production of inflammatory factors can damage the body and lead to various inflammatory diseases such as sepsis, pulmonary fibrosis, diabetes, and atherosclerosis [10,11]. Therefore, regulating the activity of secreting inflammatory factors in macrophages can inhibit pulmonary fibrosis caused by tissue inflammation to a certain extent.

Heme oxygenase 1 (HO-1) is an antioxidant-induced enzyme responsible for the breakdown of heme into biliverdin [12]. Earlier studies have identified HO-1 as having anti-inflammatory effects and the ability to modulate various immune cells, including macrophages [13–15], making HO-1 an important target in the treatment of inflammation and even pulmonary fibrosis induced by inflammation. Being similar in linear tetrapyrrole structure to that of bilirubin, phycocyanobilin (PCB) can induce the expression of HO-1. However, current bilirubin analogues are likely to cause jaundice during clinical administration, and have disadvantages such as toxic side effects and low bioavailability, which has limited the application of HO-1 targeting in anti-inflammatory and anti-fibrosis treatment.

Phycocyanin (PC) is a light-harvesting protein widely found in algae, which is soluble in water. It is composed of linear tetrapyrrole pigment PCB through thioether bond with apoprotein α and β. In PC, three PCBs are respectively connected to Cys-84 of the α chain, Cys-84, and 155 of the β chain [16] (Figure 1) and are wrapped in the hydrophobic core inside.

PC has many biological activities as an antioxidant, anti-inflammatory, and anti-pulmonary fibrosis agent [17,18]. Li et al. [18] used PC to treat mice that had bleomycin-induced pulmonary fibrosis, and found that PC could pass through the TLR2-MyD88-NF-κB signaling pathway to effectively inhibit the bleomycin-induced pulmonary fibrosis. In addition, Liu et al. [19] used PC to cure mice that had radiation-induced acute liver injury, and they found that PC significantly up-regulated the expression of Nrf2 and the downstream gene HO-1, which reduced the liver injury. Therefore, we believe that the activity of PC is in fact from PCB that is contained within PC. PCB in PC would be exposed and released after gastrointestinal digestion and be in full contact with the environment. As it features a linear tetrapyrrole structure, the active groups are the same as bilirubin to induce the expression of HO-1 against oxidation, inflammation, and pulmonary fibrosis [20,21].

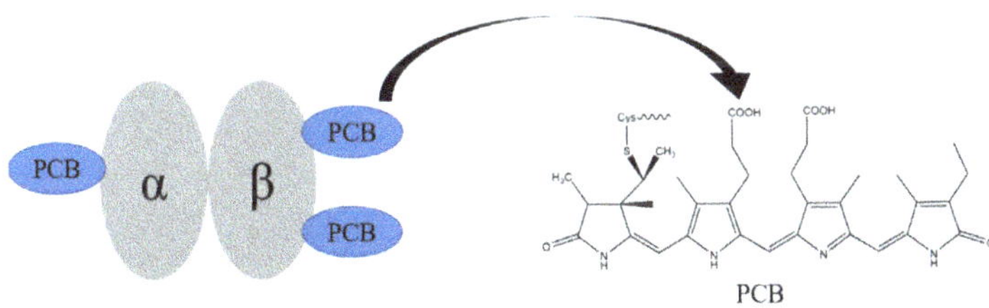

Figure 1. The structures of phycocyanin monomer and phycocyanobilin.

PC has the advantages of non-toxicity and good water solubility, and it contains PCB, which is a satisfactory material for preparing the HO-1 inducer. Therefore, we enzymatically hydrolyzed PC and decomposed it into small peptides in 10 amino acids, by which PCB that was fully exposed and phycocyanin peptide connected with PCB was formulated. In addition, the antioxidant activity, LPS-induced RAW264.7 macrophage inflammation, and TGF-β1-induced pulmonary fibrosis inhibition were determined by *in vitro* experiments.

2. Results

2.1. Enzymatic Hydrolysis of Phycocyanin and Its Absorption Spectrum and Molecular Weight Determination

PCB in PC is very sensitive when ambient pH, temperature, etc., change and the color during enzymatic hydrolysis could easily vary, thus affecting the final product. To avoid the color change after enzymatic hydrolysis and ensure the activity of phycocyanin peptide, after a large number of screening tests, we applied the composite enzyme MC101 to the enzymatic hydrolysis process of PC for the first time. The activity of MC101 was high, and the yield rate of phycocyanin peptide reached 96% in blue color. The color of the phycocyanin peptide after enzymatic hydrolysis was lighter than that of PC, showing light blue. We think that phycocyanin peptide contained a peptide fragment bound to PCB. Apparently, the PC solution was purplish blue and the solution of phycocyanin peptide after enzymatic hydrolysis turned blue. This is because phycocyanin has fluorescence characteristics, so its solution presents a purple color due to the red and blue mix. The active structure of the phycocyanin is destroyed after enzymatic hydrolysis, so the fluorescence disappears, leaving only the blue color of PCB. PCB is a fat-soluble substance and does not dissolve in water when it exists alone. However, the phycocyanin peptide after MC101 complex enzymatic hydrolysis had good water solubility, and could be quickly dissolved even at high concentrations, which broadened the application range of the phycocyanin peptide.

By configuring the *Spirulina* PC and phycocyanin peptide into a same-concentration solution and scanning in the full wavelength range of 200–800 nm, the absorption spectrum of the phycocyanin peptide was obtained (Figure 2). The absorption peaks of PC distributed mostly at 620 nm (chromophore absorption) and 280 nm (aromatic amino acid absorption), and another peak at 360 nm reflects the PCB conjugate. In the full-wavelength scan, it was shown that the absorbance peak of phycocyanin peptide at 620 nm decreased obviously, while those at 280 nm and 360 nm increased. The properties of the absorption spectrum are very similar to that of the PCB standard. Therefore, it shows that the PC had been fully decomposed into phycocyanin peptides, and the enzymatic hydrolysis was sufficient.

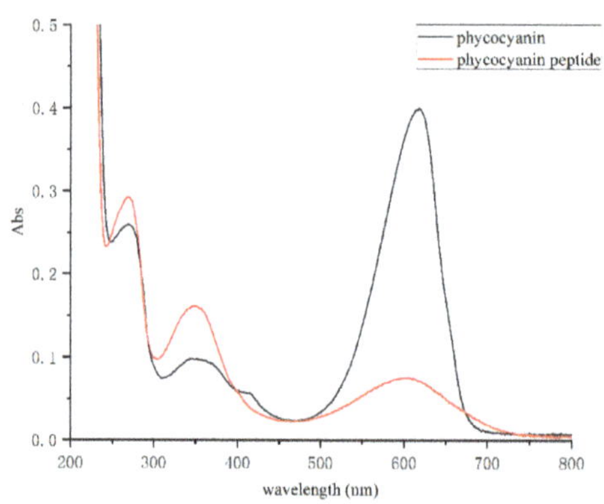

Figure 2. The absorption spectra of phycocyanin and phycocyanin peptide.

Molecular weight distribution of the phycocyanin peptide is shown in Table 1. The average molecular weight of the phycocyanin peptide is 583 Da, which is much smaller than 40 kDa of PC, while the calculated value from average molecular weight of amino acids in the protein was 110 Da. Therefore, we believe that the obtained enzymatic peptide contained 5–6 amino acids.

Table 1. Molecular weight distribution of phycocyanin peptide.

Indicators	>10,000	10,000–5000	5000–3000	3000–2000	2000–1000	1000–500	500–180	<180
Peak area percentage (%, λ220 nm)	0.72	0.39	0.58	1.81	6.83	20.56	50.14	18.96
Number average molecular weight	13,929	7536	3476	2444	1267	639	253	–
Weight average molecular weight	14,429	7826	3539	2477	1311	661	271	–
total weight average molecular weight (Mw)				583				

–: no data.

Subsequently, the phycocyanin peptide was analyzed by HPLC-MS in order to obtain information on the fragments of peptides bound with PCB. The HPLC-MS results are shown in the supplemental Table S1. From the results, it can be seen that PC, after enzymatic digestion, had 36 fragments containing PCB of different lengths, of which 9 were from the PC α subunit with lengths ranging from 9–16 amino acids and 27 were from the PC β subunit with lengths ranging from 3–21 amino acids. Considering that fragments with many amino acids would not be easily absorbed, 11 fragments with amino acid numbers less than 10 were obtained after further screening (Table 2), of which 1 was from the PC α subunit and 10 from the PC β subunit. By comparing the obtained sequences, it was shown that the PCB of the fragment from the α subunit binds to Cys84. Among the 10 fragments from the β subunit, 5 PCB bound to Cys82 and 5 PCB to Cys153 (Table 2).

Table 2. The fragments of phycocyanin peptide with PCB.

Seq Loc	Tgt Seq Mass	Sequence	Missed	Pred Mods
α(78-86)	1646.8351	QRGKDKCAR	8	PCB-α84
β(80-82)	849.3731	AAC	2	PCB-β82
β(81-84)	1047.5212	ACLR	3	PCB-β82
β(80-84)	1118.5583	AACLR	4	PCB-β82
β(81-86)	1293.5886	ACLRDM	5	PCB-β82
β(80-86)	1364.6257	AACLRDM	6	PCB-β82
β(152-155)	980.395	DCSA	3	PCB-β153
β(152-156)	1093.479	DCSAL	4	PCB-β153
β(148-155)	1348.6009	ITPGDCSA	7	PCB-β153
β(148-156)	1461.685	ITPGDCSAL	8	PCB-β153
β(147-155)	1405.6224	GITPGDCSA	8	PCB-β153

2.2. Antioxidant Activity of Phycocyanin Peptide

PC has been shown to have excellent antioxidant activity [17], but whether the enzymatic-hydrolyzed phycocyanin peptides have similar or better activity is currently unknown. Therefore, we compared the *in vitro* antioxidant activities of PC and phycocyanin peptides to understand in-depth the activity of the phycocyanin peptide.

2.2.1. Superoxide Anion Scavenging Rate

Superoxide anion (O_2^-) is a kind of reactive oxygen species (ROS) produced in organisms through respiration and other pathways [22,23]. Excessive O_2^- can cause damage to the body, produce oxidative stress, and trigger various pathological processes, such as protein denaturation, nucleotide damage, etc. [24,25]. Figure 3a shows that the O_2^- scavenging ability of PC or the phycocyanin peptide enhanced as their respective concentrations increased. Overall, the O_2^- clearance rate of the phycocyanin peptide is greater than that of that of PC. The curves were linearly fitted, and the IC50 values of PC and the phycocyanin peptide were 8.15 mg/mL and 6.65 mg/mL, respectively. The O_2^- clearance rate of the phycocyanin peptide after enzymatic hydrolysis was greatly improved from the clearance rate of PC.

2.2.2. ABTS$^+$ Scavenging Rate

Reduced ABTS is a colorless substance that can be oxidized to blue-green ABTS$^+$ free radicals by oxidants. ABTS$^+$ free radicals are very stable and have a maximum absorption peak at 734 nm. Antioxidants that are able to donate hydrogen can react with $ABTS^+$ and can turn them back into colorless ABTS. Therefore, ABTS is often used as an *in vitro* antioxidant index to judge the antioxidant capacity of a sample [26].

The scavenging ability of PC or the phycocyanin peptide of $ABTS^+$ is shown in Figure 3b. It was observed that PC and the phycocyanin peptide had different degrees of scavenging effect on $ABTS^+$. In the selected concentration range, with the increase of concentration, the clearance of PC and the phycocyanin peptide of $ABTS^+$ was logarithmically correlated. At 5 mg/mL, the scavenging rates of PC and the phycocyanin peptide of $ABTS^+$ were 88.58% and 98.02%, respectively. The IC_{50} values of PC and the phycocyanin peptide were 0.96 and 0.42 mg/mL, respectively.

2.2.3. Total Reducing Power

The reducing power refers to the color oxidation reaction that determines whether the sample can become a good electron donor. The oxidation resistance of a substance depends on the reducing power, so the oxidation resistance can be judged by measuring the reducing power of the sample. Potassium ferricyanide is an oxidizing complex. After reacting with reducing substances, it produces colorless potassium ferrocyanide. Potassium ferricyanide reacts with $FeCl_3$ to produce Prussian blue. There is a characteristic absorption peak at 700 nm. Therefore, the relative reduction power of the sample can be ascertained by an absorbance measurement at 700 nm [27].

The total reducing power of PC or the phycocyanin peptide is shown in Figure 3c. In the range of 0.5–10 mg/mL, the reducing power of PC or the phycocyanin peptide increased with the increase of its own concentration, and the reducing power was positively correlated to the concentration. The reducing power of PC and the phycocyanin peptide was 0.806 and 0.914 at a sample concentration of 10 mg/mL, respectively. In addition, the difference in total reducing power between PC and the phycocyanin peptide was not obvious at low concentrations (Figure 3c). As the concentration increased, the phycocyanin peptide enzymatically hydrolyzed with MC101 showed better reduction than that of PC.

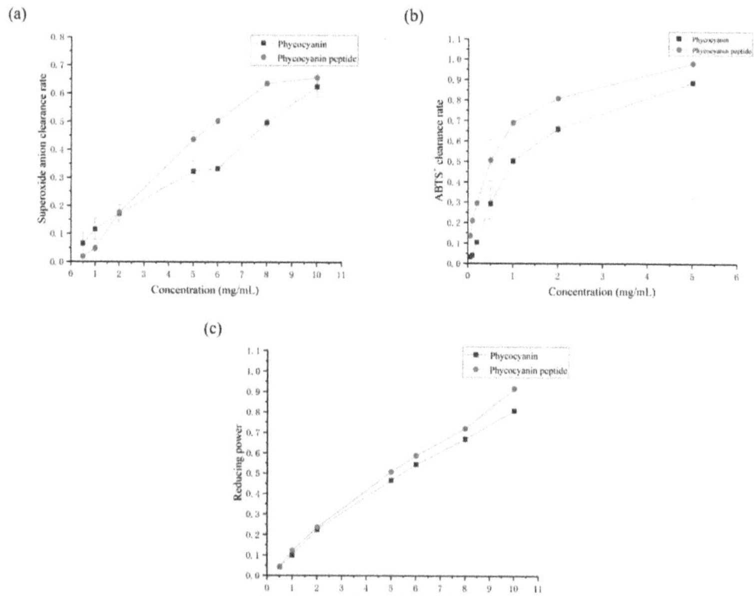

Figure 3. Antioxidant activity of phycocyanin peptide: (**a**) superoxide anion clearance; (**b**) $ABTS^+$ clearance; (**c**) total reducing power.

2.3. Anti-Inflammatory Activity of Phycocyanin Peptide

2.3.1. Effect of Phycocyanin Peptide on Survival Rate of RAW264.7 Cells

The CCK-8 assay was applied to exclude drug interference with cell status and to provide a basis for subsequent selection of the appropriate concentration. Taking the relative cell viability ≥90% as the criterion of no cytotoxicity, the phycocyanin peptide did not inhibit the growth of macrophages (Figure 4a). Overall, the phycocyanin peptide slightly promoted the proliferation of RAW264.7. The relative survival of cells at 200 μg/mL of phycocyanin peptide was approximately 138%, which proved that the phycocyanin peptide had no obvious cytotoxicity and was safe.

2.3.2. Inhibitory Effect of Phycocyanin Peptide on NO

NO is a type of cytokine which can mediate many biological functions. Macrophage-derived NO plays a major role in physiology and pathology. A proper amount of NO can promote the immune response of the body, while the over expression of NO could cause inflammation such as rheumatoid arthritis, atherosclerosis, tissue damage, etc. [28]. Therefore, the expression level of NO determines the induction or inhibitory effect of phycocyanin peptides on inflammation.

To investigate the inhibitory effect of phycocyanin peptides on NO, the effect of phycocyanin peptides on NO release from LPS-induced RAW264.7 cells was examined using the Griess method. Depending on the results of the CCK-8 experiment, the concentration of phycocyanin peptides was set at 50, 100, and 200 μg/mL. As shown in Figure 4b, the release of NO in the supernatant of RAW264.7 macrophages after LPS induction was increased dramatically compared with the control group, reaching 41.58 μM. After the intervention of phycocyanin peptides, the expression level of NO in the supernatant of RAW264.7 cells was effectively inhibited. As the concentration increased, the inhibitory effect of phycocyanin peptides on NO showed an upward trend. At a concentration of 50 μg/mL, the NO concentration was 21.75 μM, and the inhibitory rate of phycocyanin peptides on NO was 39.63%. After the concentration was raised to 200 μg/mL, the NO concentration was 20.98 μM, and the NO inhibition rate was 41.16%.

2.3.3. Inhibitory Effect of Phycocyanin Peptide on TNF-α and IL-6

As inflammatory factors secreted by macrophages, TNF-α and IL-6 overexpression often cause the development of many diseases [29,30]. Therefore, by detecting changes in the content of TNF-α or IL-6, the anti-inflammatory ability of phycocyanin peptides can be determined.

Based on the NO inhibition test, we examined the effect of phycocyanin peptides on the levels of IL-6 and TNF-α expression in the cell supernatant using the Elisa method and further validated their anti-inflammatory activity. The results are presented in Figure 4c,d. In terms of TNF-α, the induction of LPS promoted its expression, being increased from 331.29 pg/mL in the control to 454.54 pg/mL. After treatment with phycocyanin peptides, the expression of TNF-α was inhibited. Compared with the model group, which was added with 100 ng/mL LPS, the inhibitory effect of phycocyanin peptides on TNF-α was enhanced with the increase of the concentration. At 50 μg/mL, the TNF-α decreased to 362.94 pg/mL, and the relative inhibition rate was 74.32%. At 200 μg/mL, TNF-α returned to the level of the control group, or even slightly lower (304.68 pg/mL), with a relative inhibition rate of 100%. The effect of phycocyanin peptides on IL-6 expression in macrophages was similar to that of TNF-α, and the inhibitory effect continued to increase with the concentration increase. The concentration of IL-6 in the 50 μg/mL low-dose group decreased from 32.87 to 28.49 pg/mL, and the relative inhibition rate was 30.44%. The inhibitory effect was additionally enhanced at a high concentration of 200 μg/mL; after that, the IL-6 concentration was reduced to 21.95 pg/mL and the relative inhibition rate was 75.76%. The results show that phycocyanin peptides effectively suppressed the TNF-α and IL-6 expressions, and the suppression effect was gradually enhanced with increasing concentration. At a high concentration of 200 μg/mL, phycocyanin peptides restored TNF-α to normal levels. Although the recovery effect of IL-6 was not as great as that of TNF-α, the inhibition rate of IL-6 reached 75.76%, which proved that the obtained phycocyanin peptides played an anti-inflammatory role *in vitro* and had a good effect.

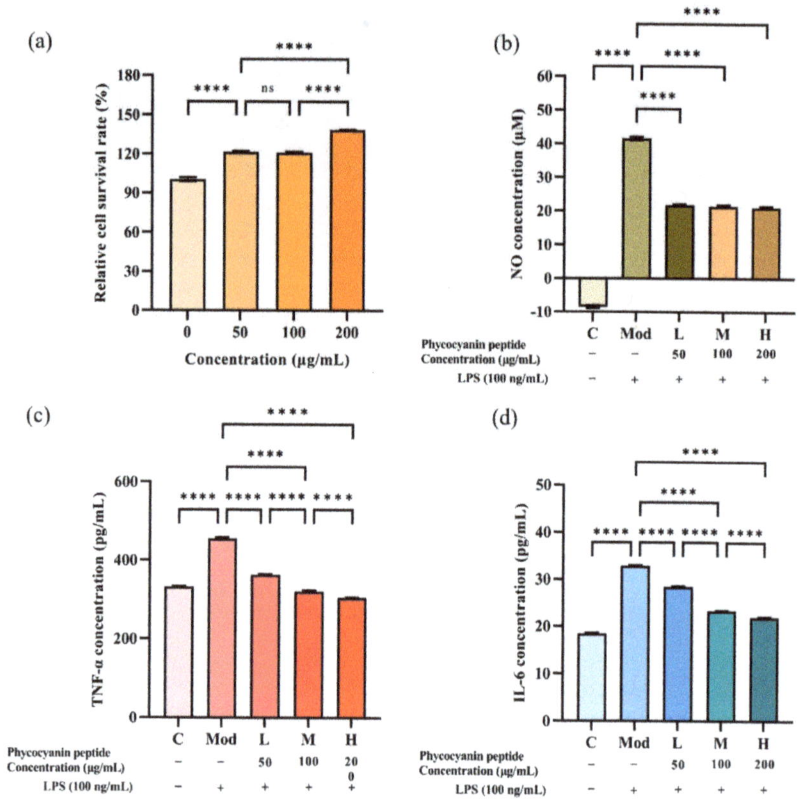

Figure 4. Anti-inflammatory activity of phycocyanin peptide: (**a**) relative cell survival rate of RAW264.7 macrophages; (**b**) NO scavenging ability; (**c**) TNF-α scavenging ability; (**d**) IL-6 scavenging ability. C: Control; Mod: Model; L: Low-dose (50 μg/mL); M: Medium-dose (100 μg/mL); H: High dose (200 μg/mL). n = 3, mean ± SD. **** $p < 0.0001$; ns $p > 0.05$.

2.4. Anti-Pulmonary Fibrosis Activity of Phycocyanin Peptide

2.4.1. A549 Cell Morphology and Collagen I Expression

Under the induction of TGF-β1, human lung epithelial cells (A549) can be polarized. The cells present a relatively slender spindle in shape, and the epithelial-mesenchymal transition (EMT) would occur, showing fibrotic activity. As Figure 5a–d show, the morphology of A549 cells after TGF-β1 significantly changed from original cobblestone-like cells to obvious spindle-shaped cells, and the cells were elongated. After treatment with phycocyanin peptides, the morphological recovery of A549 cells at low doses was not obvious, and most of the cells remained in spindle shape, while at high doses, the morphological recovery of A549 cells was more obvious, being similar to the control group.

A very significant symptom of IPF is the excessive deposition of extracellular matrix between tissues. Excessive deposition of ECM could aggravate the development of IPF. Collagen I, as the main component of ECM, plays a critical role in the excessive deposition of ECM. Collagen is not easy to be degraded as its conversion rate is very slow and has strong resistance to common proteases. Therefore, once collagen is produced, it is difficult to be removed from the human body, thereby continuously aggravating IPF. Therefore, Collagen I is an important indicator of IPF.

The effect of phycocyanin peptides is shown in Figure 5e–h in which the blue part was the nucleus of A549 cells, and the green, fluorescent part was Collagen I. After 72 h of TGF-β1 induction, the Collagen I expression of A549 cells increased significantly. Low-dose phycocyanin peptide had a certain alleviating effect on Collagen I expression, but the effect was not obvious. At high doses, phycocyanin peptides presented a better inhibitory effect on Collagen I expression, while the green fluorescence faded in general.

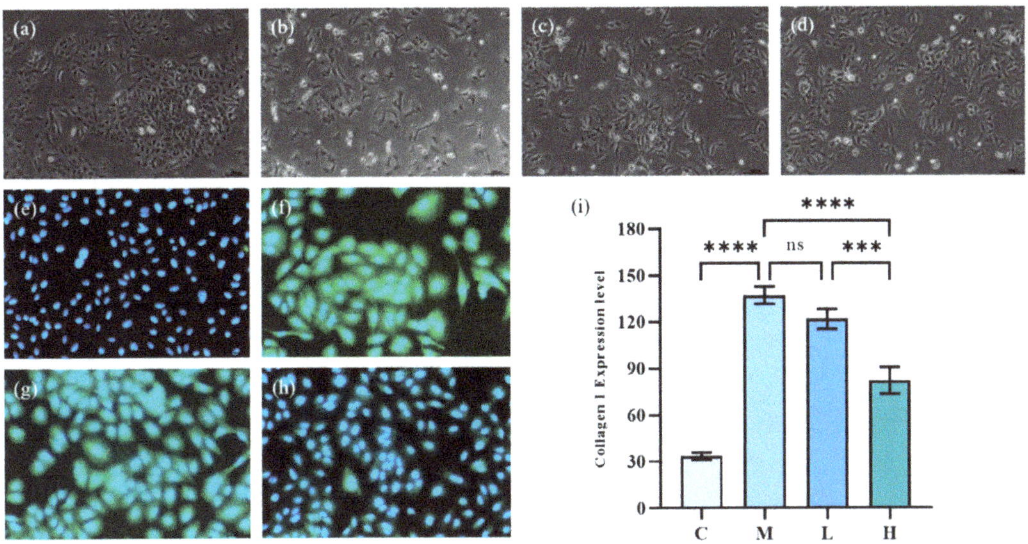

Figure 5. The effect of phycocyanin peptides on the morphology and Collagen I expression of A549 cells induced by TGF-β1: (**a**) cell morphology of control group; (**b**) cell morphology of model group; (**c**) cell morphology of low-dose group; (**d**) cell morphology of high-dose group; (**e**) expression of Collagen I in control group; (**f**) Collagen I of model group expression; (**g**) Collagen I expression in the low-dose group (10 μg/mL); (**h**) Collagen I expression in the high-dose group (30 μg/mL); (**i**) Collagen I expression level. C: the control group; M: the model group; L: the low-dose group (10 μg/mL); H: the high dose group (30 μg/mL). $n = 3$, mean ± SD. *** $p < 0.001$; **** $p < 0.0001$; ns $p > 0.05$.

2.4.2. Nrf2, NQO1 and HO-1 Expression

It has been proven that HO-1 has excellent antioxidant and anti-inflammatory activities and can modulate the activity of various immune cells. The induction of HO-1 is inseparable from the metabolism of bilirubin. Nam et al. demonstrated that bilirubin could induce the expression of HO-1 through the Keap1-Nrf2 pathway [21]. PCB in phycocyanin peptide is very similar in structure to bilirubin, so we therefore hypothesize that the anti-inflammatory activity of the peptides derives from the modulation of HO-1.

NAD(P)H: quinone oxidoreductase 1 (NQO1) is also an essential factor in the Keap1-Nrf2 signaling pathway. Through the reduction reaction, NQO1 can catalyze the reduction of quinones into hydroquinones and promote the metabolism of quinones, thereby avoiding the production of ROS by semihydroquinones and playing an antioxidant role in cells [31].

After induction by TGF-β1, the levels of NQO1, Nrf2, and HO-1 in A549 cells were significantly decreased (Figure 6, $p < 0.001$), and the cells produced an inflammatory response. After treatment with 10 and 30 μg/mL of phycocyanin peptides, the three indexes all increased to different degrees, and they all recovered to more than 60% of the control group. Statistical analysis showed that the recovery of Nrf2, NQO1, and HO-1 was related to the intervention of phycocyanin peptides ($p < 0.05$); however, with the exception of NQO1, the increase of Nrf2 and HO-1 did not appear to be related to the concentration of phycocyanin peptides ($p > 0.05$).

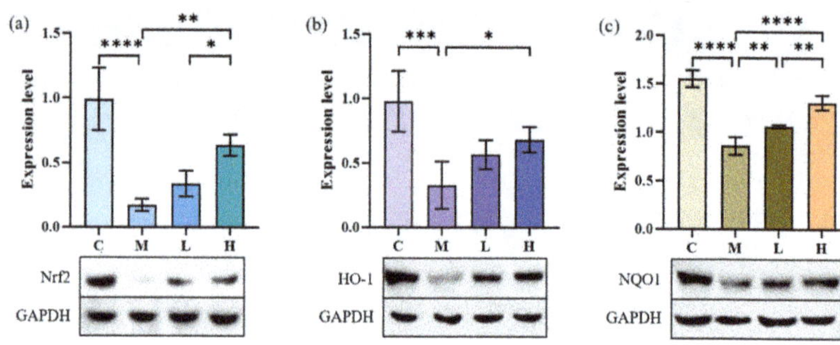

Figure 6. The effect of phycocyanin peptides on the expression of Nrf2, HO-1, and NQO1 in A549 cells induced by TGF-β1: (**a**) Nrf2 expression; (**b**) HO-1 expression; (**c**) NQO1 expression. C: the control group; M: the model group; L: the low-dose group (10 μg/mL); H: the high dose group (30 μg/mL). n = 3, mean ± SD. * $p < 0.05$; ** $p < 0.01$; *** $p < 0.001$; **** $p < 0.0001$.

2.4.3. EMT-Related Proteins Expression

EMT refers to the process of epithelial cells that are transformed into mesenchyme cells and plays a critical role in tissue inflammation and fibrosis. TGF-β1 can induce the EMT process of A549 cells through the NF-κB pathway by expressing landmark proteins such as α-SMA, vimentin, N-cadherin, and E-cadherin. The effect of phycocyanin peptides on pulmonary fibrosis can be studied by measuring the expression of several marker proteins in the cell supernatant. The expressions of α-SMA, N-cadherin, and vimentin can advance the development of IPF during the occurrence of pulmonary fibrosis. E-cadherin can maintain the connection among epithelial cells. When E-cadherin levels are down-regulated, the connection among epithelial cells would be destroyed and the adhesion lost, which would greatly increase the migration ability of cells and changes their morphology, eventually leading to the occurrence of fibrosis.

Figure 7 shows that the α-SMA content in A549 cells after induction of TGF-β1 increased from 0.31 to 0.88, that of N-cadherin increased from 0.43 to 0.86, and the expression of vimentin increased from 0.41 to 0.89. Meanwhile, the expression of E-cadherin decreased from 1.08 to 0.58, indicating that the EMT process occurred in A549 cells. After treatment with low-dose phycocyanin peptide, the expressions of α-SMA (0.80) and N-cadherin (0.85) were slightly down-regulated, that of E-cadherin (0.80) was significantly recovered, while that of vimentin (1.02) slightly increased. The expressions of α-SMA (0.61), N-cadherin (0.59), and vimentin (0.77) at high concentrations decreased largely, while that of E-cadherin (0.92) further increased, indicating that phycocyanin peptides could affect cell EMT *in vitro* with a significant reversal effect.

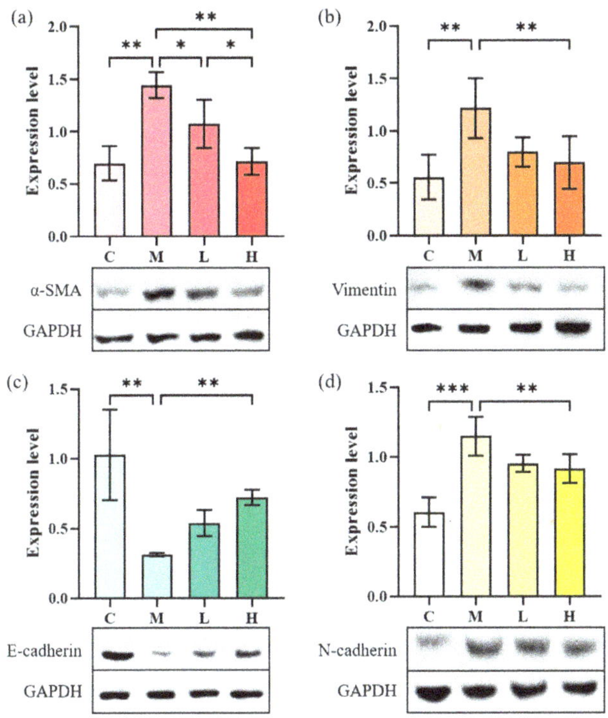

Figure 7. The effect of phycocyanin peptides on the expression of EMT-related proteins in A549 cells induced by TGF-β1: (**a**) α-SMA expression; (**b**) Vimentin expression; (**c**) N-cadherin expression; (**d**) E-cadherin expression. C: the control group; M: the model group; L: the low-dose group (10 μg/mL); H: the high dose group (30 μg/mL). n = 3, mean ± SD. * $p < 0.05$; ** $p < 0.01$; *** $p < 0.001$.

2.4.4. HFL-1 Cell α-SMA Expression

During the development of IPF, fibroblasts can be transformed into myofibroblasts, involving the process of tissue damage repair and fibrosis. Whether there is oxidative stress injury or inflammatory injury, it is often through the activation of myofibroblasts that leads to pulmonary fibrosis. The Collagen I and α-SMA expression levels are two important indicators of the fibrosis process. As an essential component of the ECM, high levels of Collagen I expression often indicate the onset of fibrosis. In 2021, Li et al. [32] obtained a 20-amino-acid peptide from PC (eicosapeptide) and investigated its anti-fibrotic effect on lung fibrosis. Their results demonstrate that eicosapeptide could dramatically depress the increase of Collagen I in HFL-1 cells induced by TGF-β1 and played an anti-fibrotic role. α-SMA is an important indicator of myofibroblast activation, and the increase in its expression signifies the activation of myofibroblasts and the development of pulmonary fibrosis [33]. Therefore, we investigated the inhibitory activity of phycocyanin peptides on the expression of α-SMA in HFL-1 cells as a complement to their anti-pulmonary fibrosis activity.

The results of immunofluorescence staining of α-SMA induced by TGF-β1 are shown in Figure 8, in which the green fluorescent part indicates the expression of α-SMA. The model group expressed a large amount of α-SMA after 72 h of TGF-β1 induction, and the cells were gradually transformed from fibroblasts to myofibroblasts. The green fluorescence of HFL-1 cells treated with phycocyanin peptides was reduced remarkably. At low doses, phycocyanin peptides could reduce the expression of α-SMA, while at high concentrations of phycocyanin peptides, α-SMA expression only appeared slightly, and returned to a normal level overall.

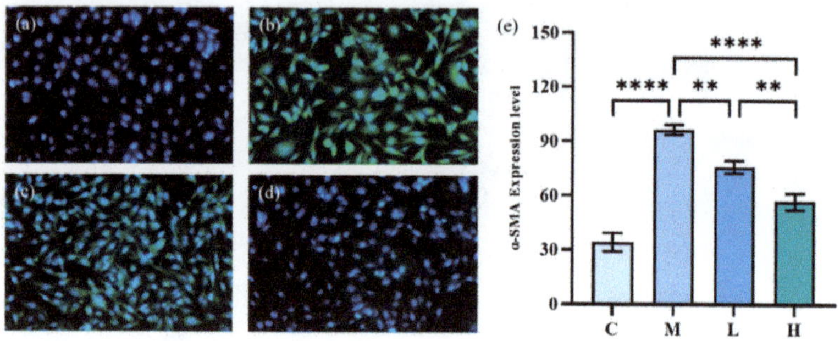

Figure 8. The effect of phycocyanin peptides on the expression of α-SMA in HFL-1 cells induced by TGF-β1: (**a**) the control; (**b**) the model; (**c**) low-dose at 10 μg/mL; (**d**) high-dose at 30 μg/mL (**e**) α-SMA expression level. C: the control group; M: the model group; L: the low-dose group (10 μg/mL); H: the high dose group (30 μg/mL). $n = 3$, mean ± SD. ** $p < 0.01$; **** $p < 0.0001$.

2.5. Covalent Docking

Britanin, CDDO, isoxazoline-based_inhibitor, DMF, and sAIM_TX64063 are all known small molecule inhibitors of the Keap1 BTB domain. Although the mechanisms of action of these small-molecule inhibitors vary, they can be grouped largely into three categories: (1) isoxazoline-based_inhibitor and DMF using α, β-unsaturated carbonyl groups as electrophilic receptors, (2) triterpenoids (CDDO and sAIM_Tx64063) taking nitrile group as electrophilic receptor, and (3) Britanin that uses a halogen as a reactive group and can suppress Keap1 activity by creating a covalent bond with the sulfur atom in Cys151 of the Keap1 BTB domain (Figure 9). PCB has α, β-unsaturated carbonyl groups, so we thought that PCB might also be able to inhibit the activity of Keap1 through the Michael addition reaction. DMF is considered to protect mouse neurons and astrocytes from oxidative stress; this process is thought to occur as DMF interacts with the Keap1 BTB domain to activate Nrf2 transcription, which in turn mediates the cellular antioxidant activity process [34]. Britanin is an herbal derivative that is thought to be capable of reversing hypoxia-glucose deprivation by inhibiting Keap1 [35]. As a triterpenoid, CDDO is thought to be a good inhibitor of various inflammatory diseases such as emphysema and COPD [36]. Isoxazoline-based–inhibitor is isoxazolinyl electrophile capable of reacting with cysteine residues, which has been used in the design of different covalent inhibitors [37]. Similar in structure to CDDO, sAIM_TX64063 is also a triterpenoid and has been demonstrated having an excellent activation effect on Nrf2, and this activity is also associated with the inhibition of Keap1 [38]. Therefore, these five small molecule compounds and PCB were selected for covalent docking simulations, and the possible anti-inflammatory and anti-fibrotic mechanisms of PCB were discussed by comparing the results after covalent docking.

Affinity (kcal/mol) shows the binding energy during covalent docking, and the lower the energy, the more effective the binding. It can be seen from the results that among the six small molecule compounds, PCB and Keap1 have the lowest binding energy of −3.8 kcal/mol, so the binding effect is the best, followed by Britanin with an affinity of −3.4 kcal/mol. CDDO and DMF are relatively poor, with an affinity of only −1.9 and −1.5 kcal/mol. The specific docking results are outlined in Table 3. The covalent docking map (Figure 10) also corroborates the results in Table 3. All of the six compounds were able to generate thioether bond with the Cys151 residue in the BTB domain of Keap1 to form a covalent bond. Among them, the steric hindrance between the conformations of PCB and Britanin and Keap1 is the smallest, so the relative affinity is low. However, CDDO and DMF interfere greatly with Keap1, resulting in high binding energy.

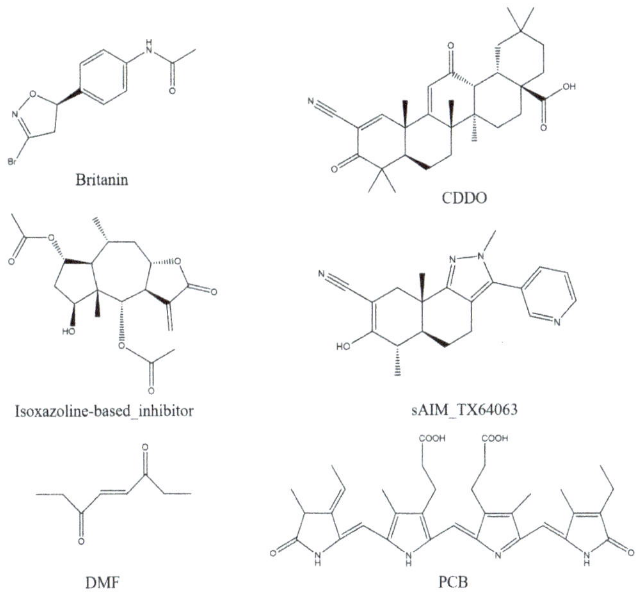

Figure 9. Keap1 small molecule inhibitor.

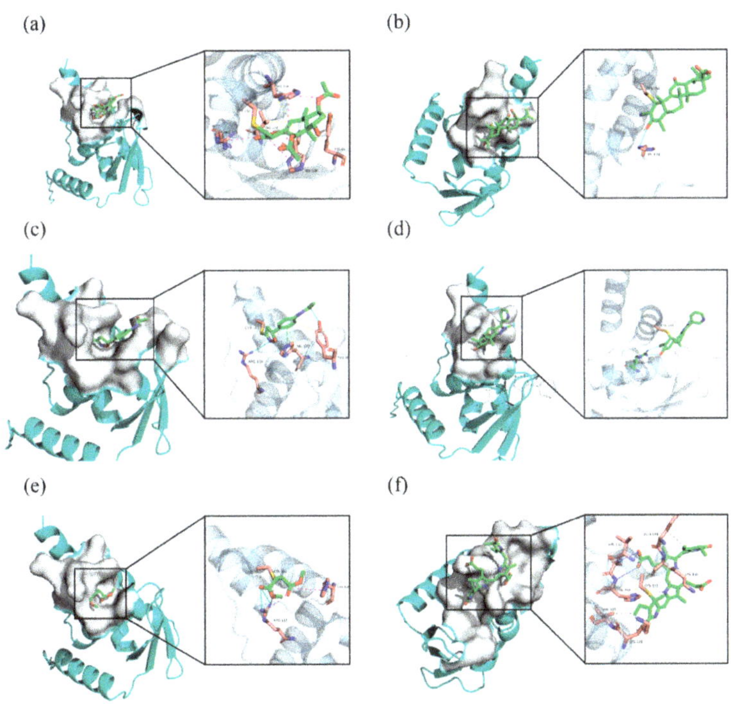

Figure 10. Small-molecule inhibitors covalent docking to the Keap1 BTB domain: (**a**) Britanin; (**b**) CDDO; (**c**) Isoxazoline-based–inhibitor; (**d**) sAIM_TX64063; (**e**) DMF; (**f**) PCB. The yellow dashed lines are hydrophobic interactions, the magenta dashed lines are salt bridges, the cyan solid lines are hydrogen bonds, and the green dashed lines are π-stacking.

Table 3. Covalent docking results.

Mode	Affinity (kcal/mol)	Clust. RMSD	Ref. RMSD	Clust. Size	RMSD Stdv	Energy Stdv	Best Run
Britanin	−3.4	0.0	−1.0	7	0.4	0.2	003
CDDO	−1.9	0.0	−1.0	8	0.0	0.0	007
Isoxazoline-based–inhibitor	−3.1	0.0	−1.0	7	0.5	0.1	004
DMF	−1.5	0.0	−1.0	8	0.4	0.1	002
sAIM_TX64063	−2.1	0.0	−1.0	8	0.2	0.1	001
PCB	−3.8	0.0	−1.0	1	NA	NA	002

By comparing the covalent docking results of five known Keap1 inhibitors with PCB, it is not difficult to see that Britanin and PCB show more interactions with the amino acid residues of Keap1, so their binding energy is lower. As shown in Figure 10, Britanin can hydrophobically interact with TRY85, HIS129, VAL132, and HIS154 of Keap1, and form salt bridges with HIS129, ARG135, and HIS154. PCB can form six hydrophobic interactions with LYS131, VAL132, GLU149, and LYS150, generate salt bridges with LYS150 and HIS154, and form three hydrogen bonds with VAL152, LEU153, and HIS154. The isoxazoline-based–inhibitor can form two hydrogen bonds with TRY85 and ARG135 and generate π-stacking with the imidazole group of HIS129. The abundant interaction force reduces the binding energy of Britanin, PCB, and the isoxazoline-based–inhibitor when they are covalently bound to the receptor, so it is more favorable for their binding to Keap1. In contrast, CDDO forms only one hydrogen bond with HIS129, sAIM_TX64063 forms two hydrogen bonds with ARG135, and DMF forms salt bridges and hydrogen bonds with HIS129 and ARG135. Furthermore, the structure of CDDO would produce steric hindrance with the Keap1 protein, which enhances its binding energy. In summary, through the simulation of the docking of PCB and Keap1 molecules, it can be confirmed that PCB can activate Nrf2 by covalently binding to Keap1.

3. Discussion

Microalgae have been widely used for food and traditional medicine in the history of China. In 284–364 AD, Chinese alchemist Ge Hong discovered a filamentous blue algae and named it Gexianmi (*Nostoc sphaeroides*). Gexianmi is rich in nutrients, and can regulate immunity with anti-inflammatory and anti-tumor effects [39]. As a kind of microalgae, *Spirulina* has similar physiological activities as *Nostoc sphaeroides*, and the isolated PC from *Spirulina* is good for having antioxidant and anti-inflammatory activities.

As PCB in PC is sensitive to temperature, pH, and other ambient conditions, its color tends to change during the enzymatic hydrolysis process. To reduce the PCB loss of phycocyanin peptide as much as possible, we hydrolyzed PC with complex protease MC101 and controlled the reaction conditions to obtain pure blue phycocyanin peptide, and the yield was high, up to 96%. Compared with the commonly used phycocyanin peptides after enzymatic hydrolysis in the market, it was found that the colors of other enzymatic peptides mostly show purple brown and more blue-loss, while the phycocyanin peptides prepared with MC101 enzyme show bright blue-green, with good color retention, and the PCB contained still presented good activity.

The anti-inflammatory effect of PC was first proposed by Romay back in 1998, and they found that C-PC could inhibit liver microsomal lipid peroxidation [17]. Numerous experimental data have confirmed the usefulness of PC in various inflammation-induced diseases, such as acute lung injury [40], pulmonary fibrosis [18], allergic airway inflammation [41], alcoholic liver [42], liver injury [19], etc., showing certain degrees of relief and recovery effects.

As a pigment protein from algae, PC contains a natural blue pigment called PCB. PCB is composed of four pyrrole rings connected by methylene pairs. It is very close to biliverdin in structure, with only the side chain substituents being slightly different, and its principal functional groups are consistent. In algae, the synthesis of PCB is through the ferrous heme metabolic pathway. Heme is oxidized to biliverdin IXα under the action of heme oxygenase (HO). It is different from biliverdin IXα in animals in that heme is reduced to bilirubin by biliverdin reductase. Biliverdin IXα in algae can be catalyzed with PCB reductase to form PCB [43], and then further connected with apoprotein to form PC. The structure of PCB is overall the same as that of bilirubin, except for slight alterations in the substituents of C3, C10, and C18; however, these differences did not affect the α, β-unsaturated carbonyl groups in the PCB (Figure 11).

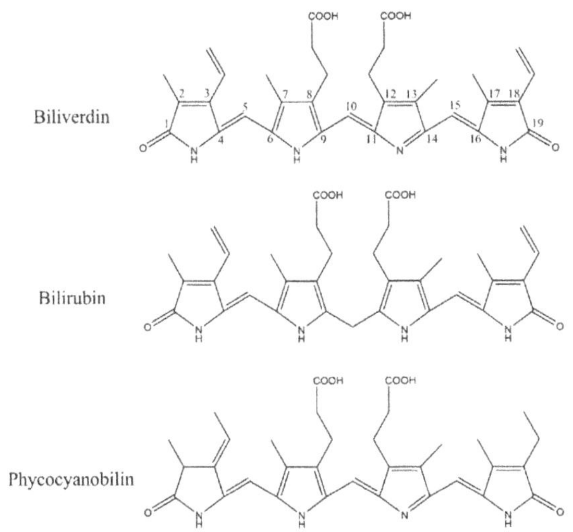

Figure 11. Structure comparison of biliverdin, bilirubin, and phycocyanobilin.

HO-1 possess antioxidant and anti-inflammatory properties to modulate a variety of immune cells [20]. If the anti-inflammatory activity of HO-1 could be used and its expression be induced, it would provide us new ideas for the development of anti-inflammatory drugs. A traditional HO-1 inducer is metalloporphyrin, and it can significantly up-regulate the expression of HO-1; however, its toxicity cannot be ignored as it greatly limits its clinical application [20]. The reaction products of HO-1 can lead to the induction of HO-1 expression. Nam et al. [21] showed that bilirubin can induce the expression of Nrf2 and further induce the export of HO-1 located downstream of Nrf2 by binding to Keap1 through an electrophilic reaction with the α, β-unsaturated carbonyl group in the structure, which inhibits the binding of Keap1 and Nrf2. Similar to metalloporphyrin, bilirubin has a series of issues for clinical application. First, bilirubin can form intramolecular hydrogen bonds, resulting in the decrease in water solubility and becoming fat-soluble, which is not conducive to the dissolution when used as a drug. In addition, taking large amounts of bilirubin would increase the concentration of bilirubin in the body, causing hyperbilirubinemia (jaundice) and adversely affecting liver function [44].

Based on published scientific research, we believe that the antioxidant, anti-inflammatory, and anti-pulmonary fibrosis activities of phycocyanin peptides are related to the structure of the PCB bound in the phycocyanin peptide. As shown in Figure 11, PCB and bilirubin are only slightly different on the substituents of C3, C10, and C18. Therefore, PCB may have a similar effect on bilirubin *in vivo*, and the phycocyanin peptide could be quickly absorbed in the gastrointestinal tract because of its small molecular weight, after entering the body. The PCB bound in the phycocyanin peptide has a linear tetrapyrrole structure. They are polypeptides, and the resulting steric hindrance prevents the formation of internal hydrogen bonds between the propionic acid group and carbonyl group, which could increase the water solubility of PCB. After being absorbed by cells, the α, β-unsaturated carbonyl group on PCB act as electrophiles to bind the thiol of the cysteine residue in Keap1, release and activate Nrf2, transfer it to the nucleus, bind with ARE, and further activate the HO-1 gene. The activated HO-1 gene expresses HO-1, regulates immune cells [45], and inhibits the expression of inflammatory pathway NF-κB, which is under a direct anti-inflammatory effect (Figure 12). In addition, elevated HO-1 has antioxidant effects to eliminate ROS together with a direct antioxidant effect of the phycocyanin peptide to reduce oxidative stress damage in the tissue.

LPS can lead to the secretion of TGF-β1 by macrophages, and elevated TGF-β1 increases the level of ROS in cells [46] and induces lung fibroblasts to secrete ECMs, such as collagen, α-SMA, etc.; the cells develop to the state of myofibroblasts, and the tissues undergo fibrosis reactions. High levels of ROS activate the NF-κB pathway, mediate the secretion of cytokines, and aggravate inflammation and fibrosis among tissues [47]. HO-1 activated by PCB can directly eliminate ROS in the body, block the NF-κB pathway, and indirectly reduce tissue inflammation. A previous study of ours on the expression of related genes after phycocyanin peptide treatment shows that PC gavage of mouse could significantly increase the export of Nrf2 and the downstream gene HO-1, thereby reducing liver damage caused by radiation [19]. Meanwhile, Strasky et al. [48] showed that PCB could suppress the development of atherosclerosis by inducing the HO-1 expression, which supports the above assumptions to a certain extent.

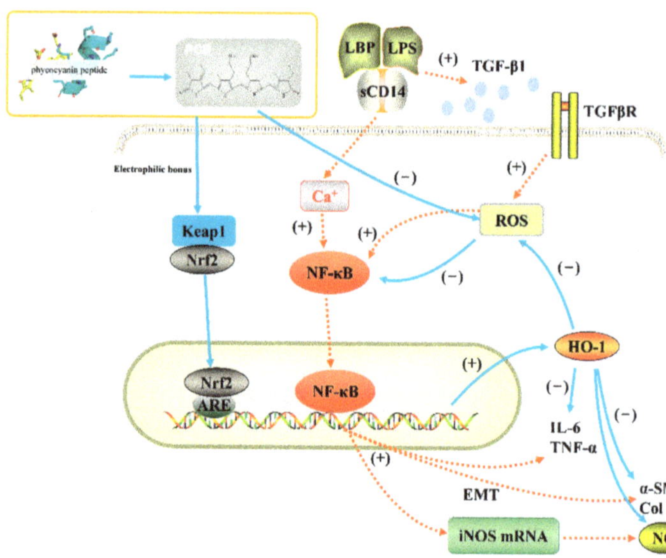

Figure 12. Anti-inflammatory pathway of phycocyanin peptide. The orange dotted line is the process of pulmonary fibrosis, and the blue solid line is the anti-pulmonary fibrosis mechanism of phycocyanin peptide.

4. Materials and Methods

4.1. Reagents

Phycocyanin was purchased from Zhejiang Binmei Biotechnology Co., Ltd., China. The compound protease MC101 was purchased from Yantai Maitel Biotechnology Co., Ltd., China (200,000 μ/g). RAW264.7 cells, CCK-8, and ELisa kits were purchased from Beijing Hua er bo si biology technology Co., Ltd., China. The Griess kits were purchased from Biyotime Biotechnology Co., Ltd., China. Immunofluorescence and the Western Blot antibodies were purchased from Beijing Hua er bo si biology technology Co., Ltd., and other reagents were of analytical grade.

4.2. Enzymatic Hydrolysis of Phycocyanin

Phycocyanin was hydrolyzed with compound protease MC101, and the enzyme addition was 5%. Other conditions were temperature 57 °C, pH natural, the ratio of material to liquid 1:80, and the reaction time 4 h. After digestion, the protease was removed by heating and centrifugation. The supernatant was suction-filtered with white diatomite, and the filtrate was vacuum freeze-dried.

4.3. Absorption Spectrum and Molecular Weight Distribution Measurement

Ultraviolet-visible spectrophotometer was used to measure the absorption spectrum of the phycocyanin peptide at 200–800 nm, and the distribution of the molecular weight of the phycocyanin peptide was determined using HPLC. The data were processed using GPC software. We used Agilent 1290 infinity II/6545 QTOF HPLC-MS to determine the fragments containing PCB in phycocyanin peptides and used Bioconfirm 8 to process the data.

4.4. Antioxidant Activity of Phycocyanin Peptide

4.4.1. Superoxide Anion Clearance

As per Zhang and Meng et al. [49,50] with modification, phycocyanin peptide was configured in different concentrations with H_2O. Tris-HCl of 0.05 M and pH 8.2, and 15 mM pyrogallic acid were kept in a water bath at 25 °C, and then 3.7 mL Tris-HCl, 1 mL phycocyanin peptide solution, and 0.3 mL pyrogallic acid were added. We shook the solution well, placed it into the spectrophotometer immediately, and measured the absorbance value every 30 s at a wavelength of 420 nm for a total of 4 min. We plotted the curve to find the slope A_i.

$$\text{Clearance rate I } (\%) = \frac{A_c - A_i}{A_c} \times 100\% \quad (1)$$

where A_i—absorbance of reaction group; and A_c—the absorbance of control group with deionized water instead of the phycocyanin peptide.

4.4.2. ABTS+ Clearance

Refer to the experimental program of Mei et al. with modification [51].

Mix 38.40 mg ABTS with 6.623 mg $K_2S_2O_8$, dissolve mixture with H_2O and dilute to 10 mL, allow to react at room temperature in darkness for 12–16 h, dilute with 0.2 M, pH 7.4 PBS buffer to the absorbance of 0.70 at 734 nm ± 0.02 or so. We took 50 μL phycocyanin peptide solution and 150 μL $ABTS^+$ solution and reacted in the dark for 6 min, then absorbance was measured at 734 nm.

4.4.3. Reducing Power

Refer to the experimental program of Klomong et al. with modification [52].

Take 1 mL phycocyanin peptide, add 2.5 mL for each of 1% potassium ferricyanide solution and 0.2 M, pH 7.4 PBS buffer, water bath at 50 °C for 20 min, then add 2.5 mL 10% $C_2HCl_3O_2$, mix well after centrifugation at 3000 rpm for 10 min. Take 2.5 mL of the supernatant, add 2.5 mL H_2O and 0.5 mL 0.1% $FeCl_3$ solution, stand still for 10 min, and detect the absorbance A_{700} at 700 nm and the value of A_{700} was used to reflect the magnitude of the reducing power.

4.5. Anti-Inflammatory Activity of Phycocyanin Peptide

4.5.1. Cell Viability

Refer to Gao et al. [53]. Use DMEM medium to dilute the phycocyanin peptide to 50, 100, 200 μg/mL; take RAW264.7 cells at logarithmic growth, adjust the cell concentration to about 5 × 10^4 cells/mL, and inoculate them in several 96-well plates, with 100 μL in each well; continue to incubate for 24 h at 37 °C in 5% CO_2. After 24 h, renew the medium containing each group of drugs and continue to culture for 24 h; after adding the drugs for 24 h, renew the medium, add 100 μL to each well containing 10 μL CCK-8.

Prepare a medium reaction solution containing no cells as the zero well, incubate it for an appropriate time and measure the absorbance at 450 nm.

4.5.2. Cytokines

Add 50 μL of 100 ng/mL LPS solutions to the treatment group and LPS group, respectively, and add the same volume of DMEM culture medium to the control group [54–56]. In the treatment group, different concentrations of phycocyanin peptide solution were added. After 24 h of culture, take 100 μL of the supernatant, use the NO detection kit to determine the NO content, and use the Elisa detection kit to determine the TNF-α and IL-6.

4.6. Determination of Anti-Pulmonary Fibrosis Activity of Phycocyanin Peptide

4.6.1. Observation of A549 Cell Morphology

Select A549 cells as the cell model, and the cell culture conditions and groupings are shown in Table 4.

Table 4. Cell culture conditions and grouping.

	Control	Model	Low Dose	High Dose
Medium	1640	1640	1640	1640
FBS	10%	10%	10%	10%
Temperature	37 °C	37 °C	37 °C	37 °C
CO_2	5%	5%	5%	5%
TGF-β1	–	10 ng/mL	10 ng/mL	10 ng/mL
Phycocyanin peptide	–	–	10 μg/mL	30 μg/mL
Sample size	100 μL	100 μL	100 μL	100 μL
Time	72 h	72 h	72 h	72 h

–: no data.

4.6.2. Immunofluorescence

The expression of A549 cells Collagen I and HFL-1 cells α-SMA was measured by immunofluorescence.

After 72 h of continuous culture, 4% paraformaldehyde fixed, 1% TritonX-100 for 5 min and 3% H_2O_2 for 10 min. Secondary antibodies incubate for 2 h. DAPI nuclei stain for 5–10 min and photographed with a fluorescence microscope.

Use ImageJ software to analyze the gray value of immunofluorescence image and quantify its fluorescence intensity by its gray value.

4.6.3. Western Blot

After the cells adhered to the wall, they were digested with 0.25% trypsin, seeded on a 6-well culture plate, and administered after the density reached 60%. The cells were observed after 72 h, washed with PBS, and scraped off with a cell scraper by adding deionized water. Cells were lysed, pipetted back and forth to mix, centrifuged to take supernatant, added with 5×loading buffer in proportion, and heated to 95 °C in a metal bath. Ten min later, the Western Blotting was performed to detect Nrf2, HO-1, NQO1, and EMT-related marker proteins.

4.7. Covalent Docking Analyze

The 3D structure of the Keap1 protein (BTB domain) was downloaded from https://www.rcsb.org, and the PDB ID is 7EXI. PyMol was used to remove water molecules. Five known small molecule inhibitors of Keap1 were selected: sAIM_TX64063 (PDB ID: 5DAF); 2-cyano-3,12-dioxooleana-1,9-dien-28-oic-acid (CDDO) (PDB ID: 4CXT); britanin (PDB ID: 5GIT); isoxazoline-based–inhibitor (PDB ID: 6FFM); dimethyl fumarate (DMF) (CAS NO. 624-49-7) and PCB (PubChem CID: 6438349) as ligands for covalent docking. Covalent docking of receptors and ligands was performed using the AutoDockFR program, and their conformations were optimized using a genetic algorithm based on the principle of minimum docking energy [57]. Interactions between receptors and ligands were analyzed using PLIP [58]. Results were visualized using PyMol.

4.8. Statistical Analysis

All experiments in this study were performed in triplicate, and results are shown as mean ± SD. Analyses were performed using SPSS Statistics 23 and graphs were constructed using GraphPad Prism 9.0. Data from two or more groups were analyzed using one-way ANOVA ($p < 0.05$ was deemed statistically significant).

5. Conclusions

The emergence of phycocyanin peptides has solved modern issues of HO-1 inducers and has the advantages of lower cytotoxicity and better water solubility. PCB in PC is similar in structure to bilirubin, so its related antioxidant, anti-inflammatory, and anti-pulmonary fibrosis activities were substantiated. Being bound to Keap1, PCBs can release Nrf2 and further induce HO-1 expression. Therefore, phycocyanin peptides can be used as a new HO-1 inducer to develop different types of anti-inflammatory and anti-pulmonary fibrosis drugs. However, the absorption and metabolism of PC, especially PCB, need to be further clarified. Future studies on the activity and bioavailability of PC will contribute to the industrial development of PC and phycocyanin peptides.

Supplementary Materials: The following supporting information can be downloaded at: https://www.mdpi.com/article/10.3390/md20110696/s1, Table S1: Results of phycocyanin peptide HPLC-MS.

Author Contributions: S.Q.: Conceptualization, Methodology, Supervision; W.-J.L.: Conceptualization, Methodology, Writing—Review & Editing, Supervision; R.-Z.L.: Methodology, Validation, Writing—Original Draft; Y.L.: Resources; J.-J.Z.: Resources; C.L.: Resources; Z.-Y.L.: Supervision. All authors have read and agreed to the published version of the manuscript.

Funding: This research was funded by the National key research and development program of China (2021YFC2103900).

Acknowledgments: We thank Yantai Jiahui Biotech Co., Ltd. for the help in the preparation of phycocyanin peptide samples, Jiangnan University for the help in high performance liquid chromatography, Hui Wang from Agilent Technologies (China) Co., Ltd. for the assistance during LC-MS analysis, and Beijing Hua Erbosi Biology Technology Co., Ltd. for the help in cell experiments.

Conflicts of Interest: The authors declare no conflict of interest.

References

1. Xin, W.; Jing, W.; Guichuan, H.; Yishi, L.; Shuliang, G. miR-320a-3P alleviates the epithelial-mesenchymal transition of A549 cells by activation of STAT3/SMAD3 signaling in a pulmonary fibrosis model. *Mol. Med. Rep.* **2021**, *23*, 357.
2. Elko, E.; Anathy, V.; van der Velden, J.; van der Vliet, A.; Janssen-Heininger, Y. Oxidation state of peroxiredoxin 4 in lungs from patients with pulmonary fibrosis and mouse models of fibrosis. *Free Radic. Biol. Med.* **2018**, *128*, S25. [CrossRef]
3. Qingbo, S.; Bingjing, L.; Hanmei, X.; Jialiang, H. Pathogenesis of Pulmonary Fibrosis and Progress in Drug Development. *Prog. Pharm. Sci.* **2018**, *42*, 868–873.
4. Zhenhua, L.; Shouchun, P.; Jian, K.; Xianming, H.; Runjiang, Y. Efect of corticosteroids upon the prognosis of idiopathic pulmonary fibrosis. *Natl. Med. J. China* **2010**, *90*, 804–807.
5. Antonio Rodriguez-Portal, J. Efficacy and Safety of Nintedanib for the Treatment of Idiopathic Pulmonary Fibrosis: An Update. *Drugs RD* **2018**, *18*, 19–25. [CrossRef] [PubMed]
6. Skold, C.M.; Bendstrup, E.; Myllarniemi, M.; Gudmundsson, G.; Sjaheim, T.; Hilberg, O.; Altraja, A.; Kaarteenaho, R.; Ferrara, G. Treatment of idiopathic pulmonary fibrosis: A position paper from a Nordic expert group. *J. Intern. Med.* **2017**, *281*, 149–166. [CrossRef] [PubMed]
7. Peltzer, N.; Walczak, H. Cell Death and Inflammation—A Vital but Dangerous Liaison. *Trends Immunol.* **2019**, *40*, 387–402. [CrossRef]
8. Ndiaye, F.; Vuong, T.; Duarte, J.; Aluko, R.E.; Matar, C. Anti-oxidant, anti-inflammatory and immunomodulating properties of an enzymatic protein hydrolysate from yellow field pea seeds. *Eur. J. Nutr.* **2012**, *51*, 29–37. [CrossRef]
9. Mao, X.Y.; Cheng, X.; Wang, X.; Wu, S.J. Free-radical-scavenging and anti-inflammatory effect of yak milk casein before and after enzymatic hydrolysis. *Food Chem.* **2011**, *126*, 484–490. [CrossRef]

10. Murray, P.J.; Allen, J.E.; Biswas, S.K.; Fisher, E.A.; Gilroy, D.W.; Goerdt, S.; Gordon, S.; Hamilton, J.A.; Ivashkiv, L.B.; Lawrence, T.; et al. Macrophage Activation and Polarization: Nomenclature and Experimental Guidelines. *Immunity* **2014**, *41*, 14–20. [CrossRef]
11. Turner, M.D.; Nedjai, B.; Hurst, T.; Pennington, D.J. Cytokines and chemokines: At the crossroads of cell signalling and inflammatory disease. *Biochim. Et Biophys. Acta-Mol. Cell Res.* **2014**, *1843*, 2563–2582. [CrossRef] [PubMed]
12. Tenhunen, R.; Marver, H.S.; Schmid, R. The Enzymatic Conversion of Heme to Bilirubin by Microsomal Heme Oxygenase. *Proc. Natl. Acad. Sci. USA* **1968**, *61*, 748–755. [CrossRef] [PubMed]
13. Jingjing, W.; Hui, Z.; Jianqiang, L. The anti-inflammatory effects of heme oxygenase-1 on emphysema model. *Chin. J. Tuberc. Respir. Dis.* **2015**, *38*, 379–383.
14. Otterbein, L.E.; Kolls, J.K.; Mantell, L.L.; Cook, J.L.; Alam, J.; Choi, A.M. Exogenous administration of heme oxygenase-1 by gene transfer provides protection against hyperoxia-induced lung injury. *J. Clin. Investig.* **1999**, *103*, 1047–1054. [CrossRef] [PubMed]
15. Otterbein, L.; Sylvester, S.L.; Choi, A.M. Hemoglobin provides protection against lethal endotoxemia in rats: The role of heme oxygenase-1. *Am. J. Respir. Cell Mol. Biol.* **1995**, *13*, 595–601. [CrossRef] [PubMed]
16. Shixiang, Z. Studies on the Anti-Hepatic Fibrosis Mechanism of Phycocyanin from Spirulina *In Vitro* and *In Vivo*. Master's Thesis, University of Chinese Academy of Sciences, Beijing, China, 2020.
17. Romay, C.; Armesto, J.; Remirez, D.; Gonzalez, R.; Ledon, N.; Garcia, I. Antioxidant and anti-inflammatory properties of C-phycocyanin from blue-green algae. *Inflamm. Res.* **1998**, *47*, 36–41. [CrossRef] [PubMed]
18. Li, C.; Yu, Y.; Li, W.; Liu, B.; Jiao, X.; Song, X.; Lv, C.; Qin, S. Phycocyanin attenuates pulmonary fibrosis via the TLR2-MyD88-NF-kappa B signaling pathway. *Sci. Rep.* **2017**, *7*, 5843. [CrossRef]
19. Liu, Q.; Li, W.; Qin, S. Therapeutic effect of phycocyanin on acute liver oxidative damage caused by X-ray. *Biomed. Pharmacother.* **2020**, *130*, 110553. [CrossRef]
20. Campbell, N.K.; Fitzgerald, H.K.; Dunne, A. Regulation of inflammation by the antioxidant haem oxygenase 1. *Nat. Rev. Immunol.* **2021**, *21*, 411–425. [CrossRef]
21. Nam, J.; Lee, Y.; Yang, Y.; Jeong, S.; Kim, W.; Yoo, J.W.; Moon, J.O.; Lee, C.; Chung, H.Y.; Kim, M.S.; et al. Is it worth expending energy to convert biliverdin into bilirubin? *Free Radic. Bio. Med.* **2018**, *124*, 232–240. [CrossRef]
22. Collin, F. Chemical Basis of Reactive Oxygen Species Reactivity and Involvement in Neurodegenerative Diseases. *Int. J. Mol. Sci.* **2019**, *20*, 2407. [CrossRef] [PubMed]
23. Deffert, C.; Cachat, J.; Krause, K.H. Phagocyte NADPH oxidase, chronic granulomatous disease and mycobacterial infections. *Cell. Microbiol.* **2014**, *16*, 1168–1178. [CrossRef]
24. Finkel, T.; Holbrook, N.J. Oxidants, oxidative stress and the biology of ageing. *Nature* **2000**, *408*, 239–247. [CrossRef] [PubMed]
25. Stadtman, E.R. Protein Oxidation and Aging. *Science* **1992**, *257*, 1220–1224. [CrossRef] [PubMed]
26. Hui, W.; Yan, Z. The Method of ABTS Assay for Screening and Evaluating Antioxidant. *Guangzhou Chem. Ind.* **2012**, *40*, 41–43.
27. Park, J.T.; Johnson, M.J. A Submicrodetermination of Glucose. *J. Biol. Chem.* **1949**, *181*, 149–151. [CrossRef]
28. MacMicking, J.; Xie, Q.W.; Nathan, C. Nitric oxide and macrophage function. *Annu. Rev. Immunol.* **1997**, *15*, 323–350. [CrossRef]
29. Kraakman, M.J.; Kammoun, H.L.; Allen, T.L.; Deswaerte, V.; Henstridge, D.C.; Estevez, E.; Matthews, V.B.; Neill, B.; White, D.A.; Murphy, A.J.; et al. Blocking IL-6 trans-Signaling Prevents High-Fat Diet-Induced Adipose Tissue Macrophage Recruitment but Does Not Improve Insulin Resistance. *Cell Metab.* **2015**, *21*, 403–416. [CrossRef]
30. Elliott, S.; Cawston, T. The clinical potential of matrix metalloproteinase inhibitors in the rheumatic disorders. *Drug Aging* **2001**, *18*, 87–99. [CrossRef]
31. Jung, K.A.; Kwak, M.K. The Nrf2 System as a Potential Target for the Development of Indirect Antioxidants. *Molecules* **2010**, *15*, 7266–7291. [CrossRef]
32. Li, Q.H.; Peng, W.; Zhang, Z.Y.; Pei, X.; Sun, Z.K.; Ou, Y. A phycocyanin derived eicosapeptide attenuates lung fibrosis development. *Eur. J. Pharmacol.* **2021**, *908*, 14. [CrossRef] [PubMed]
33. Szapiel, S.V.; Elson, N.A.; Fulmer, J.D.; Hunninghake, G.W.; Crystal, R.G. Bleomycin-induced interstitial pulmonary disease in the nude, athymic mouse. *Am. Rev. Respir. Dis.* **1979**, *120*, 893–899.
34. Linker, R.A.; Lee, D.-H.; Ryan, S.; van Dam, A.M.; Conrad, R.; Bista, P.; Zeng, W.; Hronowsky, X.; Buko, A.; Chollate, S.; et al. Fumaric acid esters exert neuroprotective effects in neuroinflammation via activation of the Nrf2 antioxidant pathway. *Brain* **2011**, *134*, 678–692. [CrossRef] [PubMed]
35. Wu, G.; Zhu, L.; Yuan, X.; Chen, H.; Xiong, R.; Zhang, S.; Cheng, H.; Shen, Y.; An, H.; Li, T.; et al. Britanin Ameliorates Cerebral Ischemia-Reperfusion Injury by Inducing the Nrf2 Protective Pathway. *Antioxid. Redox Signal.* **2017**, *27*, 754–768. [CrossRef] [PubMed]
36. Cleasby, A.; Yon, J.; Day, P.J.; Richardson, C.; Tickle, I.J.; Williams, P.A.; Callahan, J.F.; Carr, R.; Concha, N.; Kerns, J.K.; et al. Structure of the BTB Domain of Keap1 and Its Interaction with the Triterpenoid Antagonist CDDO. *PLoS ONE* **2014**, *9*, e98896. [CrossRef]
37. Pinto, A.; El Ali, Z.; Moniot, S.; Tamborini, L.; Steegborn, C.; Foresti, R.; De Micheli, C. Effects of 3-Bromo-4,5-dihydroisoxazole Derivatives on Nrf2 Activation and Heme Oxygenase-1 Expression. *Chemistryopen* **2018**, *7*, 858–864. [CrossRef]
38. Huerta, C.; Jiang, X.; Trevino, I.; Bender, C.F.; Ferguson, D.A.; Probst, B.; Swinger, K.K.; Stoll, V.S.; Thomas, P.J.; Dulubova, I.; et al. Characterization of novel small-molecule NRF2 activators: Structural and biochemical validation of stereospecific KEAP1 binding. *Biochim. Et Biophys. Acta-Gen. Subj.* **2016**, *1860*, 2537–2552. [CrossRef]

39. Li, Y.; Ma, J.; Fang, Q.; Guo, T.; Li, X. Protective effects of Nostoc sphaeroides Kütz against cyclophosphamide-induced immunosuppression and oxidative stress in mice. *Toxin Rev.* **2019**, *40*, 1118–1127. [CrossRef]
40. Leung, P.O.; Lee, H.H.; Kung, Y.C.; Tsai, M.F.; Chou, T.C. Therapeutic Effect of C-Phycocyanin Extracted from Blue Green Algae in a Rat Model of Acute Lung Injury Induced by Lipopolysaccharide. *Evid.-Based Compl. Alt. Med.* **2013**, *2013*, 916590. [CrossRef]
41. Chang, C.J.; Yang, Y.H.; Liang, Y.C.; Chiu, C.J.; Chu, K.H.; Chou, H.N.; Chiang, B.L. A Novel Phycobiliprotein Alleviates Allergic Airway Inflammation by Modulating Immune Responses. *Am. J. Resp. Crit. Care* **2011**, *183*, 15–25. [CrossRef]
42. Xia, D.; Liu, B.; Xin, W.Y.; Liu, T.S.; Sun, J.Y.; Liu, N.N.; Qin, S.; Du, Z.N. Protective effects of C-phycocyanin on alcohol-induced subacute liver injury in mice. *J. Appl. Phycol.* **2016**, *28*, 765–772. [CrossRef]
43. Okada, K. HO1 and PcyA proteins involved in phycobilin biosynthesis form a 1:2 complex with ferredoxin-1 required for photosynthesis. *FEBS Lett.* **2009**, *583*, 1251–1256. [CrossRef] [PubMed]
44. Abraham, N.G.; Kappas, A. Pharmacological and clinical aspects of heme oxygenase. *Pharmacol. Rev.* **2008**, *60*, 79–127. [CrossRef] [PubMed]
45. Yuji, N.; Tomohisa, T.; Yasuki, H. Heme oxygenase-1 and anti-inflammatory M2 macrophages. *Arch. Biochem. Biophys.* **2014**, *564*, 83–88.
46. Chanda, D.; Otoupalova, E.; Smith, S.R.; Volckaert, T.; De Langhe, S.P.; Thannickal, V.J. Developmental pathways in the pathogenesis of lung fibrosis. *Mol. Asp. Med.* **2019**, *65*, 56–69. [CrossRef]
47. Younis, T.; Khan, M.R.; Sajid, M.; Majid, M.; Zahra, Z.; Shah, N.A. Fraxinus xanthoxyloides leaves reduced the level of inflammatory mediators during *in vitro* and *in vivo* studies. *BMC Complement. Altern. Med.* **2016**, *16*, 230. [CrossRef]
48. Strasky, Z.; Zemankova, L.; Nemeckova, I.; Rathouska, J.; Wong, R.J.; Muchova, L.; Subhanova, I.; Vanikova, J.; Vanova, K.; Vitek, L.; et al. Spirulina platensis and phycocyanobilin activate atheroprotective heme oxygenase-1: A possible implication for atherogenesis. *Food Funct.* **2013**, *4*, 1586–1594. [CrossRef]
49. Qing-An, Z.; Xi, W.; Yun, S.; Xue-Hui, F.; Francisco, G.M.J. Optimization of Pyrogallol Autoxidation Conditions and Its Application in Evaluation of Superoxide Anion Radical Scavenging Capacity for Four Antioxidants. *J. AOAC Int.* **2016**, *99*, 504–511.
50. Xuelian, M.; Jia, L.; Yingying, L.; Liangchao, Z.; Chengcheng, G.; Dan, W.; Jing, L.; Changlan, C. Inhibitory effect of cordycepin on macrophage hyperactivation induced by lipopolysaccharide. *Sci. Technol. Food Ind.* **2017**, *38*, 297–301, 306.
51. Xing, M.; Guang, W.; Rui, T.; Yu, L.; Lanxin, X.; Cheng, T.; Wei, L.; Chao, C. Antioxidant Properties of Five Phycocyanins with Different Purities. *Food Sci.* **2021**, *42*, 17–23.
52. Klompong, V.; Benjakul, S.; Kantachote, D.; Shahidi, F. Antioxidative activity and functional properties of protein hydrolysate of yellow stripe trevally (Selaroides leptolepis) as influenced by the degree of hydrolysis and enzyme type. *Food Chem.* **2007**, *102*, 1317–1327. [CrossRef]
53. Gao, R.C.; Shu, W.H.; Shen, Y.; Sun, Q.C.; Bai, F.; Wang, J.L.; Li, D.J.; Li, Y.; Jin, W.G.; Yuan, L. Sturgeon protein-derived peptides exert anti-inflammatory effects in LPS-stimulated RAW264.7 macrophages via the MAPK pathway. *J. Funct. Foods* **2020**, *72*, 104044. [CrossRef]
54. Wang, Y.; Cui, Y.; Cao, F.; Qin, Y.; Li, W.; Zhang, J. Ganglioside GD1a suppresses LPS-induced pro-inflammatory cytokines in RAW264.7 macrophages by reducing MAPKs and NF-kappa B signaling pathways through TLR4. *Int. Immunopharmacol.* **2015**, *28*, 136–145. [CrossRef] [PubMed]
55. Pi, J.; Li, T.; Liu, J.; Su, X.; Wang, R.; Yang, F.; Bai, H.; Jin, H.; Cai, J. Detection of lipopolysaccharide induced inflammatory responses in RAW264.7 macrophages using atomic force microscope. *Micron* **2014**, *65*, 1–9. [CrossRef] [PubMed]
56. Qi, S.; Xin, Y.; Guo, Y.; Diao, Y.; Kou, X.; Luo, L.; Yin, Z. Ampelopsin reduces endotoxic inflammation via repressing ROS-mediated activation of PI3K/Akt/NF-kappa B signaling pathways. *Int. Immunopharmacol.* **2012**, *12*, 278–287. [CrossRef] [PubMed]
57. Bianco, G.; Forli, S.; Goodsell, D.S.; Olson, A.J. Covalent docking using autodock: Two-point attractor and flexible side chain methods. *Protein Sci.* **2016**, *25*, 295–301. [CrossRef]
58. Adasme, M.F.; Linnemann, K.L.; Bolz, S.N.; Kaiser, F.; Salentin, S.; Haupt, V.J.; Schroeder, M. PLIP 2021: Expanding the scope of the protein-ligand interaction profiler to DNA and RNA. *Nucleic Acids Res.* **2021**, *49*, W530–W534. [CrossRef]

marine drugs

Article

Two *Laminaria japonica* Fermentation Broths Alleviate Oxidative Stress and Inflammatory Response Caused by UVB Damage: Photoprotective and Reparative Effects

Qianru Sun [1,2], Jiaxuan Fang [1,2], Ziwen Wang [1,2], Zixin Song [1,2], Jiman Geng [1,2], Dongdong Wang [1,2], Changtao Wang [1,2] and Meng Li [1,2,*]

1 College of Chemistry and Materials Engineering, Beijing Technology & Business University, 11 Fucheng Road, Haidian District, Beijing 100048, China
2 Beijing Key Laboratory of Plant Resource Research and Development, Beijing Technology and Business University, 11 Fucheng Road, Beijing 100048, China
* Correspondence: limeng@btbu.edu.cn; Tel.: +86-134-2601-5179

Abstract: UVB radiation can induce oxidative stress and inflammatory response in human epidermal cells. We establish a UVB-induced damage model of human immortalized epidermal keratinocytes (HaCaT) to explore the protective and reparative effects of *Laminaria japonica* on UVB-damaged epidermal inflammation after fermentation by white *Ganoderma lucidum* (*Curtis*) *P. Karst* and *Saccharomyces cerevisiae*. Compared with unfermented *Laminaria japonica*, fermented *Laminaria japonica* possesses stronger in vitro free radical scavenging ability. *Laminaria japonica* white *Ganoderma lucidum* fermentation broth (LJ-G) and *Laminaria japonica* rice wine yeast fermentation broth (LJ-Y) can more effectively remove excess reactive oxygen species (ROS) in cells and increase the content of the intracellular antioxidant enzymes heme oxygenase-1 (HO-1) and NAD(P)H quinone oxidoreductase 1 (NQO-1). In addition, fermented *Laminaria japonica* effectively reduces the content of pro-inflammatory factors ILs, TNF-α and MMP-9 secreted by cells. The molecular research results show that fermented *Laminaria japonica* activates the Nrf2 signaling pathway, increases the synthesis of antioxidant enzymes, inhibits the gene expression levels of pro-inflammatory factors, and alleviates cellular oxidative stress and inflammatory response caused by UVB radiation. Based on the above results, we conclude that fermented *Laminaria japonica* has stronger antioxidant and anti-inflammatory activity than unfermented *Laminaria japonica*, possesses good safety, and can be developed and used as a functional inflammation reliever. Fermented *Laminaria japonica* polysaccharide has a more slender morphological structure and more rockulose, with better moisturizing and rheological properties.

Keywords: photodamage; antioxidation; inflammation; *Laminaria japonica*; fermentation technology

Citation: Sun, Q.; Fang, J.; Wang, Z.; Song, Z.; Geng, J.; Wang, D.; Wang, C.; Li, M. Two *Laminaria japonica* Fermentation Broths Alleviate Oxidative Stress and Inflammatory Response Caused by UVB Damage: Photoprotective and Reparative Effects. *Mar. Drugs* **2022**, *20*, 650. https://doi.org/10.3390/md20100650

Academic Editors: Elena Talero and Javier Ávila-Román

Received: 28 September 2022
Accepted: 18 October 2022
Published: 20 October 2022

1. Introduction

As the largest organ of the human body, the skin is in direct contact with the external environment and performs such functions as protection, excretion and body temperature regulation [1]. The epidermis is responsible for resisting toxic substances in the external environment; it can withstand mechanical friction and microbial invasion, and effectively prevent the loss of skin moisture, ions and metabolites [2]. Healthy skin has the ability to renew itself, but when damaged by external stimuli and unable to repair and renew itself in time, it can become sensitive and inflamed, and even develop pathological reactions such as atopic dermatitis and psoriasis [3]. Exposure to UV radiation, environmental changes and unhealthy cleaning practices are all contributing factors, with UV radiation being among the most significant.

Ultraviolet radiation is divided into three categories according to the length of the wavelength band: UVA (315–400 nm), UVB (280–315 nm) and UVC (100–280 nm) [4]. UVB can reach the epidermis of the skin. Excessive exposure to UVB can cause skin erythema, sunburn, and damage to intracellular macromolecules [2]. Therefore, UVB is considered to have more serious damaging effects than UVA [5]. The UVB irradiation of the skin can cause an increase in the content of reactive oxygen species (ROS) in skin cells and trigger oxidative stress response [6]. Oxidative stress can cause many skin problems, such as inflammation and barrier damage [7]. Impaired skin barrier in atopic dermatitis is strongly associated with the secretion of cytokines (inflammatory factors) by keratinocytes. Inflammatory factors can inhibit the synthesis level of filaggrin, thereby affecting the process of stratum corneum differentiation. These changes lead to increased permeability to external antigens and the gradual breakdown of the skin barrier [8]. As the hole in the ozone layer enlarges, human exposure to UV radiation increases, as well as the probability of epidermal oxidative stress, inflammation and deep skin damage caused by UVB. Thus, there is an urgent need to find effective ways to alleviate skin oxidation and inflammation caused by photodamage.

NF-E2-related factor 2 (Nrf2) is a member of the CnC subfamily of leucine-pulled transcription factors in the basic region. Under normal conditions, Nrf2 binds to the protein element Kelch-1ike ECH-associated protein 1 (Keap-1), which is in an inactive state [9]. When stimulated or damaged, intracellular ROS induces a conformational change in Keap-1 protein to promote the dissociation and termination of the ubiquitination of Nrf2. The released Nrf2 is catalyzed by the small Maf protein and translocates into the nucleus, where it activates the transcriptional activity of a further series of genes [10].

As evidenced by existing studies, Nrf2 is an anti-inflammatory pathway. Inflammation typically occurs in Nrf2-deficient chemically-induced pathologies, and Nrf2-knockout mice have a significantly increased tendency to develop multi-tissue inflammatory lesions compared with normal mice [11]. The alleviation of inflammation by Nrf2 is associated with the inhibition of pro-inflammatory cytokine production, but the molecular mechanisms underlying the interaction between the two remain largely unclear [12]. The plant-derived anti-inflammatory agent 3-hydroxyanthranilic acid (HA) can promote Nrf2 nuclear translocation and induce heme oxygenase-1(HO-1) expression, suggesting that the anti-inflammatory effect of Nrf2 is partially dependent on redox dynamic regulation [13]. Studies have shown that synthetic triterpenoids can induce the increased expression and activity of NAD(P)H quinone oxidoreductase 1(NQO-1) in mice to reduce nitric oxide synthase (iNOS) levels, and this regulatory process is strictly dependent on Nrf2 [14].

Laminaria japonica, also known as brown seaweed, is common in East Asian countries such as China, Japan and South Korea as a traditional Chinese medicinal and edible algae [15]. It is rich in active substances, such as fucoidan, fucoxanthin, dietary fiber, and carotenoids, and possesses significant antioxidant [15], anti-inflammatory [16], anti-aging and antibacterial properties [17]. Studies have confirmed that *Laminaria japonica* specifically inhibits the nuclear translocation of the STAT1 gene through the MAPK pathway, reduces skin inflammation and the expression level of cellular inflammatory chemokines, and inhibits the loss of skin moisture in mice caused by dinitrochlorobenzene (NC/Nga) sensitization [18]. Fucoidan extracted from *Laminaria japonica* has been confirmed to reduce the content of tumor necrosis factor α (TNF-α) in THP-1 cells caused by lipopolysaccharide (LPS), and it can be reduced by 50% at 50 μg/mL [19]. Research on the anti-inflammatory activity of *Laminaria japonica* is emerging, but it remains blank in terms of UVB radiation.

Fermentation technology is a kind of natural plant active substance extraction technology that is currently mature and can convert macromolecular substances into small molecular active substances to achieve detoxification and efficiency enhancement [18]. Polysaccharides extracted from *Lycium barbarum* after fermentation by *Saccharomyces cerevisiae* can significantly delay the aging process of nematodes, and possess stronger permeability [20]. Positive results have already been seen in terms of the antioxidant and disease treatment of *Laminaria japonica* treated with fermentation technology: extracts of *Laminaria japonica* fermented by *Aspergillus oryzae* have shown strong free radical scavenging effects, and some have even exceeded the same mass concentration of vitamin C [21]. The same results were obtained after the mixed fermentation of *Laminaria japonica* with two lactic acid bacteria. Fermentation treatment can not only effectively improve the extraction rate of the active substances in *Laminaria japonica*, but also promote stronger antioxidant and hypoglycemic activity [22].

In this study, traditional Chinese medicinal white *Ganoderma lucidum* and rice wine yeast are used to ferment *Laminaria japonica* in order to investigate whether the resulting fermentation broth has protective and reparative effects on skin photo-inflammation caused by UVB radiation, explore whether microbial fermentation can effectively improve the utilization rate of *Laminaria japonica* active substances and exert stronger anti-inflammatory effects, study the molecular mechanism of the effects of *Laminaria japonica* fermentation broth and provide a scientific basis for its application as an anti-inflammatory raw material.

2. Results

2.1. Changes in Active Substance Content

Laminaria japonica is rich in polysaccharides and polyphenols, which are the primary substances with antioxidant and anti-inflammatory effects [19,23].

The data showed that compared to Laminaria japonica water extract (LJ-W), the total sugar, polysaccharide, phenol and protein contents in the fermentation broth significantly increased (Table 1). The contents of total sugars, polysaccharides and proteins in Laminaria japonica white Ganoderma lucidum fermentation broth (LJ-G) and Laminaria japonica rice wine yeast fermentation broth (LJ-Y) showed an increase of 2–3 times. Although the content of total phenols in the three samples was not very high, the content after fermentation also increased by one. These experimental results confirm that the content of Laminaria japonica bioactive substances was significantly different before and after fermentation treatment. This suggests that fermentation can effectively improve the extraction rate of Laminaria japonica bioactive substances.

Table 1. Contents of total and reduction sugars and phenolic compounds analyzed by spectrophotometric methods before and after fermentation by *Laminaria japonica*. Data is represented as mean ± SEM. *** $p < 0.001$, significantly different compared to the LJ-W group.

Sample	Total Sugars (mg/g)	Polysaccharide (mg/g)	Total Phenols (mg/g)	Protein (mg/g)
LJ-W	34.56 ± 0.95	32.27 ± 2.95	0.47 ± 0.10	1.39 ± 0.19
LJ-G	81.25 ± 0.85 ***	78.99 ± 6.98 ***	0.86 ± 0.15 ***	4.66 ± 0.29 ***
LJ-Y	68.97 ± 5.19 ***	67.08 ± 5.71 ***	0.88 ± 0.06 ***	3.45 ± 0.5 ***

2.2. In Vitro Antioxidant Activity

Four types of free radical scavenging activity of *Laminaria japonica* water extract and fermentation broth were determined to evaluate the difference in antioxidant activity before and after fermentation (Figure 1A–D).

The scavenging data of the DPPH free radical test showed that the scavenging ability of the sample was weak in the concentration range of 0.25–4 mg/mL, and the scavenging effects of LJ-G and LJ-Y were worse than those of LJ-W. The clearance capacity of LJ-G was significantly improved after 4 mg/mL, and the clearance rate reached 76.75% at 8 mg/mL, which was significantly higher than that of LJ-W. The clearance rate of LJ-Y at 4–8 mg/mL was slightly improved, but there was no significant difference in the clearance between the two at the same concentration. The scavenging effects of samples on hydroxyl radicals were positively correlated with the mass concentration. The clearance of LJ-G was consistently better than that of LJ-W, reaching 86.63% and 87.09% at 4 and 8 mg/mL, respectively. The scavenging ability of LJ-Y was slightly weaker than that of LJ-G. It is worth noting that LJ-W showed the strongest scavenging ability on hydroxyl radicals at 8 mg/mL (63.51%). The same clearance was achieved by LJ-G and LJ-Y at 1 mg/mL, and the clearance rate of LJ-G was consistently higher than that of LJ-Y.

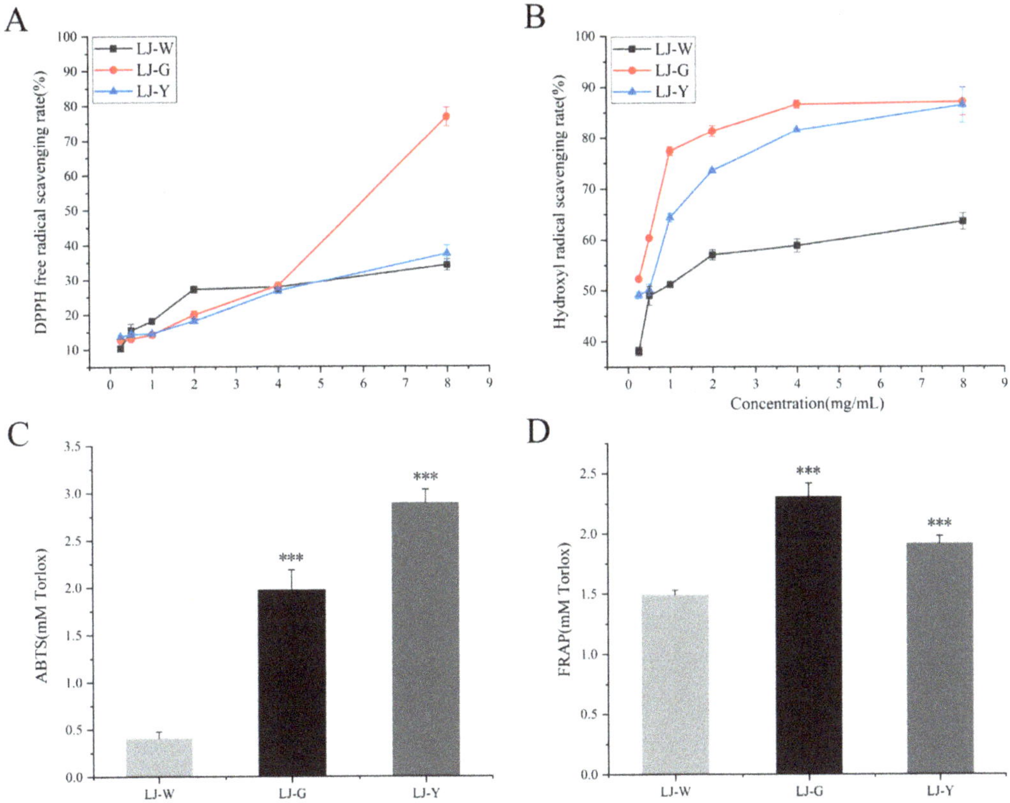

Figure 1. Antioxidant capacity of LJ-W, LJ-G and LJ-Y determined in vitro. (**A**): DPPH free radical scavenging ability; (**B**): hydroxyl radical scavenging ability; (**C**): total antioxidant capacity (ABTS method); (**D**): total antioxidant capacity (FRAP method). *** $p < 0.001$, significantly different compared to the LJ-W group.

The test results of total antioxidant capacity determination (ABTS method) showed that both LJ-G and LJ-Y had significantly ($p < 0.001$) higher ABTS+ scavenging power than LJ-W, with 2.44 and 2.81 times that of LJ-W, respectively, and LJ-Y had the best scavenging ability, reaching 1.14 mM Trolox equivalent. The test results of total antioxidant capacity determination (FRAP method) showed that compared with LJ-W, the reducing power of LJ-G and LJ-Y were significantly improved ($p < 0.001$), and the reducing ability of the two on Fe^{2+} reached 1.55 and 1.28 times that of LJ-W respectively. Unlike the total antioxidant capacity (ABTS method), LJ-G showed a stronger reduction of Fe^{2+} than LJ-Y.

Based on the above in vitro antioxidant test data, we can conclude that LJ-W, LJ-G and LJ-Y all have certain in vitro antioxidant activity, while those of LJ-G and LJ-Y are significantly stronger than that of LJ-W. The scavenging activity of LJ-G on DPPH free radicals, hydroxyl radicals, and Fe^{2+} were the strongest among the three samples.

2.3. *Effects of LJ-W, LJ-G, LJ-Y on the Viability of HaCaT Cells*

The protective and reparative effects of LJ-W, LJ-G and LJ-Y on human immortalized epidermal keratinocytes (HaCaT) cells damaged by UVB are shown in Figure 2. UVB of 10 mJ/cm^2 was selected as the experimental dose (Figure S1). In the protection group (Figure 2A), the cell viability of the three groups of samples was significantly improved compared with the model group. Cell viability increased gradually in the LJ-W group, reaching its highest level (114.17%) at 4 mg/mL. Cell viability in the LJ-G group was always greater than 100%, reaching its highest level (121.47%) at 2 mg/mL. Cell viability in the LJ-Y group was higher than that of the model group at a concentration of 0.125–2 mg/mL, and gradually increased with the concentration. At 4 mg/mL, the survival rate of HaCaT cells decreased to 73.70%, which was lower than that of the model group. The results of the repair group test (Figure 2B) showed that the cell survival rate after UVB irradiation was 72.39% (model group). The three samples with different concentrations showed similar trends in the repair of UVB-damaged HaCaT cells. With the increase in mass concentration, the cell viability of all three groups gradually increased, reaching a peak at 0.5 mg/mL, then gradually decreasing (0.5–2 mg/mL), but still remaining higher than that of the model group. At 4 mg/mL, the cell viability of the LJ-G group was still higher than that of the model group, but those of LJ-W and LJ-Y were lower.

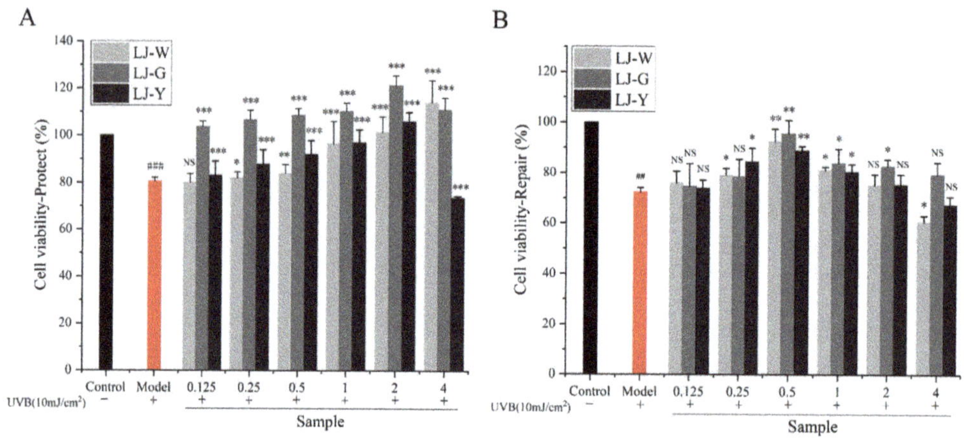

Figure 2. Effects of LJ-W, LJ-G and LJ-Y on the survival state of UVB-damaged HaCaT cells. The model group is the cell survival after UVB damage (red columns). (**A**): Protective effects of LJ-W, LJ-G and LJ-Y before UVB damage; (**B**): Reparative effects of LJ-W, LJ-G and LJ-Y after UVB damage. ^{NS} $p > 0.05$, * $p < 0.05$, ** $p < 0.01$, *** $p < 0.001$ compared with the model group, ## $p < 0.01$, ### $p < 0.001$ compared with the control group.

The above experimental data shows that *Laminaria japonica* fermentation broth can effectively alleviate the damage and death of HaCaT cells caused by UVB, and promote the proliferation of epidermal cells, but high concentrations of LJ-Y are slightly toxic to cells. In the UVB protection group, the cell survival rate of *Laminaria japonica* fermentation broth reached its highest levels at 2 mg/mL, namely, 121.47% (LJ-G) and 106.08% (LJ-Y). The fermentation broth of the UVB repair group reached maximum cell viability at 0.5 mg/mL, namely, 95.63% (LJ-G) and 88.89% (LJ-Y). Therefore, we chose 2 mg/mL and 0.5 mg/mL as the sample concentrations for subsequent protection and repair experiments. For proper expression, the concentration unit of subsequent test samples was modified to "μg/mL".

2.4. Effects of LJ-W, LJ-G and LJ-Y on ROS Inhibition and Scavenging

Excessive oxygen free radicals (ROS) will cross the cell membrane and react with most biomolecules (DNA, proteins, lipids, etc.), causing oxidative stress in the body and excessive levels of intracellular inflammatory factors, and eventually leading to the appearance of inflammatory skin problems [24]. Removing excess ROS is the guarantee for maintaining the normal functioning of the skin cells and body.

The effects of UVB damage and the protective and reparative effects of the samples on ROS content in HaCaT cells are shown in Figure 3. After UVB irradiation, the intracellular ROS content was significantly increased in the model group and significantly decreased in the protection and repair groups ($p < 0.001$). As shown by the intracellular ROS assay fluorescence values (Figure 3B), in the damage protection group, LJ-W, LJ-G and LJ-Y significantly reduced intracellular ROS contents, while the ROS contents of the LJ-G and LJ-Y groups were lower than those of the LJ-W and control groups. In the damage repair group, the intracellular ROS contents of the LJ-W and LJ-Y groups were lower than that of the control group. The scavenging effect of LJ-G was weaker than those of LJ-Y, but still significant ($p < 0.001$).

The above experimental data indicates that the water extract and fermentation broth of *Laminaria japonica* can effectively remove excess ROS in damaged HaCaT cells; the scavenging effects of *Laminaria japonica* fermentation broth are better than that of *Laminaria japonica* water extract; and LJ-G and LJ-Y in the protection group and LJ-Y in the repair group can reduce ROS in damaged cells to normal levels. This reveals that LJ-Y possesses enhanced physiological activity in ROS scavenging.

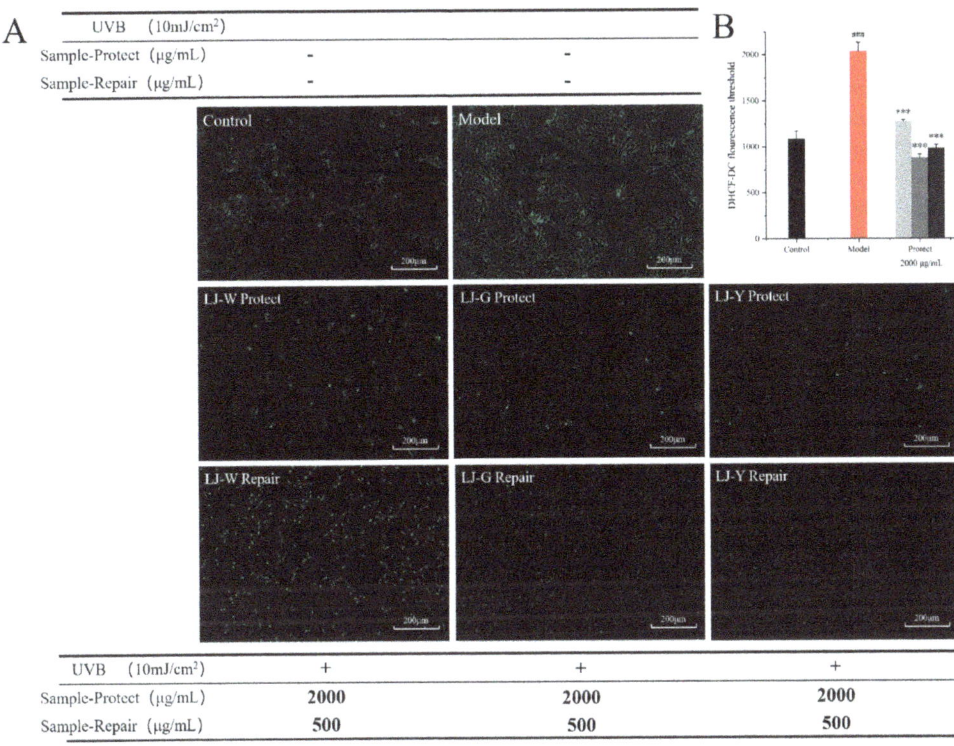

Figure 3. Effects of UVB, LJ-W, LJ-G and LJ-Y on intracellular ROS content. The model group is the intracellular ROS fluorescence values after UVB damage. (**A**): Cell ROS fluorescence intensity; (**B**): Cell fluorescence intensity value. *** $p < 0.001$ compared with the model group, ### $p < 0.001$ compared with the control group.

2.5. *Effects of LJ-W, LJ-G, LJ-Y on Antioxidant Capacity*

Heme oxygenase-1 (HO-1) is an antioxidant cytoprotective enzyme that plays a key role in oxidative stress and inflammation [25]. NQO-1 belongs to the DMES in the Phase II detoxification stage of cells and can protect the cytoplasmic membrane from ROS damage [14].

Figure 4 shows the effects of LJ-W, LJ-G and LJY on HO-1 and NQO-1 contents in cells. After UVB irradiation, compared with the blank group, the contents of HO-1 and NQO-1 in HaCaT cells were significantly decreased.

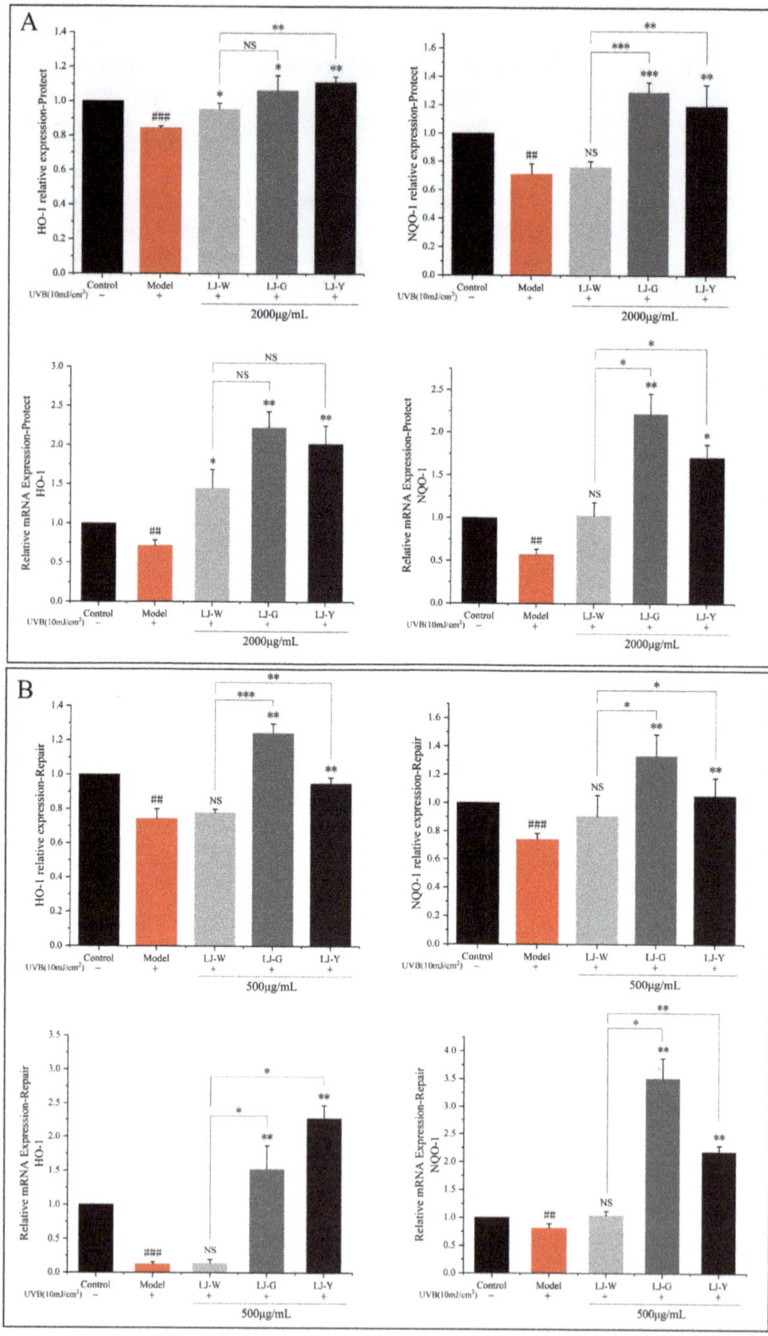

Figure 4. Effects of LJ-W, LJ-G and LJ-Y on the activity of antioxidant and detoxification enzymes in HaCaT cells. The model group was the relative expression content of intracellular inflammatory factors after UVB damage. (**A**): Differences in expression of HO-1, NQO-1 and related mRNAs in UVB-damaged protection group; (**B**): Differences in expression of HO-1, NQO-1 and related mRNAs in UVB-damaged repair group. ^{NS} $p > 0.05$, * $p < 0.05$, ** $p < 0.01$, *** $p < 0.001$ compared with the model and LJ-W groups, ## $p < 0.01$, ### $p < 0.001$ compared with the control group.

In the protection group, the effects of the three samples on HO-1 enzyme content in HaCaT cells were similar. However, after LJ-G treatment, the expression of HO-1 mRNA in cells was significantly upregulated, which was higher than LJ-Y. The LJ-G and LJ-Y groups significantly affected NQO-1 in HaCaT cells, showing excellent protective effects. Especially in LJ-G-treated HaCaT cells, the intracellular NQO-1 enzyme content was significantly increased, and its mRNA expression was four times that of the UVB-damaged model group. In the repair group, the range of antioxidant enzymes in the LJ-W group increased slightly but not significantly. LJ-G and LJ-Y could dramatically increase the content and transcriptional activity of HO-1 and NQO-1, and LJ-G was stronger.

To sum up, after UVB irradiation, while the three samples showed certain protective and reparative effects on cell oxidative stress damage, LJ-G exhibited stronger regulatory activity than LJ-Y, which will affect the activity of intracellular antioxidant enzymes and detoxification enzymes.

2.6. Effects of LJ-W, LJ-G, LJ-Y on the Expression Levels of Inflammatory Factors

Keratinocytes can secrete inflammatory chemokines (interleukins, tumor necrosis factor, etc.) and cytokines to maintain skin homeostasis. Elevated ROS levels caused by the UVB irradiation of epidermal cells can trigger the excessive secretion of inflammatory chemokines, which in turn causes the infiltration of T cells and neutrophils into the epidermis, leading to inflammation and skin lesions [18]. The elevated secretion of pro-inflammatory factors (interleukins and tumor necrosis factor) marks the beginning of cellular inflammatory response [26].

MMPs are also mediators of the vicious cycle of inflammation which is responsible for degrading various components in the extracellular matrix [27]. Studies have shown that Nrf2-knockout skin ulcer wounds heal slowly, which is attributed to the high expression of MMP-9 in the cells [28]. After Nrf2-knockout, mice are irradiated with UVB, the activity of MMPs in the skin cells significantly increases, indicating that Nrf2 can play a photoprotective role by inhibiting MMPs [29].

Laminaria japonica water extract and fermentation broth have significant effects on inflammatory chemokines interleukin-1β (IL-1β), interleukin-8, IL-8 (IL-8), tumor necrosis factor-α (TNF-α) and matrix metalloproteinase-9 (MMP-9). The effects of expression level are shown in Figure 5. After UVB irradiation, the contents and gene transcription levels of IL-1β, IL-8, TNF-α and MMP-9 in the protection and repair groups were significantly increased. In the protection group (Figure 5A), the levels of TNF-α and MMP-9 in the LJ-W group were significantly lower than those in the model group, but there was no significant difference in the levels of IL-1β and IL-8. In the LJ-G and LJ-Y groups, the contents and transcription levels of the four pro-inflammatory factors were significantly lower than those in the model group and LJ-W group. In the repair group (Figure 5B), LJ-W was able to reduce IL-1β and MMP-9 content but had no significant effect on the other inflammatory factors. The contents of the above pro-inflammatory factors were significantly decreased in the LJ-G and LJ-Y groups, and the contents of IL-1β and TNF-α in the LJ-G group were even lower than those in the LJ-Y and control group.

The above results indicate that UVB irradiation can induce the excessive secretion of intracellular inflammatory chemokines, leading to cellular inflammation. *Laminaria japonica* has a certain anti-inflammatory activity which is amplified by fermentation technology. It is worth noting that the inhibitory effects of the two *Laminaria japonica* fermentation broths on inflammatory factors were different in terms of protection and repair. In the protection group, LJ-Y showed a better ability to inhibit inflammation, while LJ-G showed a better ability in the repair group. Compared to LJ-Y, LJ-G could reduce the levels of inflammatory factors even more than undamaged cells. In conclusion, both the water extract and fermentation broth of *Laminaria japonica* have anti-inflammatory activity, and that of the fermentation broth is better than that of the water extract.

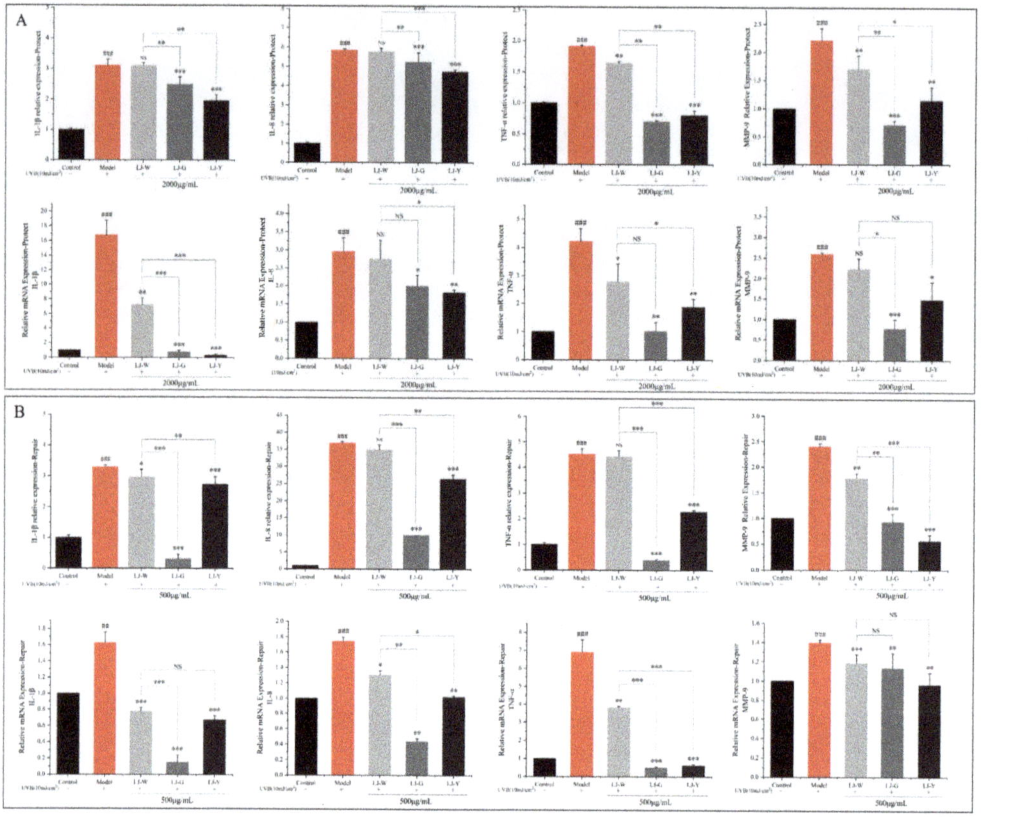

Figure 5. Effects of LJ-W, LJ-G and LJ-Y on the secretion of inflammatory factors in UVB-damaged HaCaT cells. (**A**): Differences in the content of IL-1β, IL-8, TNF-α and MMP-9 in HaCaT cells in UVB-damaged protection group; (**B**): Differences in the content of IL-1β, IL-8, TNF-α and MMP-9 in HaCaT cells in UVB-damaged repair group. * $p < 0.05$, ** $p < 0.01$, *** $p < 0.001$ compared with the model and LJ-W groups, ^{NS} $p > 0.05$, ## $p < 0.01$, ### $p < 0.001$ compared with the control group.

2.7. Effects of LJ-W, LJ-G, LJ-Y on Keratinocyte Differentiation and Skin Barrier Function

The stratum corneum of the skin is rich in functional proteins and active enzymes, such as aquaporin 3 (AQP3), filaggrin (FLG), kallikrein-7 (KLK-7) and Caspase-14. AQP3 is located on the cell membrane and is responsible for tightly bonding skin keratin [30]. FLG forms a tight physical barrier in the epidermis to prevent water loss and the invasion of external irritants [31]. KLK-7 is responsible for severing the desmosomes that connect cells, causing skin cell dysfunction and accelerating cell shedding [30]. Caspase-14 is active in the damaged epidermis and can hydrolyze FLG into the intercellular natural moisturizing factor (NMF) to maintain the moisturizing function of the skin [32].

The levels of AQP3 and FLG in the cells were significantly decreased after UVB irradiation, while the levels of KLK-7 and Caspase-14 were significantly increased (Figure 6). This indicates that UVB irradiation accelerates intracellular water loss and leads to damaged epidermal cell shedding.

After the effect of the samples, the content of water retention protein in the cells was significantly increased and the enzymatic activities of KLK-7 and Caspase-14 were significantly decreased. The protection and repair groups showed similar results. Interestingly, the barrier repair activity of LJ-G was higher than that of LJ-Y in the repair group, whereas LJ-Y was better in the protection group.

In summary, it is likely that LJ-W, LJ-G and LJ-Y can maintain the normal connections between cells. On the premise of maintaining the stability of the extracellular matrix and NMF contents, the reinforcement of filaggrin in the skin barrier is enhanced to ensure the normal function of the stratum corneum barrier, thereby effectively protecting the epidermis from UVB radiation damage.

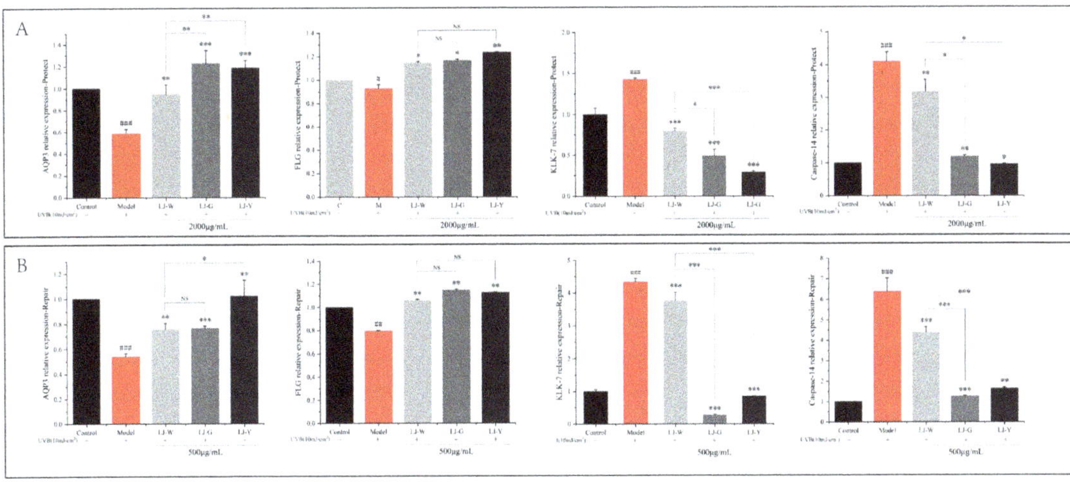

Figure 6. Effects of LJ-W, LJ-G and LJ-Y on the secretion of skin barrier factors in UVB-damaged HaCaT cells. (**A**): Differences in the content of AQP3, FLG, KLK-7 and Caspase-14 in HaCaT cells in UVB-damaged protection group; (**B**): Differences in the content of AQP3, FLG, KLK-7 and Caspase-14 in HaCaT cells in UVB-damaged repair group. * $p < 0.05$, ** $p < 0.01$, *** $p < 0.001$ compared with the model and LJ-W groups, ^{NS} $p > 0.05$, # $p < 0.05$, ## $p < 0.01$, ### $p < 0.001$ compared with the control group.

2.8. Effects of LJ-W, LJ-G, LJ-Y on Gene Expression of Nrf2 Signaling Pathway

The Nrf2 signaling pathway can regulate cellular inflammatory response [27]. The Nrf2 gene bound to the Keap-1 element is silent. At this time, the transcription of downstream genes p38, JNK1 and AP-1 is active, and can promote the expression of IL-1β, TNF-α and MMP-9. After separation from Keap-1, Nrf2 transcriptional activity is activated and translocated into the nucleus, where it regulates the downstream genes to synthesize antioxidant enzymes and inhibit the expression of pro-inflammatory factors.

We measured the transcriptional activity of the Nrf2 signaling pathway-related node genes Keap-1, p38, JNK1 and AP-1 in damaged cells before and after treatment by the samples (Figure 7). The data showed that the transcriptional activity of Keap-1 in UVB-irradiated cells was increased, the transcriptional level of the Nrf2 pathway was decreased and the transcriptional levels of its downstream genes p38, JNK1 and AP-1 were significantly increased ($p < 0.001$). After the samples were applied, the transcriptional activity of Keap-1 decreased, the transcriptional level of Nrf2 increased and the transcriptional activity of p38, JNK1 and AP-1 decreased. In the protection group (Figure 7A), the transcriptional activity of Nrf2 was significantly higher in the LJ-G and LJ-Y groups than in the LJ-W group ($p < 0.05$). In addition, LJ-G and LJ-Y had stronger inhibitory effects on p38, JNK1 and AP-1 than LJ-W. The repair group (Figure 8B) data also showed the same trend as the protection group. Notably, in the protection and repair group, the regulatory activity of LJ-G was higher than that of LJ-Y for most node genes.

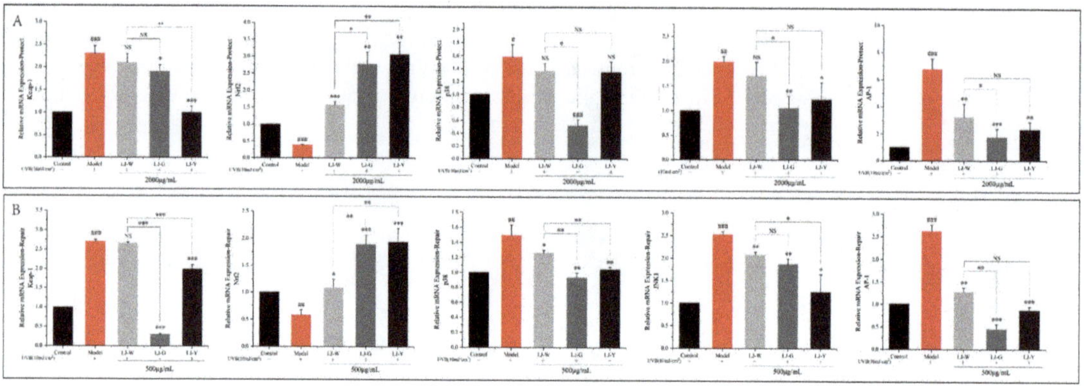

Figure 7. Effects of LJ-W, LJ-G and LJ-Y on Keap-1, Nrf2, p38, JNK1 and AP-1 genes in the Nrf2 signaling pathway. (**A**): UVB-damaged protection group; (**B**): UVB-damaged repair group. * $p < 0.05$, ** $p < 0.01$, *** $p < 0.001$, compared with the model and LJ-W groups, ^{NS} $p > 0.05$, $^{\#}$ $p < 0.05$, $^{\#\#}$ $p < 0.01$, $^{\#\#\#}$ $p < 0.001$ compared with the control group.

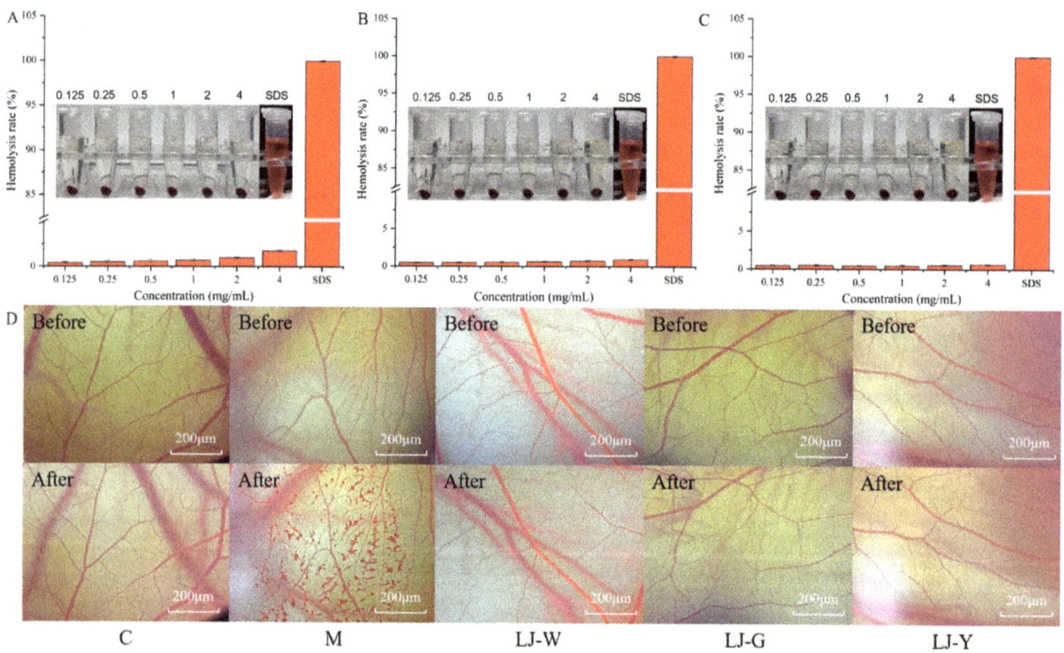

Figure 8. LJ-W, LJ-G and LJ-Y safety evaluation. (**A**): Effects of LJ-W on the hemolysis of rabbit erythrocytes; (**B**): Effects of LJ-G on the hemolysis degree of rabbit erythrocytes; (**C**): Effects of LJ-Y on the hemolysis degree of rabbit erythrocytes; (**D**): Effects of LJ-W, LJ-G and LJ-Y on the vascular damage of chick embryo chorioallantoic membrane.

The above experimental results showed that UVB irradiation could inhibit the translocation of the Nrf2 signal into the nucleus in HaCaT cells, reduce the body's antioxidant and detoxification process, and prompt cells to produce a large number of inflammatory chemokines, causing an inflammatory response. *Laminaria japonica* improved the antioxidant effects of cells by increasing the transcription level of Nrf2 and promoting its nuclear translocation. *Laminaria japonica* also reduced the transcriptional activity of AP-1, which, in turn, suppressed the high expression levels of IL-1β, TNF-α and MMP-9 induced by UVB irradiation.

2.9. LJ-W, LJ-G, LJ-Y Safety Evaluation

The chick chorioallantoic membrane eye irritation test (CAM) and erythrocyte hemolysis test can determine whether a tested substance is an irritant through the detection of vascular bleeding and hemolysis [33]. Thus, CAM and erythrocyte hemolysis tests were performed on LJ-W, LJ-G and LJ-Y to evaluate their safety (Figure 8A–D). Sodium dodecyl sulfate (SDS) was used as a positive control. The results showed that samples with different mass concentrations had little effect on the hemolysis of rabbit erythrocytes (Figure 8A–C). At the same concentration, the hemolysis degrees of LJ-G and LJ-Y were lower than that of LJ-W. 0.9% NaCl and 0.1 mol/L NaOH were used in control group and model group controls, respectively (Table 2). After the NaOH (model group) was used for 3 min, the blood vessels on the membrane showed obvious hemorrhage, vascular coagulation and vascular dissolution caused by severe stimulation, while the sample group and negative control group did not show the above phenomena. According to these results, we believe that *Laminaria japonica* fermented freeze-dried powder is non-irritating to the eyes, has high safety and can be added to skin care products as a functional raw material.

Table 2. Classification of types of hemorrhagic effects of *Laminaria japonica* water extract and fermentation broths on CAM.

Sample	Type of Bleeding	Rating/Point						ES/Point
		1	2	3	4	5	6	
Negative control (0.9% NaCl solution)	Bleeding	0	0	0	0	0	0	0
	Blood vessel coagulation	0	0	0	0	0	0	
	Angiolysis	0	0	0	0	0	0	
Positive control (0.1 mol/L NaOH)	Bleeding	2	2	2	2	2	2	12
	Blood vessel coagulation	2	1	1	1	2	1	
	Angiolysis	1	1	1	1	1	1	
LJ-W	Bleeding	0	0	0	0	0	0	0
	Blood vessel coagulation	0	0	0	0	0	0	
	Angiolysis	0	0	0	0	0	0	
LJ-G	Bleeding	0	0	0	0	0	0	0
	Blood vessel coagulation	0	0	0	0	0	0	
	Angiolysis	0	0	0	0	0	0	
LJ-Y	Bleeding	0	0	0	0	0	0	0
	Blood vessel coagulation	0	0	0	0	0	0	
	Angiolysis	0	0	0	0	0	0	

2.10. Physical Properties of LJ-Wp, LJ-Gp and LJ-Yp

The FT-IR pattern of *Laminaria japonica* is shown in Figure 9A. The absorption peaks of LJ-Wp (a), LJ-Gp (b) and LJ-Yp (c) at 3350 cm^{-1} were caused by the C-H triple bond (alkynyl) and hydroxyl stretching vibration. It is speculated that there may be a multi-molecular association structure. The absorption peaks at 3033 and 2923 cm^{-1} are caused by the stretching vibration of the carbon-hydrogen single bond. The sharp absorption peak at 1612 cm^{-1} was caused by carbonyl stretching and benzene ring skeleton vibration. The absorption peaks at 1408, 1244 and 1053 cm^{-1} are caused by the stretching of carbon single bond and carbon-hydrogen bond, and the absorption peak at 1408 cm^{-1} suggests the existence of halogen. The smaller sharp absorption peak at 826 cm^{-1} was caused by the disubstituted benzene ring, which was confirmed by the absorption peak at 1612 cm^{-1}. Comparing the FI-IR results of LJ-Wp, LJ-Gp and LJ-Yp, we found that LJ-Gp has no obvious absorption peak at 826 cm^{-1}, and the absorption peak area at 1612 cm^{-1} is correspondingly reduced. It is speculated that white *Ganoderma lucidum* may have changed the original benzene ring structure of *Laminaria japonica* during the fermentation process.

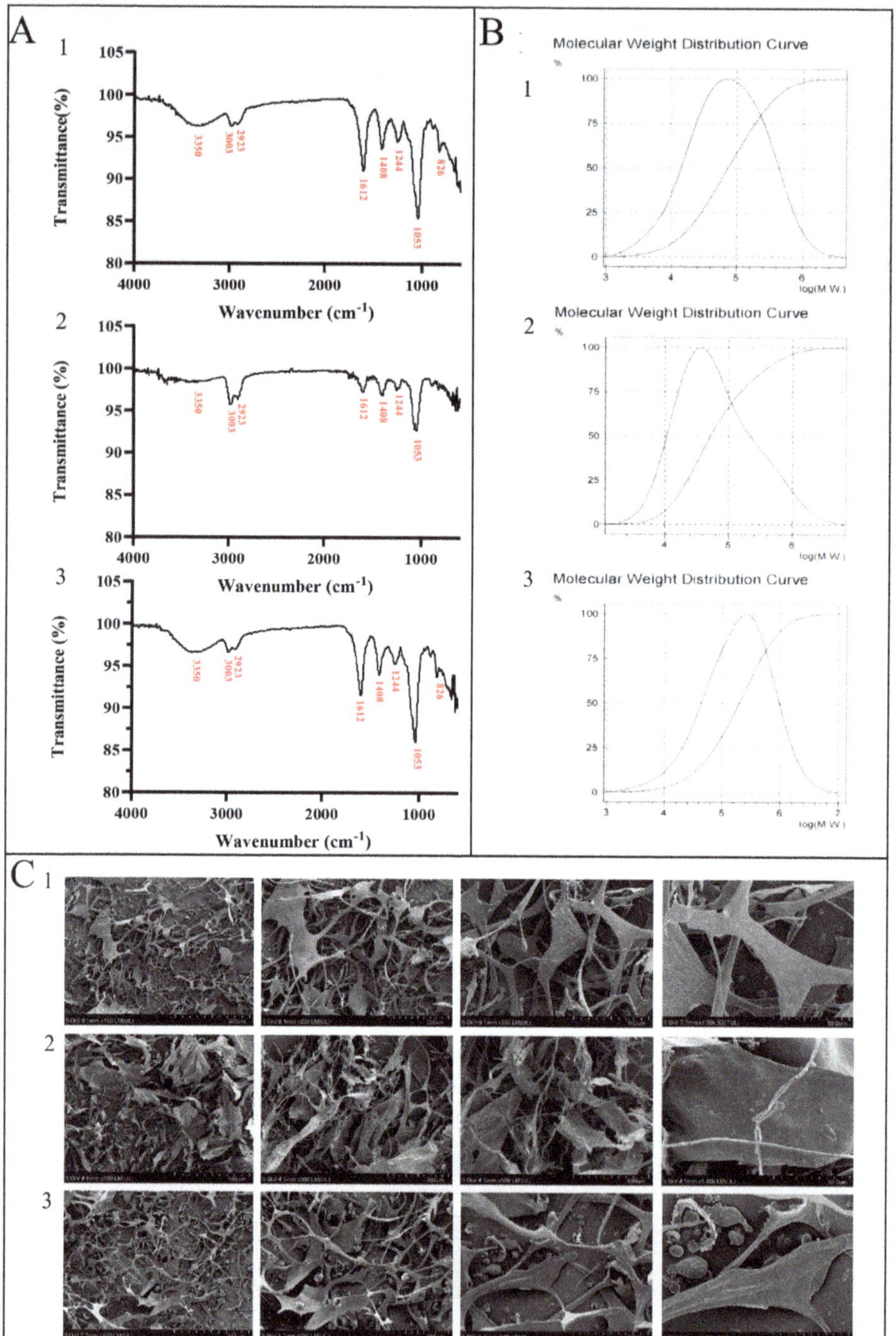

Figure 9. Morphological and structural characterization of unfermented and fermented *Laminaria japonica*. (**A**): IR spectrum of LJ-Wp (1), LJ-Gp (2) and LJ-Yp (3); (**B**): GPC Chromatogram of LJ-Wp (1), LJ-Gp (2) and LJ-Yp (3); (**C**): SEM image of LJ-Wp (1), LJ-Gp (2) and LJ-Yp (3) in 500, 200, 100 and 50 nm.

The relative molecular weights of LJ-Wp, LJ-Gp and LJ-Yp were determined using gel permeation chromatography (GPC). The elution curve is shown in Figure 9B. The measured weight average molecular weights (Mw) of LJ-Wp (a), LJ-Gp (b) and LJ-Yp (c) were 16.1232 × 10^4, 18.1159 × 10^4 and 39.6117 × 10^4 kDa, respectively (Table 3). The molecular weight of LJ-Gp and LJ-Yp was higher than that of LJ-Wp, indicating that fermentation technology can improve the *Laminaria japonica* wall-breaking rate and promote the release of macromolecular polysaccharide active substances.

Table 3. LJ-Wp, LJ-Gp and LJ-Yp molecular weight determinations.

Name	LJ-Wp	LJ-Gp	LJ-Yp
Mn (Da)	2.9466 × 10^4	2.8208 × 10^4	6.2539 × 10^4
Mw (Da)	16.1232 × 10^4	18.1159 × 10^4	39.6117 × 10^4
Mz (Da)	54.8664 × 10^4	102.6565 × 10^4	302.8553 × 10^4
Mw/Mn	5.47178	6.42222	6.33390
Mz/Mn	3.40296	5.66667	3.36240

We used scanning electron microscopy (SEM) to observe the structure and morphology of *Laminaria japonica*. Figure 9C showed the topographic features under different observation accuracies. LJ-Wp (a), LJ-Gp (b) and LJ-Yp (c) all showed irregular loose sheet-like or filamentous structures. Compared with LJ-Wp, LJ-Gp has more filamentous structure, and LJ-Yp structure is smaller and less sheet-like structure. The microstructure of fermented *Laminaria japonica* polysaccharide showed loose pores and smaller structural morphology. It is speculated that LJ-Gp and LJ-Yp have better rheological properties and absorption.

2.11. Monosaccharide Components of LJ-Wp, LJ-Gp and LJ-Yp

Table 4 and Figure 10 shows the monosaccharide composition of LJ-Wp, LJ-Gp and LJ-Yp. The highest content was found in rockulose, followed by galactose, glucuronide and mannose. The contents of fucose in LJ-Wp, LJ-Gp and LJ-Yp were 45.467, 39.620 and 44.675%, and the contents of galactose were 22.094, 21.020 and 26.090%. The content of glucuronic acid and mannose is about 10–13%. Fucoidan is widely used as a moisturizer in skin care products, and mannose can regulate the micro-ecological balance of the skin [34,35]. It is presumed that *Laminaria japonica* polysaccharide has better moisturizing effect and skin conditioning activity.

Table 4. Monosaccharide component of LJ-Wp, LJ-Gp and LJ-Yp.

Monosaccharide Ratio (%)	Sample		
	LJ-Wp	LJ-Gp	LJ-Yp
Mannose	11.789	13.903	9.696
Riboose	1.198	0.903	1.200
Rhamnose	0.749	0.541	0.583
Glucuronic acid	10.272	13.940	10.044
Galacturonic acid	0.422	0.533	0.371
Glucose	3.230	4.815	3.114
Galactose	22.094	21.020	26.090
Xylose	2.722	2.959	2.808
Arabinose	2.057	1.766	1.419
Fucose	45.467	39.620	44.675
Total	100	100	100

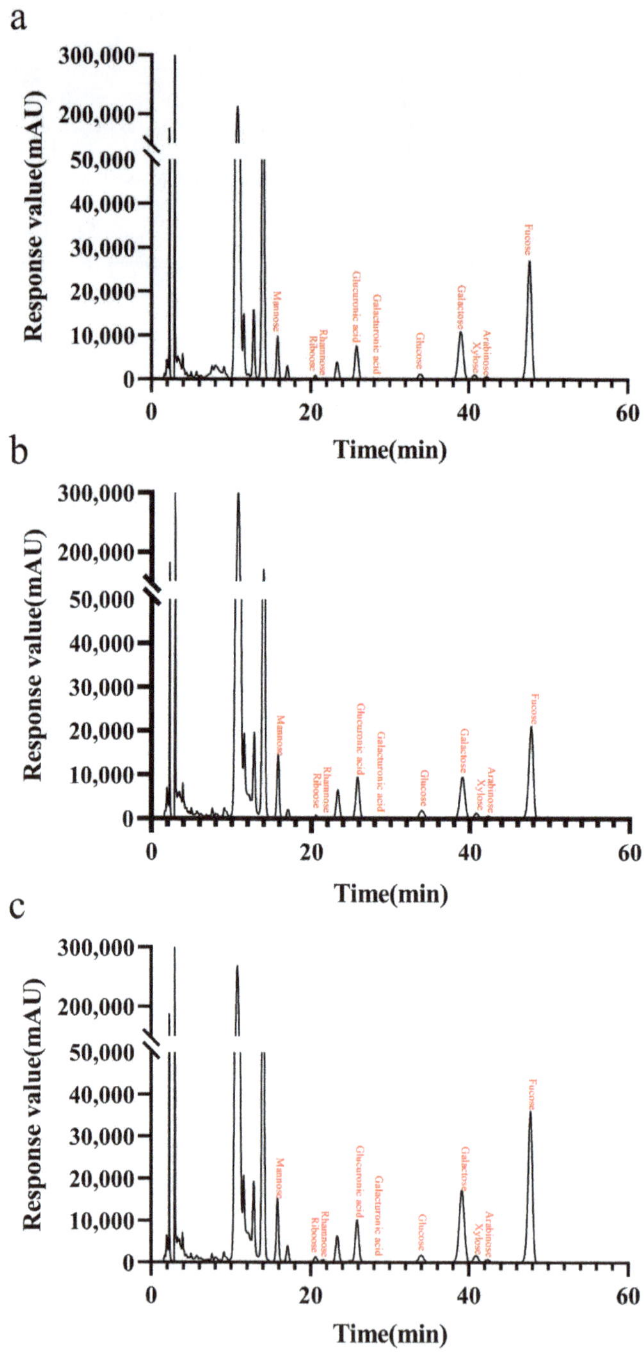

Figure 10. Monosaccharide components of LJ-Wp (**a**), LJ-Gp (**b**) and LJ-Yp (**c**). Detection wavelength is 250 nm, and detection interval is 500 ms.

3. Discussion

Inflammation is associated with many skin problems and physical diseases, and sun exposure is one factor that causes skin inflammation. UVB in sunlight can reach the epidermis of the skin, stimulate cells to produce an inflammatory response and cause other skin problems. Fucoidan extracted from *Laminaria japonica* exhibits positive in vitro anti-inflammatory activity, confirming that *Laminaria japonica* has significant anti-inflammatory effects [19], but our understanding of the role of *Laminaria japonica* in combating photo-inflammation is unclear.

The generation of photo-inflammation is complex and multidimensional. Sunlight directly stimulates skin epidermal cells to secrete pro-inflammatory factors [36]. Epidermal cells exposed to sunlight undergo oxidative stress, and excess ROS stimulates the cells to express more inflammatory chemokines [37]. The production of inflammation can lead to the destruction of the tight junction function of the epidermis, and damage to the barrier will increase the permeability of the skin to external stimuli, leading to a vicious circle of inflammation accumulation [38]. We established a UVB radiation damage system and used HaCaT cells to simulate human epidermal cells under sunlight in order to explore the protective and reparative effects of *Laminaria japonica* on epidermal inflammation at the in vitro, cellular and molecular levels. The results showed that the UVB irradiation of HaCaT cells caused changes in the contents of various substances. UVB irradiation produces a large amount of ROS in cells, which accelerates the consumption of antioxidant enzymes stored in cells and reduces their original antioxidant capacity. In addition, UVB radiation aggravates the expression of inflammatory factors, including interleukin, tumor necrosis factor, and MMPs in cells, which directly leads to the generation of epidermal inflammation.

It has been confirmed that *Laminaria japonica* possesses excellent antioxidant and anti-inflammatory properties and is rich in a variety of biologically active substances. Studies have shown that macrocystis-derived fucoidan has strong anti-inflammatory effects at low concentrations, while *Laminaria japonica*-derived macromolecular fucoidan requires five times the dose to achieve the same level [19]. However, they have very similar anti-inflammatory activity at the same molecular weight. In addition, the tough cell walls of the seaweed are a factor that hinders active substance extraction, but fermentation technology can use enzymes produced by microorganisms to increase cell wall permeability and carry out the small molecular modification of the precipitated active substances [39]. Therefore, we chose to ferment *Laminaria japonica* in order to compare and explore the effects of fermentation on its anti-inflammatory activity.

In the study, we found that all *Laminaria japonica* samples showed good antioxidant and anti-inflammatory activity. Among them, the effects of LJ-G and LJ-Y were better than that of LJ-W. In terms of antioxidation, LJ-G and LJ-Y in the protection and repair groups could promote the expression of antioxidant enzymes HO-1 and NQO-1 in the body to different degrees, while LJ-G showed higher physiological activity. In terms of inflammation inhibition, we chose to measure the most typical pro-inflammatory factors (interleukin and tumor necrosis factor). The results showed that LJ-G and LJ-Y had significant inhibitory effects on the expression levels of inflammatory factors in HaCaT cells. Interestingly, in the protection group, LJ-Y had a better inhibitory effect on interleukin, while LJ-G had stronger inhibitory activity on TNF-α and MMP-9. In the repair group, the comprehensive anti-inflammatory activity of LJ-G was stronger. Whether the differences between LJ-G and LJ-Y in the protection and repair groups were related to different bacteria fermentation methods or caused by differences in the anti-inflammatory activity of *Laminaria japonica* itself are questions that remain to be explored in the future.

Nrf2 can establish a glutathione-mediated UVB detoxification mechanism in the epidermis, alleviate the toxic effects of UVB on epidermal cells and inhibit the apoptosis of mouse epidermal cells [40]. Malonic acid extracted from Korean pine can increase the levels of antioxidant enzymes HO-1 and SOD by regulating the Nrf2 signaling pathway, and increase the survival rate of HaCaT cells by inhibiting ROS, confirming that it can inhibit skin inflammation [41]. In addition, Nrf2 affects the downstream AP-1 gene, which in turn affects the expression of genes such as IL-1β, TNF-α and MMPs, and regulates the body's inflammatory response process [27]. The results of this study show that *Laminaria japonica* can affect the expression levels of key node genes in the Nrf2 signaling pathway, promote its activation and nuclear translocation, inhibit the expression level of downstream genes, and exert its anti-inflammatory activity. Among the five node genes we explored, the LJ-G group had the most significant regulatory activity on Nrf2 and AP-1 in the protection and repair process, which are the target genes that directly regulate the inflammatory response. Combined with the research results at the cellular level, it can be considered that LJ-G has stronger anti-inflammatory activity.

In this study, fermentation technology was used to treat *Laminaria japonica*, and the damage effects of UVB on HaCaT cells and the protection and repair activity of *Laminaria japonica* on damaged cells were explored at the in vitro, cellular, and molecular levels. The results showed that after fermentation, the total sugar, polysaccharide, protein and total phenol contents in *Laminaria japonica* extracts increased significantly. This indicates that fermentation can effectively improve the extraction rate of its active substances. The in vitro antioxidant test results showed that LJ-G and LJ-Y had better in vitro oxygen-free radical scavenging activity than LJ-W, and that of LJ-G was better. In this study, a UVA-irradiated HaCaT cell inflammatory damage model was established, and a suitable concentration of *Laminaria japonica* extract was screened to investigate its anti-inflammatory activity at the cellular and molecular levels. The results showed that *Laminaria japonica* fermentation broth could activate the Nrf2 signaling pathway and promote the synthesis of antioxidant enzymes by promoting the dissociation of Nrf2 and Keap-1. In addition, *Laminaria japonica* fermentation broth regulated the transcriptional activity of Nrf2 downstream genes p38, JNK1, and AP-1, and inhibited the production of pro-inflammatory factors IL-1β, IL-8, TNF-α, and MMP-9, effectively alleviating the epidermal inflammatory response caused by UVB irradiation. The interaction of the high expression of pro-inflammatory factors and barrier disruption can lead to the aggravation of epidermal inflammation. The measurement results of skin barrier differentiation markers proved that *Laminaria japonica* fermentation broth alleviated the degree of damage to the skin barrier. This is supported by the fermentation broth's inhibition of the decomposition of the extracellular matrix by MMP-9. In this study, the eye irritation risk of *Laminaria japonica* extract was assessed using the erythrocyte hemolysis and CAM tests. The results showed that *Laminaria japonica* extract is safe and can be used as an anti-inflammatory raw material for the development and research of products related to skin inflammation. In addition, the loose filamentous structure of fermented *Laminaria japonica* polysaccharide and the monosaccharide fraction with rockulose as the main component also implied that it had good moisturizing and absorption effects.

This study used fermentation technology to treat *Laminaria japonica* and confirmed that fermentation can improve the utilization of active substances and increase anti-inflammatory activity. We compared *Laminaria japonica* alcoholic extracts with anti-inflammatory activity and found that LJ-G and LJ-Y could exert better anti-inflammatory and skin barrier repair effects [16]. Compared with the alcoholic extraction process, the *Laminaria japonica* extracts obtained using the fermentation method have a high extraction rate, high safety and high eco-friendliness that traditional organic solvent extraction lacks. However, we did not quantify the landmark anti-inflammatory active substances in the fermentation broth. We did not consider the effects of the fermented strain's own secretions on the physiological activity of *Laminaria japonica* fermentation broth. In the future, further physiological activity studies should be carried out on the anti-inflammatory active substances in the fermentation broth. In addition, does *Laminaria japonica* fermentation broth positively affect UVB-induced epidermal differentiation and defense mechanisms? Does *Laminaria japonica* fermentation relieve inflammatory skin diseases? In the future, more detailed and in-depth scientific research should be conducted on the photoprotection and repair mechanism of *Laminaria japonica* on the skin in order to further explore the biological properties of this abundant algae.

4. Materials and Methods

4.1. Preparation of Laminaria japonica Water Extract and Two Fermentation Broths

Laminaria japonica (Dalian, China) was crushed and passed through a 50-mesh sieve, dissolved in deionized water at a solid-liquid ratio of 1:40, extracted in a water bath at 80 °C for 5 h, and centrifuged at 4800 rpm for 30 min, and the supernatant was taken to obtain a *Laminaria japonica* water extract (named LJ-W) which was lyophilized through a membrane for use.

A single colony of rice wine yeast (Culture Collection Center of the Beijing Institute of Food and Brewing, Beijing, China) on YPD medium was placed in a 28 °C incubator for 48 h at 180 rpm to activate the strain. White *Ganoderma lucidum* (laboratory collection, deposit number: CGMCC No. 17789) was inoculated on a solid PDA medium and cultured at 28 °C for 12 d. The mycelium of white *Ganoderma lucidum* was picked and inoculated into a liquid PDA medium for 7 d to obtain mycelium balls.

Pretreatment was the same as that for LJ-W. *Laminaria japonica* powder was dissolved in deionized water and inserted into white *Ganoderma lucidum* mycelium balls and rice wine yeast at a material-to-liquid ratio of 1:20, then placed in an incubator and incubated at 180 rpm and 28 °C for 48 h. It was then centrifuged at 4800 rpm for 30 min, and the supernatant was taken to obtain white *Ganoderma lucidum Laminaria japonica* fermentation broth (LJ-G) and rice wine yeast *Laminaria japonica* fermentation broth (LJ-Y), which were lyophilized through a membrane for use.

4.2. Determination of Changes in Active Substance Content

Determination of total sugar content: a BC2710 Total Sugar Kit (Solarbio Biotechnology Co., Ltd., Beijing, China) was used to determine the total sugar content of LJ-W, LJ-G and LJ-Y, and the operation steps in the kit's instructions were followed.

Determination of reducing sugar content: glucose standard solution was prepared for use, 1 mg/mL of standard solution at different concentrations were taken and 2 mg/mL of DNS reagent (Carnos Technology Co., Ltd., Wuhan, China) was added. After cooling in a boiling water bath for 2 min, it was diluted five times with deionized water, the OD540 was measured, and a reducing sugar standard curve was drawn. The above steps were repeated to determine the reducing sugar content of LJ-W, LJ-G and LJ-Y.

Determination of total phenolic content: 1 mL of the standard solution of pyrogallic acid at different concentrations was prepared, 1 mL of deionized water, 0.5 mL of 2-fold diluted folin-phenol, and 1.5 mL of Na_2CO_3 solution with a mass fraction of 26.7% were added, and it was diluted 2.5 times at room temperature. After reacting for 2 h, the OD760 was determined, and a total phenol standard curve was drawn.

Determination of protein content: a BCA Protein Concentration Assay Kit (Biorigin Biotechnology Co., Ltd., Beijing, China) was used to determine the protein content of LJ-W, LJ-G and LJ-Y, and the operation steps in the kit's instructions were followed.

4.3. In Vitro Antioxidant Activity Assay

The experimental steps of DPPH and hydroxyl radical scavenging were implemented with reference to the literature [30].

Total antioxidant capacity determination: the total antioxidant capacity of the samples was determined using a Total Antioxidant Capacity Detection kit (ABTS and FRAP methods) (Beyotime Biotechnology Co., Ltd., Shanghai, China). According to the kit's instructions, Trolox and $FeSO_4 \bullet 7H_2O$ standard solution was configured, a standard curve was drawn and the total antioxidant capacity and Fe^{2+} reducing the power of LJ-W, LJ-G and LJ-Y was measured.

4.4. Cell Culture, UVB Irradiation Model Establishment and CCK8

HaCaT cells were purchased from the Cell Resource Centre (Beijing Union Medical College, Beijing, China). Cells were grown in a DMEM medium (Gibco, San Francisco, CA, USA) supplemented with 2% penicillin-streptomycin sulfate (Thermo Fisher Scientific

Co., Ltd., Shanghai, China) and 10% fetal bovine serum (Gibco, San Francisco, CA, USA) at 37 °C in a humidified environment with 5% CO_2. HaCaT cells were inoculated in T75 culture plates (Corning, Corning, NY, USA) at an inoculation density of 1×10^5/mL (the total volume of culture solution in the bottle is about 16 mL), adhered to the bottom of the plate and grew 24 h.

HaCaT cell viability was determined using the CCK8 colorimetric method. One hundred microliters of cell suspension (1×10^4/well) was inoculated into each well of a 96-well plate and it was placed in a 37 °C CO_2 incubator for 24 h. The supernatant was aspirated, 100 µL of PBS was added, different doses (0, 5, 10, 20, 30, 40 mJ/cm^2) of UVB were irradiated by UV crosslinker SCIENTZ03-II (SCIENTZ Biotechnology Co., Ltd., Ningbo, China). The total UVB energy was set at 40 mJ/cm^2 and a double layer of tinfoil was used to cover the 96-well plate, with 100 µL PBS added in the middle of each column of cells to reduce interference. The tinfoil position was adjusted during the irradiation process until the energy was depleted. PBS was aspirated and replaced by basal medium, 10 µL of CCK8 was added to each well and the OD450 was measured after incubation for 2 h. Then the cell viability was calculated, where the cell survival rate was calculated using the ratio of UVB damage group to control group multiplied by 100%.

The sample concentration selection procedure was the same as above. The difference is that the samples (100 µL/well) were added after UVB damage and continued to incubate for 12 h, control and model groups used equal amounts of basal medium instead of sample.

4.5. ROS Assay

To damage HaCaT cells, 10 mJ/cm^2 UVB was used. References [42,43] performed ROS content determination on damaged HaCaT cells that cultured in a 6-well plate. A ROS Detection Kit was selected for the experiment, the intracellular ROS fluorescence content was observed under a fluorescence inverted microscope (CKX53, Olympus LS, Tokyo, Japan) and the fluorescence intensity was recorded. The treatment of the control group and model group was carried out simultaneously with that of the sample group. During the process of damage and sample treatment in the sample group, the control and model groups were given the same amount of serum-free DMEM for culture.

4.6. Determination of Inflammation and Glial Differentiation Factor Secretion by ELISA

HaCaT cells (30×10^4/well) were cultured in a 6-well plate, the culture supernatant was collected and centrifuged at 4 °C and 6950 g for 10 min, and the supernatant was taken. ELISA kits (Wuhan Huamei Biological Engineering Co., Ltd., Wuhan, China) were used to detect the pro-inflammatory factors IL-1β, IL-8, TNF-α, MMP-9, AQP3, KLK-7, FLG, and Caspase-14.

4.7. qRT-PCR

HaCaT cells (30×10^4/well) were cultured for 24 h in a 6-well plate in order to make its cell count about 100×10^4/well, and the final volume was 2 mL per well. Total RNA extraction and reverse transcription were used by Trizol and First Stand cDNA. cDNA obtained by further reverse transcription was tested using Fast Super EvaGreen® qPCR Master Mix and qRT-PCR (QuantStudio 3, Thermoscientific, Shanghai, China). The relevant primer sequences are shown in Table 5. The primer sequences are shown in Table 6. And the qRT-PCR reaction system is shown in Table 7.

Table 5. Reverse transcription system.

Reagent Name	Volume (μL)
Total RNA	2.0
Anchored Oligo(Dt)18 Primer	1.0
2 × ES Reaction Mix	10.0
EasyScript RT/RI Enzyme Mix	1.0
Gdna Remover	1.0
Rnase-free Water	5.0

Table 6. Primer sequences for Real-Time PCR.

Gene	Direction	Primer Pair Sequence (5′→3′)
HO-1	F	CAAGCGCTATGTTCAGCGAC
	R	GCTTGAACTTGGTGGCACTG
IL-1β	F	CCTGAGCTCGCCAGTGAAA
	R	GTGGTGGTCGGAGATTCGTA
TNF-α	F	CACAGTGAAGTGCTGGCAAC
	R	AGGAAGGCCTAAGGTCCACT
Keap-1	F	GGAGGCGGAGCCCGA
	R	GATGCCCTCAATGGACACCA
AP-1	F	TCTCAACATGGGTGGTCTGT
	R	AAATGCTTCATGCGGCGAAG
IL-8	F	AAGATGTGAAGCTGACGCAGA
	R	AGAATTGAGCTGAGCCTTGG
MMP-9	F	GTACTCGACCTGTACCAGCG
	R	AGAAGCCCCACTTCTTGTCG
JNK1	F	CTGTGTGGAATCAAGCACCTTCA
	R	CTGGCCAGACCGAAGTCAAGA
p38	F	TTAACAGGATGCCAAGCCATGA
	R	GGCACCAATAAATACATTCGCAAAG
Nrf2	F	CAACTCAGCACCTTGTATC
	R	TTCTTAGTATCTGGCTTCTT
NQO1	F	CAGCCAATCAGCGTTCGGTA
	R	CTTCATGGCGTAGTTGAATGATGTC

F: forward primer; R: reverse primer.

Table 7. Reagents and dosage.

Reagent Name	Volume (μL)
Template	1.5
Forward Primer (10 μM)	0.4
Reverse Primer (10 μM)	0.4
2 × TransStart® Top Green qPCR SuperMix	10.0
Passive Reference Dye (50×)	0.4
Nuclease-free Water	7.3

4.8. Extraction of Polysaccharides from Laminaria japonica Fermentation Broth

The polysaccharide in Laminaria japonica fermentation broth was extracted by alcohol precipitation method. Dissolve the lyophilized powder prepared in 4.1 in deionized water, add three times the volume of absolute ethanol, and store at 4 °C overnight. Reconstitute the pellet with 100 mL water to the original volume and add three times the volume of Savage reagent to elute the protein. Using the alcohol precipitation method again, the precipitate was collected. The obtained precipitates were dissolved in ultrapure water and dialyzed (with 10 kDa MW cut-off) for 48 h, then freeze-dried to obtain crude polysaccharides, named "LJ-Wp, LJ-Gp, and LJ-Yp".

4.9. FT-IR, GPC and SEM

LJ-Wp, LJ-Gp, and LJ-Yp were analyzed by infrared spectroscopy using the potassium bromide tablet method, and FI-IR was analyzed by Vertex 70 FT-infrared spectroscopy in transmission mode. The wavelength range is 400–4000 cm^{-1}, and the resolution is 1 cm^{-1}.

The molecular weights of LJ-Wp, LJ-Gp, and LJ-Yp were determined by GPC-LS-IR, and the morphological characteristics of them were determined using an FEI Nova Nano SEM 450 instrument. The specific experimental method refer to Zhang et al. [43].

4.10. Monosaccharide Composition Analysis

Precisely weighed reference standards such as mannose, rhamnose and glucose, dissolved in water and dilute to 50 μg each in 1 mL of liquid. Precisely weighed the sample into a 10 mL ampoule bottle, added 3.0 mL of 2 mol/L TFA, filled with nitrogen, seal the tube, and hydrolyzed it with acid at 120 °C for 4 h. Took out and added methanol nitrogen to dry TFA, added 3.0 mL of water to reconstitute, and calculate the concentration of 50 μg/mL.

Drew the mixed standard solution, 0.6 mol/L NaOH, 0.4 mol/LPMP-methanol, the volume ratio was 1:1:2, and reacted at 70 °C for 1 h. Cooled in cold water for 10 min; added 0.3 mol/L HCl to neutralize, then added 1 mL of chloroform, vortex for 1 min, centrifuge at 3000 r/min for 10 min, and extracted three times. The supernatant was used for HPLC. The sample derivatization process repeats the above steps, except that the mixed standard solution is replaced by the sample solution.

The HPLC measurement used Xtimate C18 4.6 × 200 mm 5 μm chromatographic column, the mobile phase was composed of 83% 0.05 M potassium dihydrogen phosphate solution and 13% acetonitrile, the flow rate was 1.0 mL/min, and the injection volume was 20 μL.

4.11. Data Analysis

SPSS software (SPSS, Version 17.0, IBM Inc., Armonk, NY, USA) was used for data analysis. Origin 2018 was used for data visualization. All experiments were performed in triplicate, and data is presented as mean ± standard deviation. T-test was used to determine the significance of differences between the groups (^{NS} $p > 0.05$, $^{\#,*}$ $p < 0.05$; $^{\#\#,**}$ $p < 0.01$, $^{\#\#\#,***}$ $p < 0.001$). When $p < 0.05$, the difference is considered statistically significant.

5. Conclusions

Laminaria japonica has anti-inflammatory and skin barrier repair activity. We have successfully demonstrated that *Laminaria japonica* fermentation broth can effectively modulate UVB-induced epidermal inflammation and barrier damage. In this study, *Laminaria japonica* fermentation broths were shown for the first time to inhibit the production of inflammatory chemokines, which are the main triggers of inflammation. *Laminaria japonica* fermentation broth has also been shown to repair UVB-induced epidermal cell detachment and barrier breakage. The fermented *Laminaria japonica* polysaccharide had a more sparse and porous structure. Further studies will provide more insight into the activity of fermented polysaccharides from *Laminaria japonica* to test potential applications as a component against photo-inflammation.

Supplementary Materials: The following supporting information can be downloaded at: https://www.mdpi.com/article/10.3390/md20100650/s1, The following supporting information including Figure S1: Effects of different doses of UVB on the survival rate of HaCaT cells.

Author Contributions: Conceptualization, Q.S.; methodology, M.L.; software, J.F. and Z.S.; validation, C.W. and D.W.; formal analysis, Q.S.; investigation, Z.W. and J.G.; resources, D.W.; data curation and visualization, C.W. and D.W.; writing of the manuscript, Q.S.; writing—review and editing, Q.S.; supervision, M.L.; project administration, J.F. and C.W. All authors have read and agreed to the published version of the manuscript.

Funding: This research received no external funding.

Institutional Review Board Statement: Not applicable.

Informed Consent Statement: Not applicable.

Data Availability Statement: Such information is available from the corresponding author upon reasonable request.

Acknowledgments: We would like to thank the members of the panel for their guidance and contribution to this study.

Conflicts of Interest: The authors declare no conflict of interest.

References

1. Rodríguez-Luna, A.; Ávila-Román, J.; González-Rodríguez, M.L.; Cózar, M.J.; Rabasco, A.M.; Motilva, V.; Talero, E. Fucoxanthin-Containing Cream Prevents Epidermal Hyperplasia and UVB-Induced Skin Erythema in Mice. *Mar. Drugs* **2018**, *16*, 378. [CrossRef] [PubMed]
2. Zheng, X.; Feng, M.; Wan, J.; Shi, Y.; Xie, X.; Pan, W.; Hu, B.; Wang, Y.; Wen, H.; Wang, K.; et al. Anti-Damage Effect of Theaflavin-3′-Gallate from Black Tea on UVB-Irradiated HaCaT Cells by Photoprotection and Maintaining Cell Homeostasis. *J. Photochem. Photobiol. B Biol.* **2021**, *224*, 112304. [CrossRef] [PubMed]
3. Colombo, I.; Sangiovanni, E.; Maggio, R.; Mattozzi, C.; Zava, S.; Corbett, Y.; Fumagalli, M.; Carlino, C.; Corsetto, P.A.; Scaccabarozzi, D.; et al. HaCaT Cells as a Reliable in Vitro Differentiation Model to Dissect the Inflammatory/Repair Response of Human Keratinocytes. *Mediat. Inflamm.* **2017**, *2017*, 7435621. [CrossRef] [PubMed]
4. Kageyama, H.; Waditee-Sirisattha, R. Antioxidative, Anti-Inflammatory, and Anti-Aging Properties of Mycosporine-like Amino Acids: Molecular and Cellular Mechanisms in the Protection of Skin-Aging. *Mar. Drugs* **2019**, *17*, 222. [CrossRef] [PubMed]
5. Mapoung, S.; Umsumarng, S.; Semmarath, W.; Arjsri, P.; Srisawad, K.; Thippraphan, P.; Yodkeeree, S.; Dejkriengkraikul, P. Photoprotective Effects of a Hyperoside-Enriched Fraction Prepared from Houttuynia Cordata Thunb. On Ultraviolet B-Induced Skin Aging in Human Fibroblasts through the MAPK Signaling Pathway. *Plants* **2021**, *10*, 2628. [CrossRef] [PubMed]
6. Tang, S.C.; Liao, P.Y.; Hung, S.J.; Ge, J.S.; Chen, S.M.; Lai, J.C.; Hsiao, Y.P.; Yang, J.H. Topical Application of Glycolic Acid Suppresses the UVB Induced IL-6, IL-8, MCP-1 and COX-2 Inflammation by Modulating NF-KB Signaling Pathway in Keratinocytes and Mice Skin. *J. Dermatol. Sci.* **2017**, *86*, 238–248. [CrossRef]
7. Zhang, J.; Zheng, Y.; Hong, B.; Ma, L.; Zhao, Y.; Zhang, S.; Sun, S.; Ding, Q.; Wang, Y.; Liu, W.; et al. Dihydroquercetin Composite Nanofibrous Membrane Prevents UVA Radiation-Mediated Inflammation, Apoptosis and Oxidative Stress by Modulating MAPKs/Nrf2 Signaling in Human Epidermal Keratinocytes. *Biomed. Pharmacother.* **2022**, *155*, 113727. [CrossRef] [PubMed]
8. Vestergaard, C.; Hvid, M.; Johansen, C.; Kemp, K.; Deleuran, B.; Deleuran, M. Inflammation-Induced Alterations in the Skin Barrier Function: Implications in Atopic Dermatitis. *Chem. Immunol. Allergy* **2012**, *96*, 77–80. [CrossRef]
9. Yue, Q.; Wang, Z.; Yu, F.; Tang, X.; Su, L.; Zhang, S.; Sun, X.; Li, K.; Zhao, C.; Zhao, L. Changes in Metabolite Profiles and Antioxidant and Hypoglycemic Activities of Laminaria Japonica after Fermentation. *LWT* **2022**, *158*, 113122. [CrossRef]
10. Niture, S.K.; Khatri, R.; Jaiswal, A.K. Regulation of Nrf2-An Update. *NIH Public Access.* **2014**, *66*, 36–44. [CrossRef]
11. Itoh, K.; Wakabayashi, N.; Katoh, Y.; Ishii, T.; Igarashi, K.; Engel, J.D.; Yamamoto, M. Keap1 Represses Nuclear Activation of Antioxidant Responsive Elements by Nrf2 through Binding to the Amino-Terminal Neh2 Domain. *Genes Dev.* **1999**, *13*, 76–86. [CrossRef] [PubMed]
12. Ma, Q.; Battelli, L.; Hubbs, A.F. Multiorgan Autoimmune Inflammation, Enhanced Lymphoproliferation, and Impaired Homeostasis of Reactive Oxygen Species in Mice Lacking the Antioxidant-Activated Transcription Factor Nrf2. *Am. J. Pathol.* **2006**, *168*, 1960–1974. [CrossRef] [PubMed]
13. Ma, Q. Role of Nrf2 in Oxidative Stress and Toxicity. *Annu. Rev. Pharmacol. Toxicol.* **2013**, *53*, 401–426. [CrossRef] [PubMed]
14. Pae, H.O.; Oh, G.S.; Lee, B.S.; Rim, J.S.; Kim, Y.M.; Chung, H.T. 3-Hydroxyanthranilic Acid, One of l-Tryptophan Metabolites, Inhibits Monocyte Chemoattractant Protein-1 Secretion and Vascular Cell Adhesion Molecule-1 Expression via Heme Oxygenase-1 Induction in Human Umbilical Vein Endothelial Cells. *Atherosclerosis* **2006**, *187*, 274–284. [CrossRef] [PubMed]
15. Dinkova-Kostova, A.T.; Liby, K.T.; Stephenson, K.K.; Holtzclaw, W.D.; Gao, X.; Suh, N.; Williams, C.; Risingsong, R.; Honda, T.; Gribble, G.W.; et al. Extremely Potent Triterpenoid Inducers of the Phase 2 Response: Correlations of Protection against Oxidant and Inflammatory Stress. *Proc. Natl. Acad. Sci. USA* **2005**, *102*, 4584–4589. [CrossRef]
16. Michalak, I.; Tiwari, R.; Dhawan, M.; Alagawany, M.; Farag, M.R.; Sharun, K.; Emran, T.B.; Dhama, K. Antioxidant Effects of Seaweeds and Their Active Compounds on Animal Health and Production–a Review. *Vet. Q.* **2022**, *42*, 48–67. [CrossRef]
17. Lee, K.S.; Cho, E.; Weon, J.B.; Park, D.; Fréchet, M.; Chajra, H.; Jung, E. Inhibition of UVB-Induced Inflammation by Laminaria Japonica Extract via Regulation of Nc886-PKR Pathway. *Nutrients* **2020**, *12*, 1958. [CrossRef] [PubMed]
18. Luan, F.; Zou, J.; Rao, Z.; Ji, Y.; Lei, Z.; Peng, L.; Yang, Y.; He, X.; Zeng, N. Polysaccharides from: Laminaria Japonica: An Insight into the Current Research on Structural Features and Biological Properties. *Food Funct.* **2021**, *12*, 4254–4283. [CrossRef]
19. Hwang, Y.W. *Laminaria japonica* Suppresses the Atopic Dermatitis-Like Responses in NC/Nga Mice And Inflamed HaCaT Keratinocytes via the Downregulation of STAT1. *Nutrients* **2020**, *12*, 3238. [CrossRef]
20. Ahmad, T.; Eapen, M.S.; Ishaq, M.; Park, A.Y.; Karpiniec, S.S.; Stringer, D.N.; Sohal, S.S.; Fitton, J.H.; Guven, N.; Caruso, V.; et al. Anti-Inflammatory Activity of Fucoidan Extracts in Vitro. *Mar. Drugs* **2021**, *19*, 702. [CrossRef]

21. Wang, Z.; Sun, Q.; Fang, J.; Wang, C.; Wang, D.; Li, M. The Anti-Aging Activity of Lycium Barbarum Polysaccharide Extracted by Yeast Fermentation: In Vivo and in Vitro Studies. *Int. J. Biol. Macromol.* **2022**, *209*, 2032–2041. [CrossRef] [PubMed]
22. Park, M.-J.; Han, J.-S. Radical Scavenging and Antioxidant Activities of Fermented Laminaria Japonica Extracts. *Prev. Nutr. Food Sci.* **2006**, *11*, 10–16. [CrossRef]
23. Yi, L.; Wang, Q.; Luo, H.; Lei, D.; Tang, Z.; Lei, S.; Xiao, H. Inhibitory Effects of Polyphenols-Rich Components From Three Edible Seaweeds on Inflammation and Colon Cancer in Vitro. *Front. Nutr.* **2022**, *9*, 856273. [CrossRef] [PubMed]
24. Wang, S.H.; Chen, Y.S.; Lai, K.H.; Lu, C.K.; Chang, H.S.; Wu, H.C.; Yen, F.L.; Chen, L.Y.; Lee, J.C.; Yen, C.H. Prinsepiae Nux Extract Activates NRF2 Activity and Protects UVB-Induced Damage in Keratinocyte. *Antioxidants* **2022**, *11*, 1755. [CrossRef] [PubMed]
25. Piantadosi, C.A.; Carraway, M.S.; Babiker, A.; Suliman, H.B. Heme Oxygenase-1 Regulates Cardiac Mitochondrial Biogenesis via Nrf2-Mediated Transcriptional Control of Nuclear Respiratory Factor-1. *Circ. Res.* **2008**, *103*, 1232–1240. [CrossRef]
26. Zhang, Y.; Liu, P.; You, S.; Zhao, D.; An, Q.; Wang, D.; Zhang, J.; Li, M.; Wang, C. Anti-Inflammatory Effects of Opuntia Milpa Alta Polysaccharides Fermented by Lactic Acid Bacteria in Human Keratinocyte HaCaT Cells. *Chem. Biodivers.* **2022**, *19*, e202100923. [CrossRef]
27. Saha, S.; Buttari, B.; Panieri, E.; Profumo, E.; Saso, L. An Overview of Nrf2 Signaling Pathway and Its Role in Inflammation. *Molecules* **2020**, *25*, 5474. [CrossRef]
28. Long, M.; De La Vega, M.R.; Wen, Q.; Bharara, M.; Jiang, T.; Zhang, R.; Zhou, S.; Wong, P.K.; Wondrak, G.T.; Zheng, H.; et al. An Essential Role of NRF2 in Diabetic Wound Healing. *Diabetes* **2016**, *65*, 780–793. [CrossRef]
29. Saw, C.L.L.; Yang, A.Y.; Huang, M.T.; Liu, Y.; Lee, J.H.; Khor, T.O.; Su, Z.Y.; Shu, L.; Lu, Y.; Conney, A.H.; et al. Nrf2 Null Enhances UVB-Induced Skin Inflammation and Extracellular Matrix Damages. *Cell Biosci.* **2014**, *4*, 39. [CrossRef]
30. Zhang, Y.; Fu, H.; Zhang, Y.; Wang, D.; Zhao, D.; Zhang, J.; Li, M.; Wang, C. Reparative Effects of Dandelion Fermentation Broth on UVB-Induced Skin Inflammation. *Clin. Cosmet. Investig. Dermatol.* **2022**, *15*, 471–482. [CrossRef]
31. Sandilands, A.; Sutherland, C.; Irvine, A.D.; McLean, W.H.I. Filaggrin in the Frontline: Role in Skin Barrier Function and Disease. *J. Cell Sci.* **2009**, *122*, 1285–1294. [CrossRef] [PubMed]
32. Denecker, G.; Ovaere, P.; Vandenabeele, P.; Declercq, W. Caspase-14 Reveals Its Secrets. *J. Cell Biol.* **2008**, *180*, 451–458. [CrossRef] [PubMed]
33. da Nóbrega, A.M.; Alves, E.N.; de Farias Presgrave, R.; Costa, R.N.; Delgado, I.F. Determination of Eye Irritation Potential of Low-Irritant Products: Comparison of in Vitro Results with the in Vivo Draize Rabbit Test. *Brazilian Arch. Biol. Technol.* **2012**, *55*, 381–388. [CrossRef]
34. Lee, M.K.; Ryu, H.; Lee, J.Y.; Jeong, H.H.; Baek, J.; Van, J.Y.; Kim, M.J.; Jung, W.K.; Lee, B. Potential Beneficial Effects of Sargassum Spp. in Skin Aging. *Mar. Drugs* **2022**, *20*, 540. [CrossRef] [PubMed]
35. Vogel, V.; Olari, L.R.; Jachmann, M.; Reich, S.J.; Häring, M.; Kissmann, A.K.; Rosenau, F.; Riedel, C.U.; Münch, J.; Spellerberg, B. The Bacteriocin Angicin Interferes with Bacterial Membrane Integrity through Interaction with the Mannose Phosphotransferase System. *Front. Microbiol.* **2022**, *13*, 991145. [CrossRef] [PubMed]
36. Damage, K.C. Photo-Protective and Anti-Inflammatory Effects of Antidesma Thwaitesianum Müll. Arg. Fruit Extract against UVB-Induced Keratinocyte Cell Damage. *Molecules* **2022**, *27*, 5034.
37. Wang, P.W.; Cheng, Y.C.; Hung, Y.C.; Lee, C.H.; Fang, J.Y.; Li, W.T.; Wu, Y.R.; Pan, T.L. Red Raspberry Extract Protects the Skin against UVB-Induced Damage with Antioxidative and Anti-Inflammatory Properties. *Oxid. Med. Cell. Longev.* **2019**, *2019*, 9529676. [CrossRef]
38. Yokouchi, M.; Kubo, A.; Kawasaki, H.; Yoshida, K.; Ishii, K.; Furuse, M.; Amagai, M. Epidermal Tight Junction Barrier Function Is Altered by Skin Inflammation, but Not by Filaggrin-Deficient Stratum Corneum. *J. Dermatol. Sci.* **2015**, *77*, 28–36. [CrossRef]
39. Pérez-Alva, A.; MacIntosh, A.J.; Baigts-Allende, D.K.; García-Torres, R.; Ramírez-Rodrigues, M.M. Fermentation of Algae to Enhance Their Bioactive Activity: A Review. *Algal Res.* **2022**, *64*, 102684. [CrossRef]
40. Schäfer, M.; Dütsch, S.; Auf Dem Keller, U.; Navid, F.; Schwarz, A.; Johnson, D.A.; Johnson, J.A.; Werner, S. Nrf2 Establishes a Glutathione-Mediated Gradient of UVB Cytoprotection in the Epidermis. *Genes Dev.* **2010**, *24*, 1045–1058. [CrossRef] [PubMed]
41. Park, C.; Park, J.; Kim, W.J.; Kim, W.; Cheong, H.; Kim, S.J. Malonic Acid Isolated from Pinus Densiflora Inhibits Uvb-Induced Oxidative Stress and Inflammation in Hacat Keratinocytes. *Polymers* **2021**, *13*, 816. [CrossRef] [PubMed]
42. Zhang, Y.; Wang, D.; Fu, H.; Zhao, D.; Zhang, J.; Li, M.; Wang, C. Protective Effects of Extracellular Proteins of Saccharomycopsis Fibuligera on UVA-Damaged Human Skin Fibroblasts. *J. Funct. Foods* **2022**, *88*, 104897. [CrossRef]
43. Su, Y.; Zhang, Y.; Fu, H.; Yao, F.; Liu, P.; Mo, Q.; Wang, D.; Zhao, D.; Wang, C.; Li, M. Physicochemical and Anti–UVB–Induced Skin Inflammatory Properties of Lacticaseibacillus Paracasei Subsp. Paracasei SS–01 Strain Exopolysaccharide. *Fermentation* **2022**, *8*, 198. [CrossRef]

marine drugs

Article

Metabolites Produced by a New *Lactiplantibacillus plantarum* Strain BF1-13 Isolated from Deep Seawater of Izu-Akazawa Protect the Intestinal Epithelial Barrier from the Dysfunction Induced by Hydrogen Peroxide

Xiaozhen Diao [1,*], Katsuhisa Yamada [1,2], Yuji Shibata [2] and Chiaki Imada [1]

1 Applied Microbiology Lab, Course of Applied Marine Biosciences, Graduate School of Marine Science and Technology, Tokyo University of Marine Science and Technology, Tokyo 108-8477, Japan; k-yamada@dhc.co.jp (K.Y.); imada@kaiyodai.ac.jp (C.I.)
2 DSW Laboratory of DHC Co., Ltd., Tokyo 106-0047, Japan; yshibata@dhc.co.jp
* Correspondence: littletruth119@gmail.com

Abstract: This study aimed to investigate the protective effect of the metabolites produced by a new *Lactiplantibacillus plantarum* strain BF1-13, isolated from deep seawater (DSW), on the intestinal epithelial barrier against the dysfunction induced by hydrogen peroxide (H_2O_2) and to elucidate the mechanism underlying the effect. Protective effect of the metabolites by strain BF1-13 on the barrier function of the intestinal epithelial model treated with H_2O_2 was investigated by the transepithelial electrical resistance (TEER). The metabolites enhanced the Claudin-4 (CLDN-4) expression, including at the transcription level, indicated by immunofluorescence staining and quantitative RT-PCR. The metabolites also showed a suppression of aquaporin3 (AQP3) expression. Lactic acid (LA) produced by this strain of homofermentative lactic acid bacteria (LAB) had a similar enhancement on CLDN-4 expression. The metabolites of *L. plantarum* strain BF1-13 alleviated the dysfunction of intestinal epithelial barrier owing to its enhancement on the tight junctions (TJs) by LA, along with its suppression on AQP3-facilitating H_2O_2 intracellular invasion into Caco-2 cells. This is the first report on the enhancement of TJs by LA produced by LAB.

Keywords: *Lactiplantibacillus plantarum*; deep seawater; tight junctions; aquaporin3; hydrogen peroxide; lactic acid

Citation: Diao, X.; Yamada, K.; Shibata, Y.; Imada, C. Metabolites Produced by a New *Lactiplantibacillus plantarum* Strain BF1-13 Isolated from Deep Seawater of Izu-Akazawa Protect the Intestinal Epithelial Barrier from the Dysfunction Induced by Hydrogen Peroxide. *Mar. Drugs* **2022**, *20*, 87. https://doi.org/10.3390/md20020087

Academic Editors: Elena Talero and Javier Ávila-Román

Received: 2 December 2021
Accepted: 18 January 2022
Published: 20 January 2022

1. Introduction

It is known that the intestinal epithelium basically functions for the maintenance of the internal environment [1]. It absorbs nutrients and defends the exogenous pathogens and their secretion of toxins to protect the human body from diseases, including inflammatory bowel disease (IBD), celiac diseases, and diarrhea [2]. Between the intestinal epithelial cells, there exist TJs mediating the paracellular permeability to achieve the barrier function.

TJs are complex, contiguous, relatively, and dynamically impermeable junctional devices that take up the most apical surface of the intercellular space of intestinal epithelium [3]. TJs consist of several functional proteins including claudins, occludin (OCLN), tricellulin, and junctional adhesion molecules. The barrier function of the intestinal epithelium is closely related to these functional proteins' expression [4]. TEER assay with the intestinal epithelial model constructed by Caco-2 cells has been widely used to investigate the integrity of the intestinal epithelial barrier [5].

However, the function of TJs can be degraded by some reactive oxygen species, such as H_2O_2 [6]. H_2O_2 can be generated from the endogenous, such as the antibacterial defense by immune response or from the exogenous, such as unhealthy lifestyle, environmental pollution, and aging. Although the H_2O_2 at a low level can participate in cell signaling and cell proliferation and can eliminate invading pathogens, the high concentration may induce oxidative stress-related diseases, such as diabetes and inflammatory bowel disease in the case of the gastrointestinal tract [7,8]. Oxidative stress induced by different kinds of reactive oxygen species has been proven to cause the disruption of intestinal epithelial TJs [9]. It has been recently reported that the transportation of endogenous H_2O_2 facilitated by peroxiporins such as AQP3 (transcellular invasion) occurs prior to simple diffusion (paracellular invasion) [10].

The gut microbiota maintains the integrity and function of the gastrointestinal tract, where there are colonies of the huge community of LA-producing bacteria such as LABs [11]. LA was found as an impure component from sour milk in 1780, which was proven to be an isolated metabolite of microorganisms by Pasteur in 1857. In 1881, a French scientist successfully produced it by microbial fermentation, which leads to the first boom of LA industrial production. Since then, LA has been continuously among the hot spots in many industries, such as the food, probiotics, medical, cosmetic, and pharmaceutical industry. From the 2000s, its biotechnological application to form polylactic acid (PLA) has been of increasing interest [12]. Due to its advantage with lower cost, less limitation on production, high efficiency, and being eco-friendly, bacterial fermentation takes up about 90% of the industry market of LA production [13]. So far, there is no report on the effect of LA on the intestinal epithelial barrier function, although there are many about the LAB themselves.

Grishina et al. [14] proved that kefir and ayran supernatants, which contain large amounts of LA, decreased DNA damage in colon cells caused by fecal water in vitro. However, according to them, no effect on intestinal TJs was observed. Kefir & ayran (fermented milk) fermented by complicated bacterial species produce various components in the culture supernatants, including not only lactic acid but also other organic substances, such as ethanol [15]. Due to all these reasons, their results cannot be convincing evidence for the effect of LA on TJs. In contrast, Forsyth G. W. et al. [16] reported that L-LA reduced the net fluid secretion in ligated jejunal loops in vitro caused by cholera toxin, which indicated the potential effect of LA on intestinal epithelial permeability regulated by TJs.

It has been proven that the intestinal oxidative damage induced by H_2O_2 can be inhibited or released by some species of *Lactobacillus*, mostly isolated from the probiotics, [17,18] or by the fermentative products of them [19]. Although homofermentative LAB mainly refers to both *L. plantarum* and *Lactobacillus delbrueckii*, unlike the former, *L. delbrueckii*, which can convert 95% of glucose into LA, only produces D-LA [16], which was reported to be harmful by inducing acidosis and de-calcification [20]. The *L. plantarum* strain JCM11125 (also as *L. arizonensis* strain; 1474 bp; ENA accession number AJ965482) isolated from jojoba meal fermentation is a homofermentative strain with no catalase production and had a wide range of growth temperatures [21], which was used as the standard strain in this study.

This study aims to investigate the protective effect of the metabolites produced by DSW-derived *L. plantarum* strain BF1-13 on the intestinal epithelial barrier against the H_2O_2-induced dysfunction and its underlying mechanism related to the AQP-facilitating H_2O_2 intracellular invasion. Furthermore, it provides the potential application of the strain BF1-13 fermented food and its metabolites as supplements for IBD patients.

2. Results

2.1. Characteristics of Three Lactiplantibacillus Strains

The two isolated *L. plantarum* strains were confirmed by 16S rRNA gene sequencing and named as BF1-13 (1492 bp; DDBJ accession number LC666821) and H-6 (1492 bp; DDBJ accession number LC666820), respectively.

These three strains showed nearly the same characteristics. There is no difference in salt tolerance among the strains, whose maximum tolerances were 5% NaCl. Similar to the standard strain JCM11125, both of the two isolated strains were able to ferment all kinds of substrates and survived in the nutrients-limited medium only supplemented with specific carbohydrates All strains grew at a pH range of 2–9. However, strain BF1-13 failed to grow at 20 °C and showed a preference for a higher temperature (27–40 °C). (Table 1)

Table 1. Comparison of growth characteristics of the three strains.

		Strain BF1-13	Strain JCM11125	Strain H-6
Temperature (°C)	5	−	−	−
	10	−	−	−
	20	−	+	+
	27	+	+	+
	37	+	+	+
	40	+	+	+
	45	−	−	−
pH	2	+	+	-
	3	+	+	+
	4	+	+	+
	6.5	+	+	+
	9	+	+	+
	10	−	−	−
	11	−	−	−
NaCl (%, *m/v*)	0	+	+	+
	1	+	+	+
	2	+	+	+
	3	+	+	+
	5	+	+	+
	10	−	−	−
Fermentation substrate	Glucose	+	+	+
	Sorbitol	+	+	+
	Trehalose	+	+	+
	Xylose	+	+	+
	Arabinose	+	+	+
	Mannitol	+	+	+
	Lactose	+	+	+

+, growth; -, no growth (n = 3).

During the 12 h incubation (pH 6.5) at 37 °C, strain BF1-13 showed superiority in cell growth with the other two strains. Among the three strains, strain BF1-13 showed the fastest growth (1.67×10^9 CFU/mL, Figure 1a), and the most remarkable decreasing pH value (pH 4.599) (Figure 1b). From the measurement of the concentration of LA contained in each culture supernatant (CS), it was suggested that there was a little difference in the LA production during 12 h incubation in the following groups: strain BF1-13 (26.3 mM), strain H-6 (20.0 mM), the standard strain JCM11125 (19.8 mM) (Table 2). As mentioned above, strain BF1-13 was selected as the research target because it showed superiority in growth and LA accumulation. Strain JCM 1112 was prepared as the standard strain in this study.

Table 2. Characteristics of each strain after 12 h incubation.

Strain	BF1-13	JCM11125	H-6
CFU/mL	1.67×10^9	7.44×10^8	8.71×10^8
pH	4.6	4.9	4.9
Latic acid (mM)	26.3	19.8	20.0

The number of bacterial cells indicates as CFU/mL (n = 9, mean ± SD).

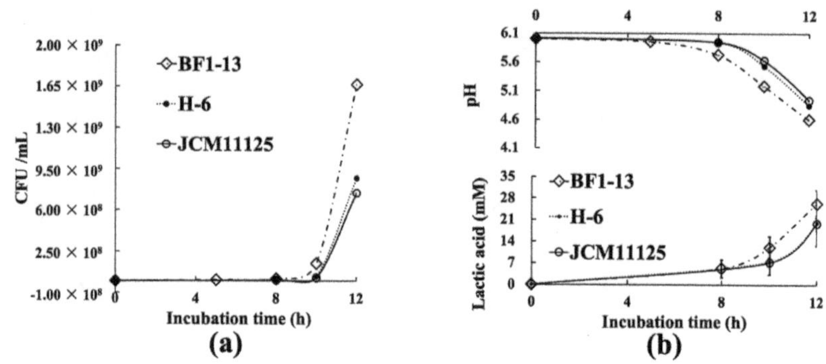

Figure 1. Comparison of each strain (BF1-13 ↓, H-6 λ, JCM11125 γ) on various characteristics. (**a**) The growth curve of the three strains. (**b**) The amounts of LA and pH changing of each strain during 12 h incubation. Data are shown as means ± SD, $n = 9$.

2.2. *Protection on the Intestinal Epithelial Barrier by the Metabolites Produced by Strain BF1-13 against the Dysfunction Caused by H_2O_2 Treatment in Caco-2 Cells*

TcTEER values decreased to 45.1% of the initial values after 6 h H_2O_2 treatment on the intestinal epithelial model using Caco-2 cells. The decreasing tcTEER value by H_2O_2 treatment was suppressed by the supplementation of 5% (*v/v*) CS of strain BF1-13. At the same supplemented concentration of 5% (*v/v*) of the CSs, strain BF1-13 (62.2%) showed slightly higher TEER values than standard strain (59.9%) (Figure 2b). However, this effect of the CS of strain BF1-13 disappeared by the 10% (*v/v*) supplementation (28.3%). (Figure 2a).

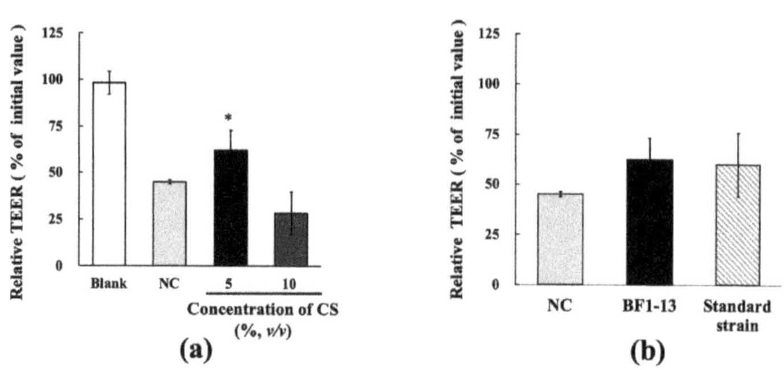

Figure 2. Effects of the metabolites of strain BF1-13 (black) and standard strain (slashes) on TJs barrier function in human Caco-2 cells. (**a**) The relationship between relative TEER value and CS supplemented concentration with 1 mM H_2O_2 (NC) or without (blank). (**b**) The comparison of the protective effect between the metabolites of the two strains on TJs barrier function. Monolayers were preincubated with 5% (*v/v*) CS of the two strains separately or without (NC) for 2 h, followed by 1 mM H_2O_2 treatment for 6 h or not (blank). Relative TEER values are means ± SD, $n = 3$. Asterisks indicate a significant difference with NC (* $p < 0.05$) determined by the Mann–Whitney U test.

2.3. *Enhancement on the TJs-Related Proteins by the Metabolites Produced by Strain BF1-13 from the Suppression Induced by H_2O_2*

2.3.1. Enhancement on CLDN-4 by the Metabolites Produced by Strain BF1-13 through Immunofluorescent Microscopy

Decreasing expression of CLDN-4 by H_2O_2 treatment was inhibited by the supplementation of the CS of the strain BF1-13. This effect was also shown on the standard strain (Figure 3). However, the expression of OCLN was different from that of CLDN-4. There was a little difference between the metabolites of the two strains. In the case of strain

BF1-13, the difference of the effects between supplemented concentration 5% and 10% was barely seen with the immunofluorescent staining microscopy (Data not shown.)

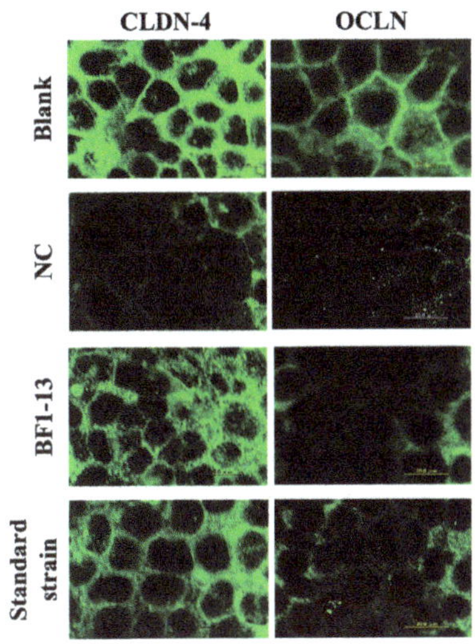

Figure 3. The comparison of the enhancements between the metabolites of the two strains on the expression of CLDN-4 and OCLN. Monolayers were preincubated with 5% (*v/v*) CS of the two strains separately or without (NC) for 2 h, followed by 1 mM H_2O_2 treatment for 2 h or not (blank). Images were observed by fluorescence microscopy. Each image is representative of 3 similar experiments. The scale bar is 20 μm.

2.3.2. Inducement on CLDN-4 mRNA by the Metabolites Produced by Strain BF1-13 through Quantitative RT-PCR

The mRNA expression of CLDN-4 was only induced by the supplementation with the CS of the strain BF1-13 in Caco-2 cells (Figure 4a). The mRNA expression of CLDN-4 with the supplementation with the CS of strain BF1-13 (fold-change of 1.75 & 1.53) was higher than that with the NC (fold-change of 1.20 & 1.15) during the 2 h H_2O_2 treatment. However, enhancement of CLDN-4 expression was not shown on that treatment with the CS of the standard strain (fold-change of 0.85 & 0.91). In the case of OCLN, the mRNA expression was not maintained by the supplementations with the CSs of both two strains (Figure 4b).

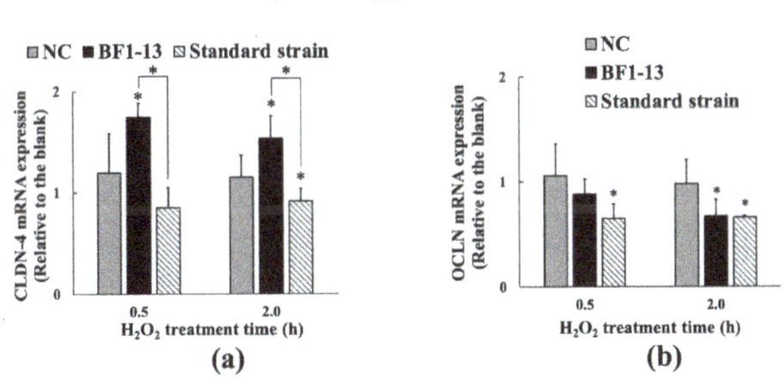

Figure 4. The inducement of the metabolites of the strain BF1-13 (black) on the expression of CLDN-4 (**a**) at the transcription level compared with the standard strain (slashes), which was compared with the expression of OCLN (**b**). Monolayers were preincubated with 5% (*v/v*) CS of the two strains separately or without (NC) for 2 h, followed by 1 mM H_2O_2 treatment for 0.5 and 2 h separately or not (blank). The mRNA expressions relative to the blank are shown as means ± SD, $n = 3$. Asterisks indicate a significant difference with the blank, or between BF1-13 and the standard strain (* $p < 0.05$) determined by the Mann–Whitney U test.

2.4. Suppression on AQP3 by the Metabolites Produced by Strain BF1-13

2.4.1. Suppression on AQP3 by the Metabolites Produced by Strain BF1-13 through Immunofluorescent Microscopy

The expression of AQP3 was enhanced by 1 mM H_2O_2 treatment compared to the blank, which was also supported by the previous report [10]. The AQP3 expression was suppressed by the supplementation with the CS of strain BF1-13. Then, the same effect was shown in the CS of the standard strain (Figure 5).

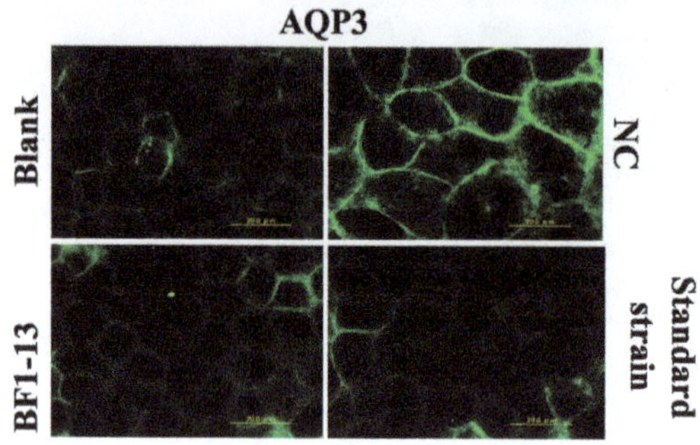

Figure 5. The suppressions of the AQP3 expression in human Caco-2 cells induced by the supplementation of the CSs. Monolayers were preincubated with 5% (*v/v*) CS of the two strains separately or without (NC) for 2 h, followed by 1 mM H_2O_2 treatment for 2 h or not (blank). Images were observed by fluorescence microscopy. Each image is representative of 3 similar experiments. The scale bar is 20 μm.

2.4.2. Suppression on AQP3 mRNA by the Metabolites Produced by Strain BF1-13 through Quantitative RT-PCR

The mRNA expression of AQP3 was significantly suppressed by the metabolites by strain BF1-13. This suppression was also shown on the metabolites by the standard strain (Figure 6). The mRNA expression of AQP3 with the supplementation with the CS of strain BF1-13 (fold-change of 0.57 & 0.52) was strongly suppressed during the 2 h H_2O_2 treatment, as well as that of the standard strain (fold-change of 0.45 & 0.45).

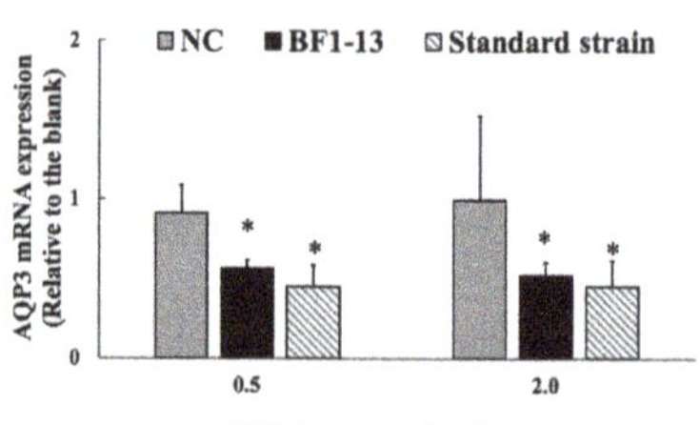

Figure 6. The suppressions of the metabolites of strain BF1-3 (black) and standard strain (slashes) on the expression of AQP3 at the transcription level. Monolayers were preincubated with 5% (*v/v*) CS of the two strains and LA (1.32 mM) separately or without (NC) for 2 h, followed by 1 mM H_2O_2 treatment for 0.5 and 2 h separately or not (blank). The mRNA expressions relative to the blank are shown as means ± SD, $n = 3$. Asterisks indicate a significant difference with the blank, or between BF1-13 and the standard strain (* $p < 0.05$) determined by the Mann–Whitney U test.

2.5. Effect by LA Contained in the Metabolites Produced by Strain BF1-13 on the Intestinal Epithelial Barrier with H_2O_2 Treatment

2.5.1. Protection on the Intestinal Epithelial Barrier by LA against the Dysfunction Caused by H_2O_2 Treatment

Equal concentration of authentic LA (1.32 mM), which was contained in the metabolites, included in the 5% (*v/v*) CS of strain BF1-13 indicated almost the same protective effect on decreasing TEER by H_2O_2 treatment on the model (Figure 7a). Supplementation of authentic LA (1.32, 1.97, and 2.63 mM) suggested the significant maintenance effect of relative TEER values (83%, 74%, 74%) on the model treated with H_2O_2 (Figure 7b). However, the protective effect of LA decreased by the higher concentration, which was more than 1.32 mM.

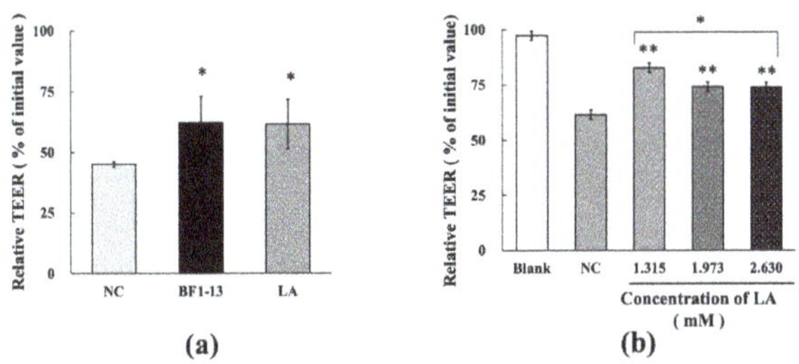

Figure 7. Protective effects of the metabolites of strain BF1-13 (black) and LA (spots) on TJs barrier function. (**a**) The comparison of the protective effect between 5% (*v/v*) CS and LA (1.32 mM) on TJs barrier function with 1 mM H_2O_2 (NC) or without (blank). (**b**) Dose-dependent protective effect of authentic LA on TJs barrier function. Monolayers were pre-incubated with 1.32, 1.97, and 2.63 mM authentic LA separately or without (NC) for 2 h, followed by 1 mM H_2O_2 treatment for 6 h or not (blank). Relative TEER values are means ± SD, n = 3 (**a**), 6 (**b**). Asterisks indicate a significant difference with NC (* $p < 0.05$, ** $p < 0.005$) determined by the Mann–Whitney U test, or among various concentrations of LA (* $p < 0.01$) determined by the Kruskal–Wallis test.

2.5.2. Enhancement on TJs-Related Proteins by LA Contained in the Metabolites Produced by Strain BF1-13

The CLDN-4 expression was maintained by the supplementation of the CS of strain BF1-13 and the equal concentration of authentic LA against H_2O_2 treatment (Figure 8). There was little difference in the effect between them. However, in the case of the equal concentration of authentic LA contained in the metabolites, the suppression of OCLN expression by the H_2O_2 treatment was relieved.

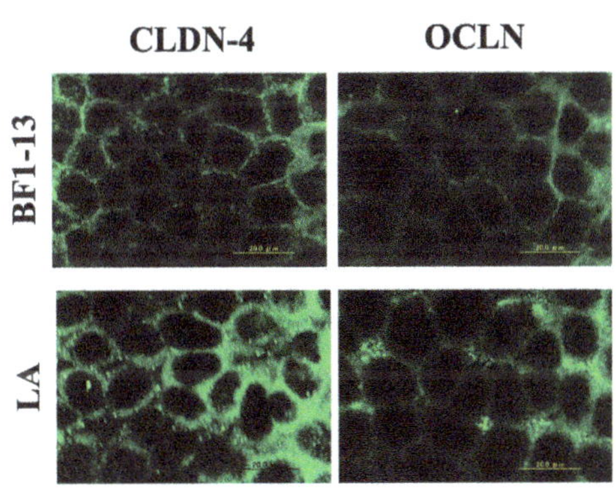

Figure 8. The comparison of the enhancements between the metabolites of strain BF1-13 and LA on the expression of CLDN-4 and OCLN. Monolayers were preincubated with 5% (*v/v*) CS of the strain and LA (1.32 mM) separately for 2 h, followed by 1 mM H_2O_2 treatment for 2 h. Images were observed by fluorescence microscopy. Each image is representative of 3 similar experiments. The scale bar is 20 μm.

The mRNA expression of CLDN-4 was induced by the supplementation with the CS of strain BF1-13. This effect was also shown in the equal concentration of authentic LA (fold-change of 1.55 & 1.26) (Figure 9a). However, the mRNA expression of OCLN was only induced by the supplementation with LA at 0.5 h (fold-change of 1.25) (Figure 9b).

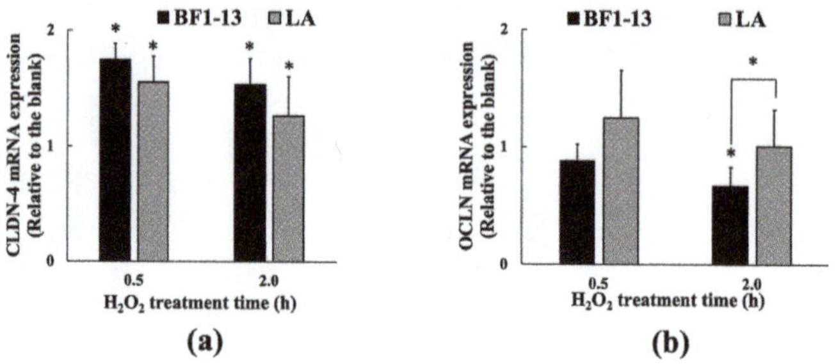

Figure 9. The comparison of the effect between the metabolites of strain BF1-13 (black) and LA (spots) on the expression of CLDN-4 (**a**) and OCLN (**b**) at the transcription level. Monolayers were preincubated with 5% (*v/v*) CS of strain BF1-13 and LA (1.32 mM) separately or without (NC) for 2 h, followed by 1 mM H_2O_2 treatment for 0.5 and 2 h separately or not (blank). The mRNA expressions relative to the blank are shown as means ± SD, $n = 3$. Asterisks indicate a significant difference with the blank, or between BF1-13 and LA (* $p < 0.05$) determined by the U-Mann-Whitney test.

2.5.3. Effect on AQP3 by LA Contained in the Metabolites Produced by Strain BF1-13

The LA (1.32 mM) contained in the metabolites of strain BF1-13 showed no suppressive effect on AQP3 expression (Figure 10) nor on its mRNA expression (Figure 11).

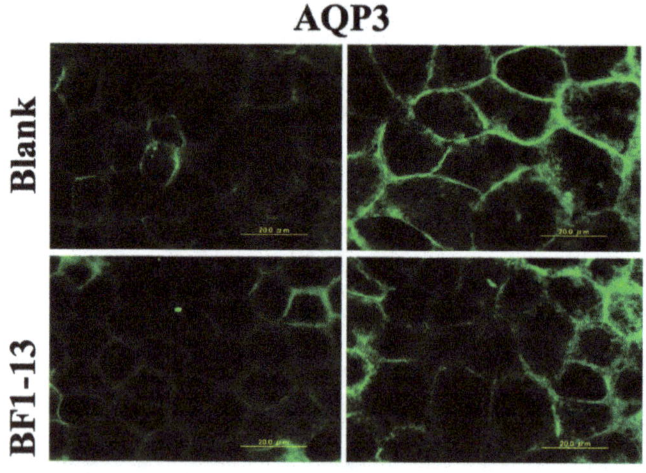

Figure 10. The effect of the metabolites of strain BF1-13 and equal concentration of authentic LA on the expression of APQ3 in human Caco-2 cells. Monolayers were preincubated with 5% (*v/v*) CS of the strain BF1-13 and LA (1.32 mM) separately or without (NC) for 2 h, followed by 1 mM H_2O_2 treatment for 2 h or not (blank). Images were observed by fluorescence microscopy. Each image is representative of 3 similar experiments. The scale bar is 20 μm.

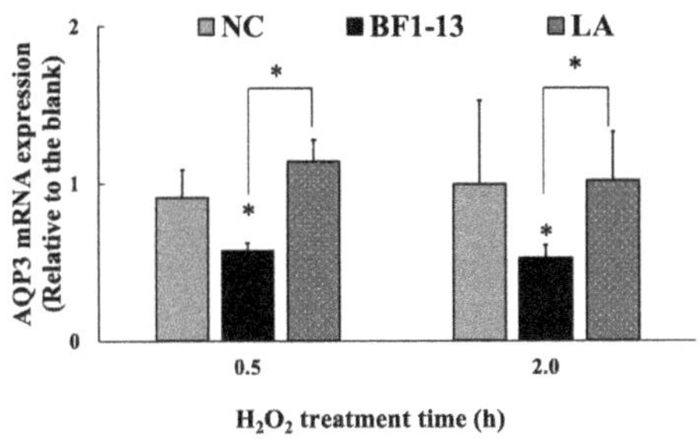

Figure 11. The effect of the metabolites of strain BF1-3 (black) and equal concentration of authentic LA (spots) on the expression of AQP3 at the transcription level. Monolayers were preincubated with 5% (*v/v*) CS of the strain BF1-13 and LA (1.32 mM) separately or without (NC) for 2 h, followed by 1 mM H_2O_2 treatment for 0.5 and 2 h separately or not (blank). The mRNA expressions relative to the blank are shown as means ± SD, *n* = 3. Asterisks indicate a significant difference with the blank, or between BF1-13 and LA (* $p < 0.05$) determined by the U-Mann-Whitney test.

3. Discussion

The *L. plantarum* strain BF1-13 grew at a wide range of temperature (27–40 °C) and pH (2–9), which also showed good tolerance to salt (up to 5% NaCl). The optimal culture conditions for *L. plantarum* strain BF1-13 were determined as the anaerobic incubation around pH 6.5 at 37 °C. The LA contained in the CS after 12 h incubation of strain JCM11125 was less than that of strain BF1-13, which is consistent with the fact it can only convert 80% of the glucose into LA [22]. Good tolerance to low pH allows the strain BF1-13 to colonize in an acidic environment such as the human intestine. Its utilizing of the single carbohydrates source overcomes food intolerances during consumption, such as lactose intolerance. In summary, strain BF1-13 was suggested to have less limitation on both production and consumption when applied to different industries.

When the mechanism on the maintenance of human health via oral intake by the metabolites produced by marine-derived strain is studied, it is more practical to combine it with the transportation routes through the epithelial membrane during their absorptions in the small intestine. The small intestine is the main place for the nutrients to be taken up from the intestinal lumen to reach the lamina propria through the intestinal epithelium. In this study, all the tested samples were supplemented from the apical side, which stands for the lumen in vivo. As an endogenous reactive oxygen species, H_2O_2 is mostly produced in the lamina propria by antibacterial defense. Therefore, the monolayers in vitro model constructed by Caco-2 cells, which imitated the intestinal epithelial barrier, was treated with H_2O_2 from the basolateral side in this study. As a result, it was suggested by the relative TEER value that the metabolites produced by *Lactiplantibacillus plantarum* strain BF1-13 protected the intestinal epithelial barrier from the dysfunction caused by H_2O_2 treatment. The equal amount of LA included in the metabolites of strain BF1-13 also showed the same protective effect on the barrier. According to the results shown by immunofluorescence staining and quantitative RT-PCR, LA (1.32 mM) contained in the CS of strain BF1-13 showed the similar enhancement on CLDN-4 expression at the transcription level with the metabolites themselves. In summary, it was suggested that LA, which was the only organic acid included in the metabolites produced by homofermentative LAB, was one of the bioactive substances to enhance the expression of TJs-related protein especially CLDN-4.

However, this enhancement on OCLN was much weaker than CLDN-4 especially by the supplementation of the CSs of two strains. This difference between CLDN-4 and OCLN was shown on both the protein expressions and the mRNA expression. To investigate why the protective effect on the barrier function of the intestinal epithelial model was achieved by the different enhancement on TJs expressions, further experiments related to the expression of AQP3, which was recently proven to facilitate H_2O_2 intracellular invasion, was prepared in this study [23]. It was known that AQP3 was expressed only on the basolateral side of the cell membrane [24]. Combining the positions of CLDN-4 and OCLN in TJs structure, it was indicated that H_2O_2 added to the basolateral side of the monolayer in the model had two different invasion routes, including intercellular invasion by simple passive diffusion and intracellular route by AQP3-facilitating diffusion. It was suggested that H_2O_2 intracellular invasion via APQ3 caused the suppression of the expression of TJs-related proteins, especially CLDN-4. It was also suggested that the suppression by the metabolites produced by the strain on AQP3 expression at the transcription level made the H_2O_2 preferentially target on OCLN via intercellular route than CLDN-4 via AQP3-facilitating intracellular invasion. The suppression was not shown by LA, mostly because, as a small molecule, it prefers the paracellular route by TJs rather than the transcellular route.

Lactiplantibacillus plantarum strain BF1-13 isolated from DSW in Izu-Akazawa was suggested to be a fast-growing strain with a higher LA accumulation compared to the same species strains isolated from surface seawater and terrestrial. According to this, strain BF1-13 was suggested to have the possibility of wide application in the LAB market. In addition, the protective effect of the metabolites produced by strain BF1-13 discovered by this study will provide the possibility of the application in the maintenance of human health against intestinal related diseases. Moreover, the relationship between the dysfunction of intestinal epithelium and H_2O_2 intracellular invasion has been revealed for the first time. This is the first time that LA has been reported to be an essential substance for the enhancement on the intestinal epithelial barrier from the dysfunction caused by H_2O_2, although many other effects of LA have been well-known. Still, the suppression of the AQP3 expression by the metabolites remains to be further researched.

4. Materials and Methods

4.1. Chemicals

High-glucose DMEM (HG-DMEM) was prepared with Dulbecco's modified Eagle's medium (DMEM; Nissui, Tokyo, Japan). De-Man Rogosa Sharpe medium (MRS broth, BD Difco^{TM}, Sparks, MD, USA) was purchased for the incubation of LAB. Rabbit anti-CLDN-4, OCLN, AQP3 antibodies and horseradish peroxidase conjugated Goat anti-Rabbit IgG H+L antibody were purchased from Abcam (Tokyo, Japan). Lactic acid and H_2O_2 was purchased from Wako (Osaka, Japan).

4.2. Bacterial Strains and 16S rRNA Sequencing

Two isolated strains were prepared in the following methods in this study. *L. plantarum* strain BF1-13 was isolated by the DSW laboratory of DHC Co. from the bag filter (pore size, 0.5 μm) at 800 m depth DSW in the pumping station in Izu-Akazawa, Shizuoka Prefecture and strain H-6 was isolated from a seaweed '*Polyopes* sp.' collected in Okinawa Prefecture. The genomic DNA used in the analysis of 16S rRNA gene sequences was extracted by Achromopeptidase (Wako, Osaka, Japan). Extracted DNA was PCR amplified by TKs Gflex DNA Polymerase (Takara Bio, Shiga, Japan) and was sequenced on ABI PRISM 3130 × 1 Genetic Analyzer System by using BigDye Terminator v3.1 Cycle Sequencing Kit (Applied Biosystems, Waltham, MA, USA). The 16S rDNA sequences were assembled using ChromasPro 2.1 (Technelysium, South Brisbane, AUS). Phylogenetic analysis was performed by using ENKI software (TechnoSuruga Laboratory, Shizuoka, Japan). Strain JCM11125 as the standard strain was obtained from RIKEN BRC (Ibaraki, Japan).

4.3. Cell Culture

Caco-2 cells (No. RCB0988, RIKEN BRC, Japan) were cultured by HG-DMEM at 37 °C in 5% CO_2. The cells at the density of 0.24×10^6 cells/cm^2 were seeded on each hanging cell culture insert (Millicell, Φ 6.5 mm, 1.0 μm pore size, PET; 0.3 cm^2 area; Merck KGaA, Darmstadt, Germany) to construct the intestinal epithelial model. The medium was changed every two days until the monolayer was completed. The cells used were between passage 11 and 25.

4.4. Bacteria Incubation and the Preparation of Metabolites Containing CSs from the Isolated Strains

Both of the two strains and the standard strain were incubated in liquid MRS medium (pH 6.5) at 37 °C for 12 h and the CSs were separately obtained from the bacterial suspension (with final bacterial concentrations of strains BF1-13 and JCM11125 as 1.67×10^9, 7.44×10^8 CFU/mL, respectively) by centrifugation (13,200× *g*, 4 °C, 5 min) followed by the sterilization through 0.2 µm filter (ADVANTEC, Tokyo, Japan). For each CS, the ability of LA production by the strain was measured at 450 nm by a microplate reader (Model 550, Bio-Rad, Hercules, CA, USA) using Lactate Assay Kit-WST (DOJINDO, Kumamoto, Japan) after pH value was measured. The comparison of the growth temperature among three strains was investigated following the incubation at 5, 10, 20, 27, 37, 40, 45 °C, by using the MRS agar medium (pH 6.5) for 2 weeks. Salt tolerance was investigated by 0~10% (m/v) NaCl supplementation to the MRS agar medium (pH 6.5). Carbohydrate fermentation ability was conducted by using the nutrient-limited medium (pH 6.5) containing different substrates (0.4%, m/v) separately, including glucose, sorbitol, trehalose, xylose, arabinose, mannitol, and lactose. Salt tolerance, pH tolerance and carbohydrate fermentation ability of the strains were determined following the 2 weeks incubation at 37 °C.

4.5. Measurement of the Intestinal TJs Barrier Function

Intestinal epithelial barrier function was evaluated as the TEER value on the intestinal epithelial model using Caco-2 cells by Millicell-ERS-2 system (Millipore, Billerica, MA, USA). Caco-2 cells achieved to form the monolayer from 7 to 14 days after seeding when temperature corrected TEER (tcTEER) values of the model became stable around 200–300 Ω cm^2 (310.15 K) [25]. In this study, the CSs of the two strains were prepared with HG-DMEM separately as the evaluation medium. Each evaluation medium was added to the apical side of the membrane and the model was cultured for 2 h (37 °C, 5% CO_2) to be habituated. After culturing, all the media in both apical and basolateral sides were replaced by HG-DMEM without FBS due to its eliminative effect of H_2O_2. Then, HG-DMEM (without FBS) containing 1 mM H_2O_2 (negative control, NC) or without H_2O_2 (blank) was added to the basolateral side. The model was incubated for 6 h (37 °C, 5% CO_2). TEER values were measured before and until 8 h of the CSs supplementation. Authentic LA (1.32, 1.97, and 2.63 mM) prepared by HG-DMEM was also administered to the apical side following the same procedure to investigate its effect on TJs barrier function.

4.6. Immunofluorescence Staining

Cell monolayer was cultured separately with various evaluation media, including 5% (v/v) CSs of the two strains or authentic LA (1.32 mM) for 2 h (37 °C, 5% CO_2) to be habituated. After culturing, all the media were replaced by HG-DMEM without FBS. Then, the model was treated without (blank) or with 1 mM H_2O_2 (NC) for 2 h (37 °C, 5% CO_2). After H_2O_2 treatment, the monolayers were rinsed with cold PBS, fixed in methanol at 4 °C for 30 min. Next, they were exposed to PBS containing 0.3% Triton-X100 for 5 min. Monolayers were blocked in PBS containing 5% (v/v) normal fetal serum for 30 min at room temperature. The monolayers were separately incubated with rabbit anti-CLDN-4, OCLN, AQP3 antibody overnight at 4 °C, followed by the 2nd antibody (horseradish peroxidase conjugated goat anti-rabbit IgG H+L antibody) for 1 h in the dark at room temperature. Each protein expression was observed by the fluorescence microscope with a magnification of 1000 (BX51N-33P, Olympus, Tokyo, Japan).

4.7. RNA Extraction and Quantitative RT-PCR

Confluent Caco-2 cells were preincubated without (blank and NC) or with 5% (v/v) CS of each strain or authentic LA (1.32 mM) for 2 h (37 °C, 5% CO_2). After preincubation, all the media were replaced by HG-DMEM without FBS. Then, cells were treated with or without it (blank) 1 mM H_2O_2 for 0.5, 1, and 2 h, separately (37 °C, 5% CO_2). After H_2O_2 treatment, total RNA was extracted using ISOGENII (NIPPON GENE, Tokyo, Japan) according to the supplier's protocol by using sterile and RNase-free tubes. Total RNA was quantified by measuring optical density at 260 nm. cDNAs were synthesized using 1 μg total RNA by PrimeScriptTM Reverse Transcriptase (TaKaRa, Shiga, Japan) following the instruction. Quantitative RT-PCR was conducted with 2 μL cDNA template in a total volume of 20 μL by TB Green® Fast qPCR Mix (TaKaRa, Shiga, Japan) using Applied Biosystems® StepOnePlusTM Real-Time PCR Systems (Thermo Fisher, Waltham, MA, USA) with specific primers of each gene (Table S1). PCR reaction for each sample was done in triplicate in 96-well plates. Gene expression was normalized to glyceraldehyde-3-phosphate dehydrogenase (GAPDH) and calculated by the comparative CT method to be expressed as fold-change compared to the blank.

4.8. Statistical Analysis

All the tcTEER values were indicated as relative TEER values by which the initial values were 100%. All values were indicated as mean ± SD. Statistical significance was determined by the U-Mann-Whitney test for the comparison between two groups and the Kruskal-Wallis test for that among groups.

5. Conclusions

In summary, metabolites produced by DSW-derived *Lactiplantibacillus plantarum* strain BF1-13 protect the intestinal epithelial barrier from the dysfunction caused by H_2O_2 treatment. It was also elucidated that the mechanism of this protective effect was achieved by both the enhancement of CLDN-4 expression and the suppression on AQP3-facililating H_2O_2 invasion. The oxidative stress induced by H_2O_2 impairs the intestinal epithelial barrier function which is mainly regulated by TJs, especially CLDN-4 [26]. The dysfunction of the intestinal epithelial barrier is correlated to the IBD, including ulcerative colitis (limited to the large intestine) and Crohn's disease (mainly affects the small intestine), which correlates with a higher risk for colorectal cancer [27]. For the people who are at risk of IBD that could be triggered by both genetic susceptibility and environmental exposure [28], the daily intake of the food fermented by the strain BF1-13 could help them with maintaining the intestinal epithelial barrier function to lower the risk. For the patients who are suffering the prolonged inflammation, metabolites produced by the strain could be the supplements to support them in the daily diet or help them recover from the surgical treatment.

Supplementary Materials: The following supporting information can be downloaded at: https://www.mdpi.com/article/10.3390/md20020087/s1, Table S1: Sequences of primers (human) used for quantitative RT-PCR.

Author Contributions: Conceptualization, X.D. and K.Y.; methodology, X.D. and K.Y.; software, X.D. and K.Y.; validation, X.D.; formal analysis, X.D., K.Y., Y.S. and C.I.; investigation, X.D. and K.Y.; resources, K.Y., Y.S. and C.I.; data curation, X.D. and K.Y.; writing—original draft preparation, X.D.; writing—review and editing, K.Y., C.I.; visualization, X.D. and K.Y.; supervision, X.D. and K.Y.; project administration, K.Y. and C.I.; funding acquisition, X.D. All authors have read and agreed to the published version of the manuscript.

Funding: This research received no external funding.

Institutional Review Board Statement: Not applicable.

Informed Consent Statement: Not applicable.

Data Availability Statement: The data that support the administration concentration of hydrogen peroxide used in this study are openly available at http://doi.org/10.2741/3223 (accessed on 1 May 2008).

Acknowledgments: Thanks to Nomura, M. and Yamamoto, T. for giving advice on the research. Also, thanks to Terahara, T. for the instruction on the quantitative RT-PCR.

Conflicts of Interest: The authors declare no conflict of interest.

References

1. König, J.; Wells, J.; Cani, P.; García-Ródenas, C.; MacDonald, T.; Mercenier, A.; Whyte, J.; Troost, F.; Brummer, R. Human Intestinal Barrier Function in Health and Disease. *Clin. Transl. Gastroenterol.* **2016**, *7*, e196. [CrossRef] [PubMed]
2. Camilleri, M.; Madsen, K.; Spiller, R.; Van Meerveld, B.; Verne, G. Intestinal Barrier Function in Health and Gastrointestinal Disease. *Neurogastroenterol. Motil.* **2012**, *24*, 503–512. [CrossRef] [PubMed]
3. Natsuga, K. Epidermal Barriers. *Cold Spring Harb. Perspect. Med.* **2014**, *4*, a018218. [CrossRef] [PubMed]
4. Liévin-Le Moal, V.; Servin, A. Pathogenesis of Human Enterovirulent Bacteria: Lessons from Cultured, Fully Differentiated Human Colon Cancer Cell Lines. *Microbiol. Mol. Biol. Rev.* **2013**, *77*, 380–439. [CrossRef]
5. Srinivasan, B.; Kolli, A.; Esch, M.; Abaci, H.; Shuler, M.; Hickman, J. TEER Measurement Techniques For In Vitro Barrier Model Systems. *J. Lab. Autom.* **2015**, *20*, 107–126. [CrossRef]
6. Birben, E.; Sahiner, U.; Sackesen, C.; Erzurum, S.; Kalayci, O. Oxidative Stress and Antioxidant Defense. *WAO J.* **2012**, *5*, 9–19. [CrossRef]
7. Saeidnia, S.; Abdollahi, M. Toxicological and Pharmacological Concerns on Oxidative Stress and Related Diseases. *Toxicol. Appl. Pharm.* **2013**, *273*, 442–455. [CrossRef]
8. Sies, H. Role of metabolic H_2O_2 generation: Redox signaling and oxidative stress. *J. Biol. Chem.* **2014**, *289*, 8735–8741. [CrossRef]
9. Rao, R. Oxidative Stress-Induced Disruption of Epithelial and Endothelial Tight Junctions. *Front. Biosci.* **2008**, *13*, 7210–7226. [CrossRef]
10. Bienert, G.; Chaumont, F. Aquaporin-Facilitated Transmembrane Diffusion of Hydrogen Peroxide. *Biochim. Biophys. Acta* **2014**, *1840*, 1596–1604. [CrossRef]
11. Jandhyala, S.M.; Talukdar, R.; Subramanyamet, C.; Vuyyuru, H.; Sasikala, M.; Reddy, D.N. Role of the normal gut microbiota. *World J. Gastroenterol.* **2015**, *21*, 87887–88803. [CrossRef] [PubMed]
12. Ghaffar, T.; Irshad, M.; Anwar, Z.; Aqil, T.; Zulifqar, Z.; Tariq, A.; Kamran, M.; Ehsan, N.; Mehmood, S. Recent Trends in Lactic Acid Biotechnology: A Brief Review on Production to Purification. *J. Radiat. Res. Appl. Sci.* **2014**, *7*, 222–229. [CrossRef]
13. Alves de Oliveira, R.; Komesu, A.; Vaz Rossell, C.; Maciel Filho, R. Challenges and Opportunities in Lactic Acid Bioprocess Design—From Economic to Production Aspects. *Biochem. Eng. J.* **2018**, *133*, 219–239. [CrossRef]
14. Grishina, A.; Kulikova, I.; Alieva, L.; Dodson, A.; Rowland, I.; Jin, J. Antigenotoxic Effect of Kefir and Ayran Supernatants on Fecal Water-Induced DNA Damage in Human Colon Cells. *Nutr. Cancer.* **2011**, *63*, 73–79. [CrossRef]
15. Bengoa, A.A.; Iraporda, C.; Garrote, G.L.; Abraham, A.G. Kefir micro-organisms: Their role in grain assembly and health properties of fermented milk. *J. Appl. Microbiol.* **2018**, *126*, 686–700. [CrossRef]
16. Forsyth, G.; Kapitany, R.; Hamilton, D. Organic Acid Proton Donors Decrease Intestinal Secretion Caused by Enterotoxins. *Am. J. Physiol. Gastrointest. Liver Physiol.* **1981**, *241*, G227–G234. [CrossRef]
17. Seth, A.; Yan, F.; Polk, D.; Rao, R. Probiotics Ameliorate the Hydrogen Peroxide-Induced Epithelial Barrier Disruption by A PKC- And MAP Kinase-Dependent Mechanism. *Am. J. Physiol. Gastrointest. Liver Physiol.* **2008**, *294*, G1060–G1069. [CrossRef]
18. Miyauchi, E.; O'Callaghan, J.; Buttó, L.; Hurley, G.; Melgar, S.; Tanabe, S.; Shanahan, F.; Nally, K.; O'Toole, P. Mechanism of Protection of Transepithelial Barrier Function By *Lactobacillus salivarius*: Strain Dependence and Attenuation by Bacteriocin Production. *Am. J. Physiol. Gastrointest. Liver Physiol.* **2012**, *303*, G1029–G1041. [CrossRef]
19. Yoda, K.; Miyazawa, K.; Hosoda, M.; Hiramatsu, M.; Yan, F.; He, F. *Lactobacillus* GG-Fermented Milk Prevents DSS-Induced Colitis and Regulates Intestinal Epithelial Homeostasis through Activation of Epidermal Growth Factor Receptor. *Eur. J. Nutr.* **2013**, *53*, 105–115. [CrossRef] [PubMed]
20. Pohanka, M. D-Lactic Acid as A Metabolite: Toxicology, Diagnosis, And Detection. *BioMed. Res. Int.* **2020**, *2020*, 3419034. [CrossRef] [PubMed]
21. Kostinek, M.; Pukall, R.; Rooney, A.; Schillinger, U.; Hertel, C.; Holzapfel, W.; Franz, C. *Lactobacillus arizonensis* Is a Later Heterotypic Synonym of *Lactobacillus plantarum*. *Int. J. Syst. Evol. Microbiol.* **2005**, *55*, 2485–2489. [CrossRef]
22. Swezey, J.; Nakamura, L.; Abbott, T.; Peterson, R. *Lactobacillus arizonensis* Sp. Nov., Isolated from Jojoba Meal. *Int. J. Syst. Evol. Microbiol.* **2000**, *50*, 1803–1809. [CrossRef]
23. Bollag, W.; Aitkens, L.; White, J.; Hyndman, K. Aquaporin-3 in the Epidermis: More Than Skin Deep. *Am. J. Physiol. Cell Physiol.* **2020**, *318*, C1144–C1153. [CrossRef] [PubMed]
24. Rai, T.; Sasaki, S.; Uchida, S. Polarized Trafficking of the Aquaporin-3 Water Channel Is Mediated by An NH_2-Terminal Sorting Signal. *Am. J. Physiol. Cell Physiol.* **2006**, *290*, C298–C304. [CrossRef] [PubMed]
25. Blume, L.F.; Denker, M.; Gieseler, F.; Kunze, T. Temperature Corrected Transepithelial Electrical Resistance (TEER) Measurement to Quantify Rapid Changes in Paracellular Permeability. *Pharmazie* **2010**, *65*, 19–24. [PubMed]
26. Landy, J.; Ronde, E.; English, N.; Clark, S.; Hart, A.; Knight, S.; Ciclitira, P.; Al-Hassi, H. Tight Junctions In Inflammatory Bowel Diseases And Inflammatory Bowel Disease Associated Colorectal Cancer. *World J. Gastroenterol.* **2016**, *22*, 3117–3126. [CrossRef] [PubMed]
27. Nadeem, M.S.; Kumar, V.; Al-Abbasi, F.A.; Kamal, M.A.; Anwar, F. Risk of Colorectal Cancer in Inflammatory Bowel Diseases. *Semin. Cancer Biol.* **2020**, *64*, 51–60. [CrossRef] [PubMed]
28. Gu, P.; Feagins, L. Dining With Inflammatory Bowel Disease: A Review of the Literature on Diet in the Pathogenesis and Management of IBD. *Inflamm. Bowel Diseases.* **2019**, *26*, 181–191. [CrossRef]

marine drugs

Review

Microalgae, Seaweeds and Aquatic Bacteria, Archaea, and Yeasts: Sources of Carotenoids with Potential Antioxidant and Anti-Inflammatory Health-Promoting Actions in the Sustainability Era

Paula Mapelli-Brahm [1,*], Patricia Gómez-Villegas [2], Mariana Lourdes Gonda [3], Antonio León-Vaz [2], Rosa León [2], Jennifer Mildenberger [4], Céline Rebours [4], Verónica Saravia [5], Silvana Vero [3], Eugenia Vila [5] and Antonio J. Meléndez-Martínez [1,*]

1 Food Colour and Quality Laboratory, Facultad de Farmacia, Universidad de Sevilla, 41012 Sevilla, Spain
2 Laboratory of Biochemistry, Faculty of Experimental Sciences, Marine International Campus of Excellence and REMSMA, University of Huelva, 21071 Huelva, Spain; patricia.gomez@dqcm.uhu.es (P.G.-V.); antonio.leon@dqcm.uhu.es (A.L.-V.); rleon@uhu.es (R.L.)
3 Área Microbiología, Departamento de Biociencias, Facultad de Química, Universidad de la República, Gral Flores 2124, Montevideo 11800, Uruguay; mgonda@fq.edu.uy (M.L.G.); svero@fq.edu.uy (S.V.)
4 Møreforsking AS, Borgundveien 340, 6009 Ålesund, Norway; jennifer.mildenberger@moreforsking.no (J.M.); celine.rebours@moreforsking.no (C.R.)
5 Departamento de Bioingeniería, Facultad de Ingeniería, Instituto de Ingeniería Química, Universidad de la República, Montevideo 11300, Uruguay; vsaravia@fing.edu.uy (V.S.); mvila@fing.edu.uy (E.V.)
* Correspondence: pmapelli@us.es (P.M.-B.); ajmelendez@us.es (A.J.M.-M.)

Abstract: Carotenoids are a large group of health-promoting compounds used in many industrial sectors, such as foods, feeds, pharmaceuticals, cosmetics, nutraceuticals, and colorants. Considering the global population growth and environmental challenges, it is essential to find new sustainable sources of carotenoids beyond those obtained from agriculture. This review focuses on the potential use of marine archaea, bacteria, algae, and yeast as biological factories of carotenoids. A wide variety of carotenoids, including novel ones, were identified in these organisms. The role of carotenoids in marine organisms and their potential health-promoting actions have also been discussed. Marine organisms have a great capacity to synthesize a wide variety of carotenoids, which can be obtained in a renewable manner without depleting natural resources. Thus, it is concluded that they represent a key sustainable source of carotenoids that could help Europe achieve its Green Deal and Recovery Plan. Additionally, the lack of standards, clinical studies, and toxicity analysis reduces the use of marine organisms as sources of traditional and novel carotenoids. Therefore, further research on the processing of marine organisms, the biosynthetic pathways, extraction procedures, and examination of their content is needed to increase carotenoid productivity, document their safety, and decrease costs for their industrial implementation.

Keywords: bioactives; natural pigments; antioxidant activity; marine organisms; human health; blue economy; agro-food; functional foods; marine resources; cosmetics

Citation: Mapelli-Brahm, P.; Gómez-Villegas, P.; Gonda, M.L.; León-Vaz, A.; León, R.; Mildenberger, J.; Rebours, C.; Saravia, V.; Vero, S.; Vila, E.; et al. Microalgae, Seaweeds and Aquatic Bacteria, Archaea, and Yeasts: Sources of Carotenoids with Potential Antioxidant and Anti-Inflammatory Health-Promoting Actions in the Sustainability Era. *Mar. Drugs* **2023**, *21*, 340. https://doi.org/10.3390/md21060340

Academic Editor: João Carlos Serafim Varela

Received: 8 May 2023
Revised: 25 May 2023
Accepted: 27 May 2023
Published: 1 June 2023

1. Introduction: Sustainability, the Blue Economy, and the Versatility of Carotenoids in Agro-Food and Health

It is estimated that 40% of the global land surface is used by agriculture and that 30% of global greenhouse gas emissions and 70% of freshwater use are related to food production. Besides, 820 million people have insufficient food, 2000 million adults are overweight or obese, and the global prevalence of diabetes has doubled since 1990 [1]. Thus, research on food science, technology, and related fields such as agriculture, aquaculture, and nutrition must be directed toward the sustainable production of health-promoting foods to combat diet-related diseases and preserve key resources for future generations. The synergistic work carried out in large related networks (CYTED-IBERCAROT, CYTED-MICROAGRO, COST-EUROCAROTEN, COST-SEAWHEAT, COST-OCEAN4BIOTECH, CaRed, etc.) is important to orchestrate efforts and produce breakthroughs.

The rapid growth of the global population requires new solutions to ensure food security, which cannot rely exclusively on land-based agriculture; hence, the importance of tapping further into the vast aquatic biodiversity. The blue economy of the EU is fundamental for the European Green Deal and the Recovery Plan for Europe, which will define the European economy for many years, as the ocean is the main climate regulator, offers clean energy, and provides us with oxygen and food, among other resources. The Blue Economy includes all those activities that are marine-based or marine-related and encompasses sectors including living and non-living resources or renewable energy, among others. Important marine species featured in the blue bioeconomy are algae (microalgae and seaweeds), bacteria, fungi, and invertebrates. Each year, hundreds of untapped new compounds derived from these organisms are discovered, including carotenoids and their derivatives. The applications of their biomasses in agro-food health range from foods, supplements, cosmetics, and feeds to biostimulants and fertilizers. They can also be used for biomaterials, bioremediation, or biofuels [2]. Regarding the carotenoids, the Action COST EUROCAROTEN (www.eurocaroten.eu (accessed on 30 May 2023)) highlighted that apart from marine sources, understudied carotenoids (the great majority, as only 10–20 carotenoids are being studied in depth) offer many possibilities for innovation in agro-food, health, cosmetics, or biomaterials [3].

Algae and microbes (as well as insects) are highlighted as matrices that can be further harnessed to produce innovations in the context of healthy diets from sustainable food systems in the EAT-Lancet Commission report [1]. Microalgae and macroalgae received much interest as biofuel feedstocks in response to the uprising energy crisis, climate change, and the depletion of natural sources. Concomitantly, high-value co-products, including carotenoids, have been produced through the extraction of a fraction of algae to improve the economics of a microalgae biorefinery. This is the process of obtaining biofuels, energy, and diverse high-value products through biomass transformation and process equipment [4,5]. The advantages of using algae biomass as feedstock to produce biofuels and co-products compared to higher plants are several: higher growth and productivity due to their all-year production capability; growth under stress conditions and lower nutritional and water requirements; no requirements of herbicides or pesticides; contribution to CO_2 sequestration and wastewater bioremediation; easier modulation of the biosynthesis of valuable co-products by modifying growth conditions [5–7]. It is no wonder that the importance of algae (either as whole cells or for the extraction of high-value components) for innovation in agro-food was highlighted in recent reports from the European Commission and the Food and Agriculture Organization (FAO) [8,9]. More than 20 genera of cyanobacteria and microalgae are currently used for food or feed applications. This number is expected to increase considerably over the next few years as research on such microalgal applications grows incessantly [8]. Regarding macroalgae, some species are exploited and used for human consumption in Europe, particularly in France, Spain, and Ireland, where several companies harvest edible seaweed. These new types of industries have recently been developed following the increasing demand from European consumers [10]. Recent reviews highlight the incorporation of algae (either whole cells or extracts) into a wide variety of foods (emulsions, vegetarian gels, dairy products, cookies, bread, and pasta) with potential health-related and even technofunctional benefits [6,11].

Being microbes, bacteria share biotechnological and sustainability advantages with microalgae. Indeed, they are similarly important for alleviating the greenhouse effect, as they can also assimilate CO_2 [12]. Bacteria have long been used for food production in diverse types of fermentation to obtain food components (including carotenoids, as discussed later) [13,14] and in the last decades, they have attracted great interest in the context of the development of probiotics and the role of the microbiome on human health [15–17]. Currently, there is a growing trend to look for innovative applications in aquaculture [18].

Marine archaea offer many opportunities for innovation as they produce unique metabolites as a result of their adaptation to extreme environments. Examples are the C45 carotenoid dihydroisopentenyldehydrorhodopin, the C50 carotenoid bacterioruberin, or the glycoside esters thermozeaxanthin and salinixanthin (Figure 1), whose structures (and therefore their physico-chemical properties) differ greatly relative to typical dietary carotenoids [19].

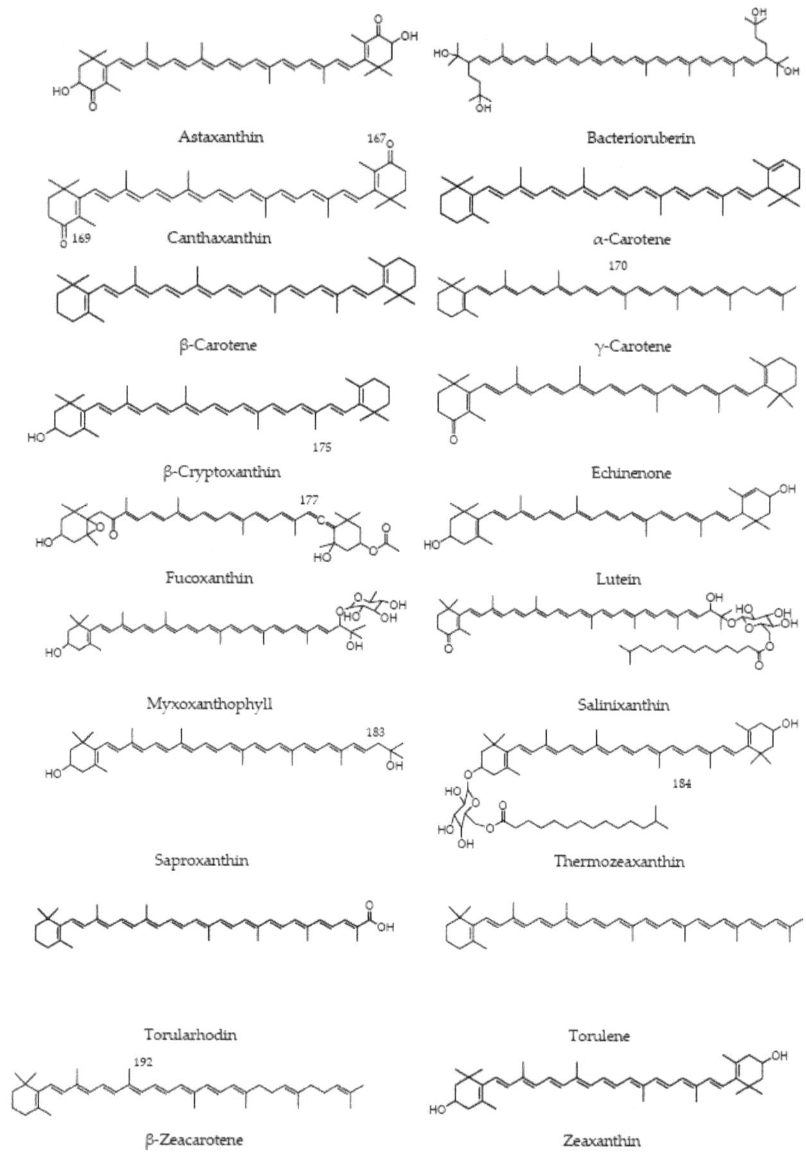

Figure 1. Chemical structures of the main carotenoids present in marine organisms.

Yeasts are ubiquitous. Marine yeasts were first discovered in the Atlantic Ocean, and since then they have been isolated from seawater, marine deposits, seaweed, fish, marine mammals, and sea birds. They have also been found in benthic animals and seafloor sediment at depths up to 11,000 m. They are more abundant in nearshore habitats (10–1000 cells/L of water), whereas in low organic surface to deep-sea oceanic regions, they are found at densities <10 cells/L [20]. Some of the many marine yeast genera are *Rhodotorula*, *Rhodosporidium*, *Candida*, *Cryptococcus*, *Torulopsis*, and *Saccharomyces* [21].

The ability of algae, bacteria, and yeasts to use diverse products for growing (especially agriculture and aquaculture wastes or even flue gases in the case of algae) makes them very attractive in the context of the circular economy [22–31].

Carotenoids are unsaturated compounds that can be classified into two types: carotenes, which are hydrocarbons, and xanthophylls, which present one or more functional groups containing oxygen [32]. Beyond their long-known roles as natural colorants and (some of them) precursors of vitamin A, a large body of evidence indicates that they (or their derivatives, commonly known as apocarotenoids) may be involved in health-promoting biological actions, contributing to reducing the risk of non-communicable diseases as well as attracting renewed interest in cosmetic products [33–35]. According to the last update of the carotenoid database on 1 November 2020 (http://carotenoiddb.jp, accessed on 30 May 2023), 680 natural carotenoids from different eukaryotic organisms have been described, including animals, plants, algae, filamentous fungi, and yeasts. Various marine organisms can accumulate carotenoids and represent a route of entry for these compounds into the food chain (Figure 1, Tables 1–5). Thus, they can be considered a natural and renewable source for carotenoid production. However, to date, the industrial production of carotenoids has been based primarily on chemical synthesis [29]. Their extraction from marine organisms cannot compete economically with chemical synthesis, due to the high costs associated with their production and with downstream extraction and purification processes. However, it should be taken into account that obtaining the same isomeric mixtures as those found in natural sources could increase the cost of the production process. For example, in the case of astaxanthin, a mixture of stereoisomers is generally obtained by chemical synthesis, while astaxanthin from natural sources is mainly represented by a single isomer and has higher antioxidant activity and stability [36]. In any case, current trends in natural products have renewed interest in the use of microbial carotenoids as an alternative to those chemically produced [37]. This preference is mainly based on safety concerns about the hazardous waste and intermediate products that could be produced in chemical synthesis. The obtention of carotenoids from marine organism cultures is considered a sustainable process by which carotenoids are produced by milder and more environmentally friendly procedures. The extraction of carotenoids from marine organisms also has advantages over the extraction from plants. They have higher growth rates than vegetables, and their production is for most of these organisms not as affected by seasons or geographic variabilities. Most of them are reasonably easy to cultivate and can be produced in large quantities within a limited space. Moreover, some marine organisms, as a source of carotenoids, can be cultivated on agro-industrial waste or by-products in processes supporting the concept of a circular economy.

This review assesses the potential use of marine organisms as a sustainable source of carotenoids, including well-known and novel carotenoids, as a fundamental part of the European Green Deal and its Recovery Plan. In addition, the different challenges that need to be addressed to obtain new carotenoids on an industrial scale are discussed. The possible antioxidant and anti-inflammatory actions of marine carotenoids are also discussed.

2. Main Marine Organisms Containing Carotenoids

2.1. *Archaea*

2.1.1. Ecological Importance

Archaea constitute a domain of prokaryotic organisms with riveting characteristics and high ecological importance. This domain was introduced by Carl Woese in 1990 [38] and originally included only extremophilic microorganisms, concretely methanogens, thermophiles, and halophiles. However, members of the domain Archaea are now known to be widespread in non-extreme habitats, including aquatic and terrestrial environments.

In marine ecosystems, planktonic archaea are classified into four distinct groups designated Marine Group (MG) from I to IV, with a growing number of phylum, class, order, and family designations [39]. *Thaumarchaeota* (MGI) have an important role in marine biogeochemistry as they perform the oxidation of ammonia to nitrite, the first step of nitrification. In addition, thaumarchaea are capable of synthesizing cobalamin, the precursor of vitamin B12, which is required by other planktonic members that are not capable of producing it [40]. The MGII archaea are a diverse group of uncultivated heterotrophic and photoheterotrophic microbes with a potential role in the ocean carbon cycle because they are able to break high-molecular-weight compounds. Finally, the MGIII and MGIV archaea are usually found in the deep sea, and so far, little information about them is available [41].

Archaeal communities are also present in freshwater environments, where their main biogeochemical role is related to methanogenesis and ammonium metabolism [42]. Methanogenic archaea, including *Methanobacteriales*, *Methanosarcinales*, and *Methanomicrobiales*, are usually found in anaerobic water and sediments, while ammonia-oxidizing archaea, such as *Thaumarchaeota* and *Nitrosotalea-like*, are often present in oligotrophic surface freshwaters. In addition, some potential genome-based roles, including fermentative metabolism or symbiosis, are being studied in *Woesearchaeota*, the main group found in boreal lakes [43].

The most studied archaea are extremophiles distributed in aquatic environments under severe conditions. Halophiles inhabit salt lakes, soda lakes, and salterns, while hyperthermophiles live in hot springs and shallow or deep hydrothermal systems. Among them, methanogens are represented in halophiles, thermophiles, psychrophiles, acidophiles, and alkaliphiles [44]. Many hyperthermophilic archaea are sulfur-chemolithoautotrophs (e.g., *Sulfolobus*), and complete denitrification is usual in some families of haloarchaea such as *Haloarculaceae* and *Haloferacaceae* [45].

2.1.2. Sustainability Pros and Cons

Archaea are relevant microorganisms for the sustainable production of carotenoids. Currently, bacterioruberin is one of the few products commercialized from archaea [46], and its bioprocess production ensures a low environmental footprint by re-using raw materials in the fermentation cycles [47]. From a biotechnological point of view, the extreme conditions in which haloarchaea proliferate reduce the risk of culture contamination by other microorganisms and allow their cultivation under non-sterile conditions with reduced cultivation costs. Additionally, given that haloarchaea undergo cell lysis at low concentrations of salt, carotenoids can be easily recovered under mild conditions with lower costs than from other microorganisms that require harsh cell disruption methods. Finally, haloarchaea show an advantage over other microorganisms as a source of bioactive products that can be used in personal care products, drugs, or foods, as they are non-pathogenic for animals and humans [19].

2.1.3. Main Species (and Compounds) in the Context of Agro-Food and Health

Archaea are a valuable source of unique carotenoids with promising applications in cosmetics, food, pharmaceuticals, and nutraceuticals. To date, 25 different carotenoids and apocarotenoids have been described in Archaea, 13 of which are only present in this domain. Most archaeal carotenoids are synthesized by haloarchaea, with bacterioruberin being the most abundant [19] (Table 1, Figure 1).

Table 1. Examples of carotenoid content in marine archaea.

Species	Carotenoids	Collection Sites	Experimental Conditions	Quantification/ Identification Methodologies	Contents	References
Halobacterium halobium	Total carotenoids (mainly bacterioruberin)	Saltern brine, Sfax, Tunisia	Liquid culture (10 g/L yeast extract, 7.5 g/L casamino acids, 250 g/L NaCl, 20 g/L $MgSO_4.7H_2O$, 2 g/L KCl, and 3 g/L trisodium citrate), 37 °C, 240 rpm, 7 days	Spectrophotometry	5.66–7.63 mg/L, FW	[48]
Haloarcula japonica	Total carotenoids (mainly bacterioruberin)	Saltern soil	Liquid culture, 37 °C, 10 days	HPLC-MS	335 µg/g, DW. Bacterioruberin was up to 68.1% of the total carotenoids (mol%)	[49]
Halorubrum sp.	Total carotenoids (mainly bacterioruberin)	Saltern brine	Liquid culture, 37 °C	Spectrophotometry	25 mg/L, FW	[50]
Haloterrigena turkmenica	Total carotenoids (mainly bacterioruberin)	Salt soil crust	Liquid culture, 37 °C, 180 rpm	Spectrophotometry	74.5 µg/g, DW	[51]
Halorubrum sp.	Total carotenoids (mainly bacterioruberin)	Seven distinct saline habitats, Algeria	Liquid culture, 37 °C	Spectrophotometry	3.68 mg/L, FW	[52]
Haloarcula sp.	Total carotenoids (mainly bacterioruberin)	Tebenquiche Lake of the Atacama Saltern, Chile	Liquid culture (5 g/L proteose-peptone, 10 g/L yeast extract, 1 g/L glucose, with 25% (*w*/*v*) total salts), 40 °C, 120 rpm, 10 days	HPLC-MS	488.88–871.53 mg/g, DW	[53]
Haloferax mediterranei strain R-4 (ATCC33500)	Total carotenoids (mainly bacterioruberin)	Salt pond	Liquid culture, 36.5 °C	Spectrophotometry	92.2 µg/mL, FW	[54]

DW, dry weight; FW, fresh weight; HPLC, high-performance liquid chromatography; MS, mass spectrometry.

Bacterioruberin is an unusual C50 carotenoid whose function is related to UV protection and membrane stability. The complete biosynthetic pathway of bacterioruberin from lycopene was elucidated in *Haloacula japonica* by Yang et al. [55], and intermediates such as monoanhydrobacterioruberin and bisanhydrobacterioruberin are usually present in different proportions in haloarchaeal extracts [49–51]. Bacterioruberin possesses 13 carbon double bonds and 4 hydroxyl groups at the terminal ends that confer it a high antioxidant potential, even greater than the reference antioxidant β-carotene under certain conditions. Several studies have demonstrated the antioxidant activity of bacterioruberin-rich extracts [51,56], and some others have also evaluated the different bioactivities of haloarchaeal pigments. In this context, carotenoids extracted from *Halobacterium halobium*, *Halogeometricum limi*, and *Haloplanus vescus* have shown antiproliferative activities in the human cancer cell line HepG2 (hepatocarcinoma) [48,57]. In addition, carotenoids from *Natrialba* have shown antiviral activity against viral hepatitis (HCV and HBV) and potent antiproliferative and apoptotic activities against colon, breast, liver, and cervical cancer cells, as well as being safer than the standard anticancer drug 5-FU for normal human cells under certain conditions [58]. Furthermore, bacterioruberin-rich extracts from various species of haloarchaea, including the genera *Haloarcula*, *Halobacterium*, *Halorubrum*, and *Haloferax*, have displayed diverse bioactive properties in vitro, such as antibacterial, neuroprotective, anti-inflammatory, antiglycemic, and antilipidemic activities [52–54,59]. Finally, bacterioruberin-enriched extracts from a genetically modified *Haloferax volcanii* strain were shown to improve the quality of cryopreserved ram sperm cells, suggesting their potential application to increase insemination yields [60].

2.1.4. Production Approaches and Production Data in Different Continents

The extreme growth conditions required for extremophilic archaea that synthesize carotenoids have made the development of laboratory strategies for their culture and the production of biocompounds difficult. Studies on the genus *Haloferax* revealed that the bacterioruberin synthesis can be enhanced by adding carbon sources (glucose or starch) to the medium [54] and also by using a culture medium with lower salt (less than 15% *w*/*v*) [61].

Thus, bacterioruberin production from *Haloferax* has been proposed in a two-step process, using a first step with a high salt concentration (20%, *w/v*) for biomass production and a second step with lower salt concentrations for the accumulation of carotenoids [62]. However, other species of the genera *Halorubrum* and *Natriabla* required a salt concentration of 20–25% *w/v* for the maximum production of carotenoids [50,52,58], and therefore the process could be performed in a single step. Moreover, many other physicochemical parameters have to be taken into account to optimize the carotenoids produced by haloarchaea. In this regard, diverse species of haloarchaea cultured with a nearly neutral pH, a temperature of 30 to 40 °C, high light exposure, and high oxygen availability have experimented with an increase in growth rate and carotenoid productivity. It has also been demonstrated that the amount of $MgSO_4$, KCl, and trace elements may influence the production of bacterioruberin [63].

2.1.5. Current and Prospective Applications

Archaeal carotenoids offer a wide range of potential applications in foods, health promotion, and cosmetics, for instance, as nutraceuticals, colorants, or food preservatives. Given the antioxidant and photoprotective properties of bacterioruberin, the possibility of using it as a natural additive for foods and personal care products is quite realistic. The haloarchaeal carotenoid bacterioruberin is commercially produced as "Halorubin" by the company Halotek (Leipzig, Germany) [46].

Despite the bioactive properties described for bacterioruberin, its chemical lability, poor water solubility, and low bioavailability limit many of its potential health-promoting uses. Therefore, the use of nanovesicles (archaeosomes or nanoarchaeosomes) and nanoparticles has been proposed to improve the stability and biological activities of bacterioruberin [64]. In the same way, the encapsulation of archaeal carotenoids into two oil-in-water dispersions has been reported as a useful system for the delivery and protection of carotenoids for food applications [65].

2.1.6. Research Needs

Carotenoids from Archaea have shown considerable potential for health-promoting effects. Nonetheless, the small number of studies on natural products from Archaea compared to Bacteria and Eukarya suggests that this field of research is still relatively underexplored. Considering the extreme environments in which extremophilic archaea reside, it can be assumed that they contain many new molecules of biotechnological interest whose identification is still pending.

The biotechnological applications of bacterioruberin contribute to the increasing demand for this carotenoid. Nowadays, commercial bacterioruberin extracted from lab-scale cultivation of *H. salinarum* can be purchased at a price point of 553 EUR/mg (CAS: 53026-44-1) [46]. Thus, more studies on the biosynthetic pathways of carotenoids and cultivation methods are needed to obtain the maximum yield of growth and pigment production, as well as the implementation of large-scale production systems with reduced costs.

Genetic engineering offers an interesting way to enhance the biosynthesis of high-value-added products. Increasing bacterioruberin biosynthesis through the genetic manipulation of haloarchaea is now possible due to recent advances in the knowledge of carotenogenesis and the sequencing of the genomes of haloarchaea belonging to different genera. However, this approach has only been tested in some species [60]. Genetic manipulation of the carotenoid biosynthetic pathway by overexpression, edition, or knockout of specific genes of the pathway should be explored in more depth to overcome the low carotenoid concentration observed in many species.

Finally, it is worth mentioning that most studies testing the bioactive properties of bacterioruberin have been performed in vitro, so it is necessary to study whether these results can be extrapolated in vivo. Moreover, the toxicity of bacterioruberin should be thoroughly analyzed in animal and human models, and the development of efficient and stable transport systems must be achieved.

2.2. Bacteria

2.2.1. Ecological Importance

The bacterial communities inhabiting marine environments vary from coastal to offshore areas, as in the water column, due to differences in physico-chemical and biological properties to which they are exposed. Depending on their energy and carbon sources, bacteria can be phototropic or non-phototropic.

Carotenoids are produced in all phototrophic bacteria (Table 2), and their biological function is to support photosynthesis as accessory pigments, protect the photosynthetic machinery from excess light that induces photodamage, and quench reactive oxygen species (ROS) [66–68]. There are four main types of bacterial phototrophs in marine microbial communities: (i) Cyanobacteria, which are mainly represented in marine environments with the genera *Synechococcus* and *Prochlorococcus* as one of the main contributors to planktonic systems [69]; (ii) anaerobic anoxygenic phototrophic bacteria such as purple sulfur bacteria, purple non-sulfur bacteria, and green sulfur bacteria, which are generally restricted to oxygen-free habitats, such as sediments and mats [67,70–72]; (iii) aerobic anoxygenic phototrophic bacteria, which includes members belonging to either α-, β-, or γ-proteobacteria, and inhabit the upper layer of the ocean, representing up to 15% of the total bacterial community [72]; and (iv) proteorhodopsin-containing bacteria. It is estimated that as much as half of the surface ocean bacteria could be part of them, but functional evidence is still scarce [69,72–74].

On the other hand, the widespread presence of carotenoids in non-phototrophic bacteria indicates their importance for their viability (Table 2). α-proteobacteria and γ-proteobacteria, together with members of the Bacteroidetes phylum, are the most abundant groups of heterotrophic bacteria in the sea [73]. Due to the absence of photosynthetic apparatus, the role of carotenoids in organoheterotrophs lies mainly in protection from photooxidative damage and in the regulation of membrane fluidity [75,76]. The production of carotenoids is proposed to be an adaptive response to certain environmental and stressor conditions, such as freeze–thaw cycles and solar radiation [77].

2.2.2. Sustainability Pros and Cons

Marine bacteria could be a promising alternative for the development of sustainable processes to produce carotenoids. As with other microorganisms, the culture systems used to grow bacteria are independent of seasonal restrictions, have low utilization of space, and do not require the use of arable land. Even when more research is needed to improve bacteria as a carotenoid source, novel strategies aligned with the development of sustainable processes could be a niche for the use of marine bacteria diversity. Second- and third-generation feedstocks require the maximum utilization of resources with the generation of multiple products, as in a biorefinery, to favor the global economic balance of the process. Microorganisms are known for their ability to use a wide variety of substrates. Bacteria present certain advantages, such as a high growth rate, metabolic diversity, and easier genetic manipulation [78]. Even though up-to-date scare reports include marine bacteria, promising research is being carried out. Kristjansdottir et al. proposed that *Rhodothermus marinus* has potential for its application in biorefineries since it is a thermophile that produces various biomass-degrading enzymes and, among other interesting products, produces carotenoids [79]. Park et al. isolated *Vibrio* sp. SP1 from a starfish for its ability to grow from alginate, and after introducing genes encoding enzymes to produce lycopene and β-carotene, it could grow on brown macroalgae *Sargassum* and produce lycopene [80].

2.2.3. Main Species (and Compounds) in the Context of Agro-Food and Health

Overall, the carotenoids most demanded for their importance in agro-food and health that are produced by bacteria are β-carotene, β-cryptoxanthin, zeaxanthin, astaxanthin, and canthaxanthin (Figure 1). In Table 2, a list of the species and main carotenoids produced is presented.

Table 2. Examples of carotenoid content in marine bacteria and cyanobacteria.

Organisms	Species	Carotenoids	Collection Sites	Experimental Conditions	Quantification/Identification Methodologies	Contents	References
Bacteria	*Rubritalea squalenifaciens* (DSM 18772[T])	Diapolycopenedioc acid xylosylesters	Miura Peninsula, Kanagawa, Japan	Liquid culture (1.0% starch, 0.4% yeast extract, and 0.2% peptone in seawater), 30 °C, 120 rpm, 2 days	Carotenoids purification by chromatography and identification of the structures by HRESI-MS and spectroscopic analyses	2.2–10.2 mg purified from cells in a 42 L culture	[81]
	Planococcus maritimus strain iso-3	Methyl 5-glucosyl-5,6-dihydro-apo-4,4′-lycopenoate	Clyde estuary, UK	Marine broth 2216), 30 °C, 120 rpm, 1 day	Carotenoids purification by chromatography and identification of the structures by HRESI-MS and spectroscopic analyses	2.5 mg purified from cells in an 18 L culture	[81]
	04OKA-13-27 (MBIC08261)	(3*R*)-saproxanthin	Coast of Okinawa Prefecture, Japan	Marine Broth 2216, 30 °C, 100 rpm, 1 day	Carotenoids purification by chromatography and identification of the structures by MS, ^{1}H-NMR, and CD analyses	0.3 mg purified from cells in a 2 L culture	[81]
	YM6-073 (MBIC06409)	(3*R*,2′*S*)-myxol	Coast of Okinawa Prefecture, Japan	Marine broth, 30 °C, 100 rpm, 1 day	Carotenoids purification by chromatography and identification of the structures by MS, ^{1}H-NMR, and CD analyses	0.5 mg purified from cells in a 2 L culture	[81]
	Erythrobacter rubeus sp. nov.	Keto-spirilloxanthin	Hamdeok Beach, Jeju Island, Republic of Korea	Marine agar 2216, 25 °C	-	-	[82]
	Flavobacterium sp. P8	Zeaxanthin	King George Island, Antarctica	2 L medium (7 g/L peptone, 7 g/L yeast extract, and 15 g/L NaCl) in a 5 L-bioreactor, 20 °C, pH 7, 1 vvm airflow, 20% pO_2, 72 h	HPLC	2.15 ± 0.15 mg/L, DW	[83]
	Brevundimonas scallop	Astaxanthin, 2,2′-dihydroxy-astaxanthin, and 2-hydroxy-astaxanthin	Nanao Island of Guangdong Province, China	100 mL LB medium, 25 °C, 150 rpm, 3 days	Spectrophotometry	104.29 ± 4.98 µg/g astaxanthin, 1068.46 ± 52.42 µg/g 2,2′-dihydroxy-astaxanthin, and 130.36 ± 6.11 µg/g 2-hydroxy-astaxanthin, DW	[84]
	Erythrobacter citreus LAMA 915	Adonixanthin, caloxanthin, canthaxanthin, erythroxanthin sulfate, nostoxanthin, and zeaxanthin	Atlantic Ocean (30° S, 36° W)	Soluble fish hydrolysate 0.5%	Spectrophotometry	1.5 mg/L total carotenoids, FW	[85]
	Erythrobacter flavus KJ5	Caloxanthin, caloxanthin sulfate, caloxanthin sulfate isomer, β-carotene, β-carotene *cis* isomer, β-cryptoxanthin, nostoxanthin sulfate, nostoxanthin isomer, zeaxanthin, zeaxanthin isomer, zeaxanthin sulfate, and zeaxanthin sulfate *cis* isomer	Hard coral Acropora nasuta, Karimunjawa Islands, Indonesia	Shioi broth, 28 °C, under dark conditions, 48 h	-	-	[86]
	Planococcus sp. Eg-Natrun	Astaxanthin and β-carotene	Umm Risha Lake, Wadi El-Natrun, south of Al-Buhayrah province, Egypt	E1 rich marine medium (C/N~3.95), 37 °C, 150 rpm, 3 days	Spectrophotometry	1024 ± 53 µg/g, DW	[87]

Table 2. *Cont.*

Organisms	Species	Carotenoids	Collection Sites	Experimental Conditions	Quantification/Identification Methodologies	Contents	References
	Brevundimonas sp. strain N-5	Astaxanthin	Shimoda Port, Shizuoka Prefecture on the Pacific coast of Middle Japan	NB broth, 150 rpm, 2 days	HPLC	601.2 μg/g total carotenoids, 364.6 μg/g astaxanthin (3*S*, 3′*S*), DW	[88]
	Paracoccus sp. strain N-81106	Astaxanthin	Iwate, Japan	OEG medium, 25 °C, 150 rpm, 18 h	HPLC	0.9 ± 0.1 mg/L astaxanthin,3.1 ± 0.2 mg/L total carotenoids, FW	[89]
	Vitellibacter sp. BW	Zeaxanthin and β-carotene	East and West coast of India (Tuticorin, Mandappam, Rameshwaram, and Mangalore)	Marine broth, 37 °C, 200 rpm, 5 days	HPLC	115.7 ± 5.0 mg/g total carotenoid, FW	[90]
	Arenibacter sp. 4W	Zeaxanthin and β-carotene	East and West coast of India (Tuticorin, Mandappam, Rameshwaram, and Mangalore)	Marine broth, 37 °C, 200 rpm, 5 days	HPLC	99.3 ± 21.9 mg/g total carotenoid, FW	[90]
	Paracoccus sp. LL1 (KP288668)	Astaxanthin	Lonar lake, India	Fed-batch culture in minimal medium supplemented with 20 g/L of glucose. Fermenter with a membrane of cell filtration, 1.5 vvm airflow, 300 rpm, DO > 20%	HPLC	8.51 ± 0.20 mg/L intracellular astaxanthin, 10.2 ± 0.24 mg/L extracellular astaxanthin, FW	[91]
	Paracoccus haeundaensis KCCM 10460	Astaxanthin	-	Co-culture with lactic acid bacteria. PMF medium, 25 °C, 160 rpm, 1 vvm airflow, 72 h	HPLC	821.09 ± 30.98 μg/g, DW	[92]
	Paracoccus zeaxanthinifaciens ATCC 21588	Zeaxanthin	Purchased from BCCM/LMG, Ghent University, Belgium	Culture media, 30 °C, 180 rpm, 72 h	TLC and spectrophotometry	11.63 mg/L, FW	[93]
	Paracoccus zeaxanthinifaciens ATCC 21588	Zeaxanthin	Purchased from BCCM/LMG, Ghent University, Belgium	Culture media in a bubble column reactor, 30 °C, 80 h	HPLC	13.76 ± 0.14 mg/L, FW	[94]
Cyanobacteria	*Anabaena* sp. PCC 7120	Canthaxanthin, β-carotene, echinenone, ketomyxol 2′-fucoside, and myxol 2′-fucoside	A gift from Waseda University, Japan	BG-11 medium, 26–28 °C, 110 rpm, white fluorescent light (30–40 μE m^{-2} s^{-1}), 2 weeks	HPLC	Relative molecular masses (%): 1 canthaxanthin, 62 β-carotene, 25 echinenone, 4 ketomyxol 2′-fucoside, and 8 myxol 2′-fucoside	[95]
	Anabaena variabilis IAM M-3 (=PCC 7118, ATCC 27893)	Canthaxanthin, β-carotene, echinenone, ketomyxol 2′-fucoside, 3′-hydroxyechinenone, and myxol 2′-fucoside	A gift from Waseda University, Japan	BG-11 medium, 26–28 °C, 110 rpm, white fluorescent light (30–40 μE m^{-2} s^{-1}), 2 weeks	HPLC	Relative molecular masses (%): 4 canthaxanthin, 38 β-carotene, 33 echinenone, 13 ketomyxol 2′-fucoside, 1 3′-hydroxyechinenone, and 11 myxol 2′-fucoside	[95]
	Nostoc punctiforme PCC 73102 (=ATCC 29133)	Canthaxanthin, β-carotene, echinenone, ketomyxol 2′-fucoside, and myxol 2′-fucoside	A gift from Kanazawa University, Japan	BG-11 medium, 26–28 °C, 110 rpm, white fluorescent light (30–40 μE m^{-2} s^{-1}), 2 weeks	HPLC	Relative molecular masses (%): 13 canthaxanthin, 45 β-carotene, 17 echinenone, 13 ketomyxol 2′-fucoside, and 11 myxol 2′-fucoside	[95]

Table 2. *Cont.*

	Anabaena variabilis ATCC 29413 (=IAM M204)	Canthaxanthin, β-carotene, echinenone, 4-hydroxymyxol, and myxol	A gift from Waseda University, Japan	BG-11 medium, 26–28 °C, 110 rpm, white fluorescent light (30–40 μE m^{-2} s^{-1}), 2 weeks	HPLC	Relative molecular masses (%): 22 canthaxanthin, 51 β-carotene, 20 echinenone, 2 4-hydroxymyxol, and 5 myxol	[96]
	Synechocystis sp. PCC 6803	β-carotene, deoxymyxol 2′-dimethyl-fucoside, echinenone, 3′-hydroxyechinenone, myxol 2′-dimethyl-fucoside, and zeaxanthin	A gift from Waseda University, Japan	BG-11 medium supplemented with 10 mM HEPES buffer (pH 7.5) in 10 L carboys, 30 °C with aeration and irradiation (40–100 E/m^2·s)	HPLC	Relative molecular masses (%): 26 β-carotene, 1 deoxymyxol 2′-dimethyl-fucoside, 18 echinenone, 4 3′-hydroxyechinenone, 36 myxol 2′-dimethyl-fucoside, and 14 zeaxanthin	[97]

CD, circular dichroism; DW, dry weight; FW, fresh weight; ^{1}H-NMR, proton nuclear magnetic resonance; HPLC, high-performance liquid chromatography; HRESI-MS, high-resolution electrospray ionization-mass spectroscopy; MS, mass spectrometry; TLC, thin-layer chromatography.

Among Bacteroidetes, Flavobacteriaceae is an important family with several species that have been reported to produce carotenoids. Marine *Flavobacterium*, *Chryseobacterim*, *Formosa*, *Vitellibacter*, and *Arenibacter* are genera with species known for producing zeaxanthin, β-cryptoxanthin, and β-carotene [98–100].

Regarding Alphaproteobacteria, the genera *Paracoccus* and *Brevindimonas* have been widely studied for their capacity to produce carotenoids. *Paracoccus haeundaensis* sp. and *Paracoccus* sp. LL1 (KP288668) were reported to produce astaxanthin [91,101], while *Paracoccus zeaxanthinifaciens* produces mainly zeaxanthin [93,102]. *Brevundimonas scallop* and *Brevundimonas* sp. N-5 produce astaxanthin [84,88].

Regarding photosynthetic bacteria, several members are reported as carotenoid producers. Members of *Anabaena*, *A. variabilis* ATCC 29413, *A. variabilis* IAM M-3, and *Anabaena* sp. PCC 7120, were reported for their ability to produce β-carotene, echinenone, and canthaxanthin (Figure 1) [103]. *Synechococcus* species were studied for the production of carotenoids: zeaxanthin, cryptoxanthin, myxoxanthophyll (myxol-2′-fucoside), echinenone, 3′-hydroxyechinenone, and synechoxanthin. *Synechococcus elongatus* PCC 7942 is a cyanobacterium that synthesizes zeaxanthin as one of the predominant carotenoids [104]. Other carotenoids-producing cyanobacteria are *Nostoc punctiforme* PCC 73102, which produces β-carotene, echinenone, and canthaxanthin; *Thermosynechococcus elongatus* BP-1 produces mainly β-carotene and nostoxanthin; *Gloeobacter violaceus* PCC 7421 produces β-carotene and echinenone; and *Prochlorococcus marinus* MED4 produces β-carotene, β-cryptoxanthin, and zeaxanthin [103].

2.2.4. Production Approaches and Production Data in Different Continents

The industrial production of bacterial carotenoids is still in development. The literature reports laboratory-scale experiments and scales up to laboratory bioreactors operated as batch cultures, with alternative approaches such as fed-batch and co-culture. Most of the research refers to aquatic bacteria, but not those of marine origin. Joshi et al. studied marine *Paracoccus zeaxanthinifaciens* as an interesting strain for zeaxanthin production [93]. These authors developed a media culture using surface response methodology and an artificial neural network that allowed them to reach an overall yield of 11.6 mg/L after 72 h of culture. Vila et al. [83] formulated a media culture to produce zeaxanthin by a marine Antarctic *Flavobacterium* sp. P8 utilizing factorial designs and later used a 5 L bioreactor in batch mode to obtain 2.15 mg/L of zeaxanthin. A strategy of co-culture was studied by Choi et al. [92] in the production of astaxanthin by *Paracoccus haeundaensis* sp. This strategy, coupled with the optimization of the media culture composition, led to an increase in astaxanthin production of 2.5 times compared to *P. haeundaensis* cultivation in Luria–Bertani broth medium. Genetic engineering has been applied to improve carotenoids yields in the last few decades. Sarnaik et al. [104] developed a genetically modified strain PCC 7942 (*Synechococcus* 79R48) by homologous recombination, which improves β-carotene flux toward zeaxanthin synthesis. The modified strain improved zeaxanthin production two times relative to the wild-type strain. In addition, heterologous production of astaxanthin from marine bacterial genes was achieved in *Synechocystis* sp. PCC 6803 [105].

2.2.5. Current and Prospective Applications

Nowadays, the only bacteria commercially exploited as a carotenoid source is *Paracoccus carotinifaciens*, which has approvals as a feed product from EFSA and as an astaxanthin-rich extract to be used in food from the FDA. Although it is not of marine origin, it shows that bacteria are a viable option for carotenoid production if they fulfill scale-up requirements and economic balance. To the authors' knowledge, there are not yet any products for food or health purposes on the market. Resistance to bacteria-based products has been presented as a possible cultural obstacle to overcome [78].

Part of the ongoing research involves modifications of the carotenoid biosynthetic pathways of the well-known *Corynebacterium glutamicum* and heterologous expression in *E. coli* to increase carotenoid production [106–108]. Besides, the application of genetic manipulation tools to increase the carotenoid yields of marine carotenogenic strains is still promising [89]. Then, marine bacteria could be a cell factory [109], a source of carotenogenic genes contributing to the improvement of tools for heterologous expression, as well as a source of novel carotenoids with potential new applications [110]. On the other hand, the inclusion of carotenoids as a high-value product in the development of biorefineries in the context of sustainable processes could be promising. The diversity of marine bacteria may contribute to providing appropriate degrading properties of different feedstocks combined with the production of carotenoids.

2.2.6. Research Needs

The existence of much more diversity than that recovered by classical culture techniques from marine environments has challenged bioprospecting strategies, and new approaches incorporating heterologous expressions of metagenomic data are possible but difficult [111]. It is still necessary to isolate and culture bacteria to study the production of bioactive metabolites such as carotenoids [112], and meanwhile, improvements in culture methods are reported [113–118].

Much of the chemical identification of new carotenoids is carried out by mass spectrometry analysis, but the lack of commercial standards (for identification and quantification) and chemical data literature slows progress [35]. Besides, among the wide variety of carotenoids produced by bacteria, only a few have proven applications. New and rare carotenoids are frequently characterized by antioxidant activity and preliminary assays with cell lines, but more research is required to evaluate their real potential application [81].

Finally, there is a lack of economic, technological, and environmental assessment research to guide the progress of the bioprocess of bacterial carotenoid production. As mentioned above, integrating the carotenoid production process into a multi-product biorefinery approach could improve competition with the actual production processes.

2.3. *Macroalgae*

2.3.1. Ecological Importance

Macroalgae play an important role in coastal ecosystems, mainly as a habitat and food source [119,120]. They require certain salinity levels, sunlight, and adequate attachment substrates, although they can also float. Macroalgae are at the basis of the food chain, providing vegetal material and nutrition to other marine organisms, partly through associated microbiota [121]. They form complex structures that serve as an important shelter, support, or breeding substrate. Similarly, macroalgae vegetation is essential for the early developmental stages of many commercially interesting fish species [122,123]. Large seaweed can even contribute to protecting the shoreline from climate-imposed changes [124,125]. Due to their large surface area, macroalgae filter surrounding water, retain excessive nutrients, and mitigate pollution for the integrity of their marine environment [126]. Macroalgae can also act as carbon sinks through the uptake of atmospheric carbon dioxide and the export of dissolved organic carbon to offshore areas [7,127]. In addition, seaweeds contribute to oxygen production during the daytime due to their photosynthetic activity. Overall, macroalgae are regarded as essential in the aquatic trophic chain web, and some of the environments to which they belong are regarded as among the most productive habitats on the planet. Other ecosystem services identified are their use as ingredients in the formulation of feeds, foods, fertilizers, drugs, and other products for human consumption [5].

The photosynthetic pigments and their combinations present in seaweeds have high taxonomic importance in their classification. Green algae are a large group, of which one class is Chlorophyceae. Chlorophyceae are the closest to terrestrial green plants and contain chlorophylls a and b and carotenoids. The Phaeophyceae, which represent one class of brown algae, contain mainly chlorophyll c and the brown carotenoid pigment fucoxanthin. Rhodophyceae, or red seaweed, is a group distinguished by the presence of pigments such as chlorophyll, carotenoids, and the phycobiliprotein phycoerythrin. Carotenoids in seaweed can function as light energy harvesters (passing on light excitation to chlorophyll) and antioxidants that inactivate the ROS formed under exposure to high light and air [128,129].

2.3.2. Sustainability Pros and Cons

Concepts including sustainability and the circular economy are a prerequisite that is now associated with food production. The agro-food system is always in search of sustainable and healthy diets for a growing population [130,131]. Today, most carotenoids are chemically synthesized or extracted from land plants. Therefore, macroalgae could serve as an alternative source of carotenoids, with the prerequisite that algae production and carotenoid extraction can be sustainable. This is well represented in the literature, with a focus on induced carotenoid synthesis in green [132] or red algae [133]. Several extraction techniques have been investigated and are suggested to be faster, more sustainable, and more efficient than traditional methods to isolate bioactive compounds from algae [134]. Finally, the sustainability of seaweed farming is also yet to be demonstrated for its production in Europe [135,136].

2.3.3. Main Species (and Compounds) in the Context of Agro-Food and Health

The main bioactive compounds in macroalgae are polysaccharides, phenolic compounds, carotenoids, vitamins, minerals, and peptides [137,138]. Carotenoids are photosynthetic and bioactive compounds found in all of the seaweed divisions (Table 3), playing critical roles in photosynthesis. Seaweed carotenoids are generally localized in the chloroplast or accumulated in vesicles, the cytoplasmic matrix, or bound to membranes and other macromolecules in the intracellular space. Carotenoids in seaweeds are synthesized from pyruvate and/or acetyl-CoA. The first carotenoid in the isoprenoid route is phytoene, from which lycopene is formed after several desaturation steps. Lycopene is further transformed sequentially into compounds including α-carotene and lutein in red and green algae or β-carotene and zeaxanthin in the three types of macroalgae. Green algae further transform zeaxanthin into violaxanthin and neoxanthin, while only brown algae finally form fucoxanthin [139–141]. Furthermore, different carotenoids are synthesized in response to exposure to light and air since they scavenge ROS with their photoprotective properties [139,142].

In the marine environment, carotenoids are widespread in seaweeds (Table 3). Green seaweeds have a carotenoid profile very similar to that of green vegetables (with lutein and β-carotene as major carotenoids) along with violaxanthin, neoxanthin, and others [35,132]. In addition to fucoxanthin, Phaeophyceae are rich sources of β-carotene and violaxanthin and also contain ε-carotene, antheraxanthin, diatoxanthin, diadinoxanthin, and neoxanthin at lower concentrations. α-carotene, β-carotene, zeaxanthin, and lutein are commonly found in Rhodophyceae [143].

One very well-studied carotenoid in seaweeds is fucoxanthin, the brown pigment that colors brown seaweeds and, in particular, kelps, as well as diatoms. Fucoxanthin is one of the most abundant carotenoids, contributing around 10% of the estimated total production of carotenoids in nature [144]. While fucoxanthin from unicellular diatoms is interesting for industrial production, brown seaweed is the only dietary source of fucoxanthin. Fucoxanthin is produced industrially using macroalgae, such as *Laminaria japonica* or *Undaria pinnatifida*, although its fucoxanthin content is low (0.2–0.6% of dry weight) [145]. A survey in Japan showed that while the intake of carotenes and other xanthophylls was covered by different food sources, fucoxanthin was mainly provided by the intake of the brown seaweed wakame [146]. The fucoxanthin content will vary between different species and genera and according to environmental factors, such as geographical location and season. Furthermore, the detected content will also be influenced by the extraction method [139].

Table 3. Examples of carotenoid content in marine macroalgae.

Organisms	Species	Carotenoids	Collection Sites	Experimental Conditions	Quantification/ Identification Methodologies	Contents	References
Macroalga (brown)	*Sargassum horneri*	Fucoxanthin	Nesaki, Hokkaido (41°45′ N, 140°49′ E) and Matsushima, Miyagi (38°23′ N, 141°04′ E), northern seashore of Japan	Mariculture of collected thalli in Usujiri, Japan	HPLC	1.35–4.5 mg/g DW	[147]
	Cystoseira hakodatensis	Fucoxanthin	Usujiri, Hokkaido (41°56′ N, 140°57′ E), Japan	Natural growth in Usujiri, Japan	HPLC	0.6–4.1 mg/g DW	[147]
	15 species of brown seaweed	Fucoxanthin	Wild harvest in intertidal zones in Shinori and Nesaki, Japan	Natural growth	HPLC	0.1–3.7 mg/g DW	[148]
Macroalga (green)	*Ulva* spp.	Total carotenoids	Wild harvest in Inter- and subtidal zones in India, Egypt, China, Spain, and Chile	Natural growth	Spectrophotometry	1.25–4.6 mg/g DW	[132]
	Ulva compressa	Lutein, neoxanthin, violaxanthin, and zeaxanthin	Coasts of Mangaluru (12°45′31.7″ N 74°51′53.2″ E to 13°06′25.9″ N 74°46′ 03.6″ E), India	Natural growth	HPLC	4.7 μg/g lutein, 3.8 μg/g neoxanthin, 4.0 μg/g violaxanthin, and 3.9 μg/g zeaxanthin, DW	[149] *
	Chaetomorpha antennia	Lutein, neoxanthin, violaxanthin, and zeaxanthin	Coasts of Mangaluru (12°45′31.7″ N 74°51′53.2″ E to 13°06′25.9″ N 74°46′ 03.6″ E), India	Natural growth	HPLC	141.3 μg/g lutein, 33.3 μg/g neoxanthin, 33.7 μg/g violaxanthin, and 34.6 μg/g zeaxanthin, DW	[149]
Macroalga (red)	*Grateloupia* sp.	Lutein and zeaxanthin	Coasts of Mangaluru (12°45′31.7″ N 74°51′53.2″ E to 13°06′25.9″ N 74°46′ 03.6″ E), India	Natural growth	HPLC	166.6 μg/g lutein and 36.3 μg/g zeaxanthin, DW	[149]
	Pyropia yezoensis	α-carotene, β-carotene, lutein, zeaxanthin	Thalli of *P. yezoensis* strain U-51, maricultured at Shichigahama, Miyagi, Japan	Mariculture	HPLC	0.7 mg/g α-carotene, 1.8 mg/g β-carotene, 1.4 mg/g lutein, and 0.15 mg/g zeaxanthin, DW	[150]

DW, dry weight; HPLC, high-performance liquid chromatography. * In this reference, more species of algae have been analyzed, in addition to those indicated in the table.

2.3.4. Production Approaches and Production Data in Different Continents

Seaweed accounts for approximately 50% of the production of marine biomass by aquaculture, which is the fastest-growing component of food production with an estimated growth of >7%/year, ~3–5-fold that of agriculture (2%/year), livestock (2.6%/year), and wild fisheries (0.1%/year) [151].

Almost all the production happened in Asian countries (especially China and Indonesia). To be sustainable, the activities associated with seaweed farming require coordinated and participatory governance, regulations, and best practices. Globally, seaweed mariculture is still marginal in Europe although is expected to grow dramatically to become important in the European Blue Growth and Bioeconomy strategies [152–154]. The main exploited algae species in Europe are *Laminaria hyperborea, Laminaria digitata*, and *Ascophyllum nodosum*, which, as kelp forests, are considered among the world's most ecologically dynamic and biologically diverse habitats [155,156]. Other seaweeds such as *Alaria esculenta*, *Pyropia yezoensis*, and *Palmaria palmata* are ingredients in many European cuisines [10].

2.3.5. Current and Prospective Applications

There is an important interest in bioactive molecules with antioxidant and anti-inflammatory effects, which makes seaweed compounds interesting ingredients for functional food and cosmetic formulations. Fucoxanthin is the most studied carotenoid in seaweed and has exhibited antioxidant and anti-inflammatory effects, which are extensively reviewed [139,157–160]. Olsthoorn et al. [160] have recently published a comprehensive review in which they identify four inflammatory principles on which brown seaweeds have an effect. Fucoxanthin is effective in two of them, namely regulating ROS and innate immune responses, thus reducing pro-inflammatory cytokines induced by both [160].

Only two studies were found on clinicaltrial.gov when searching for "fucoxanthin NOT microalgae", which plan to investigate the effect of fucoxanthin supplementation in non-fatty liver disease and metabolic syndrome.

In animal studies, the administration of fucoxanthin-rich algae extracts increased antioxidative enzymes and decreased ROS in two mouse models of diabetes and colitis, respectively, and in cisplatin-induced hamsters [161–163].

Fucoxanthin treatment ameliorated histological damage and inflammatory levels of PGE_2 in the colon of mice after induced colitis and downregulated the inflammation-related enzyme cyclooxygenase-2 (COX-2) and the transcription factor nuclear factor kappa B (NF-kB) [164]. Oral and percutaneous administration of fucoxanthin reduced ear swelling in mice and limited the induction and activity of COX-2, phospholipase A2 (PLA_2), and hyaluronidase (HA) [165]. Fucoxanthin administration also reduced inflammatory cytokines (IL-1β, IL-6, and TNF-α) in an obesity mouse model [166] and by lipopolysaccharide (LPS)-induction as well as depression-associated behavior in the LPS-stimulated mice [167]. These effects have also been reviewed in light of their potential protective role in several connected pathologies, such as liver and skin damage or cardiovascular and metabolic disease [159]. Miyashita and Hosokawa have also reviewed the potential uses of fucoxanthin in the management of obesity and diabetes, including a downregulation of inflammatory adipokines [158,168] and reduced body weight and abdominal fat in overweight Japanese subjects [169].

Yang et al. summarize the studies in cell models and mice, which show the potential of fucoxanthin for therapeutic efficacy in neurodegenerative disorders [164]. Of special interest, fucoxanthin had a neuroprotective effect against cerebral ischemic/reperfusion injury through activation of the antioxidative response pathway involving nuclear factor erythroid 2-like 2 (NFE2L2/Nrf2) and heme oxygenase 1 (HMOX-1) [170]. Fucoxanthin also reduced neuroinflammation in a mouse model of Parkinson's disease [171].

Here, we only mention some in vivo studies illustrating the anti-inflammatory and antioxidative effects of fucoxanthin, as we understand these to be most relevant for its final application. An extensive body of work in vitro has been published to evaluate this and other seaweed-derived carotenoids, with some described in Section 5 [139,172,173]. There are a few additional studies with whole seaweed and mixed extracts, which also provide valuable information on the effects of dietary intake of seaweed [174,175]. During uptake, fucoxanthin is metabolized to fucoxanthinol and amarouciaxanthin A, which were found in the adipose tissue, kidney, heart, lung, and spleen of mice [176,177]. In humans, fucoxanthin intake led to a peak of fucoxanthinol in plasma of 44.2 ± 14.9 nM after 4 h, gradually decreasing within 24 h [178,179]. Therefore, the bioavailability of fucoxanthin is lower and more transient than that of other dietary carotenoids such as β-carotene and lutein [178,180]. This underlines the importance of in vivo studies that take into account the limiting steps of metabolism and uptake.

2.3.6. Research Needs

The anti-inflammatory and antioxidant effects are among the three most commonly published pharmacological properties of fucoxanthin in recent years. However, most studies are based on cell and mouse models, with only a minority focusing on clinical interventions [181]. Thus, more clinical studies are needed to confirm and fully exploit the described anti-inflammatory and antioxidant effects of this carotenoid. Compared to fucoxanthin, there is almost no literature on the potential health effects of other carotenoids derived from green or red seaweeds, including their metabolized or degradation products, which could also have beneficial effects. For example, a carotenoid degradation product, 3-hydroxy-4,7-megastigmadien-9-one, extracted from the green seaweed *Ulva pertusa*, was shown to limit the inflammatory response to microbial products in mouse innate immune cells [173].

In light of increasing interest in carotenoids from alternative sources, the production of seaweeds with a higher content of carotenoids could be one approach. Aquaculture technology would need to be optimized with a better understanding of how environmental parameters affect seaweed physiology and therefore their carotenoids content.

Finally, the economic and environmental sustainability as well as the safety of the whole value chain from seaweed to the carotenoid final product still need to be validated and documented according to the regulatory framework to ensure sustainable and safe use of this ingredient.

2.4. *Microalgae*

2.4.1. Ecological Importance

Microalgae are a heterogeneous group of photosynthetic microorganisms of high ecological importance. They are responsible for approximately half of the global CO_2 fixation, which makes them "cell factories" capable of taking up the main greenhouse gas in the atmosphere [182]. Their cultivation does not require arable land, and microalgal biomass can be produced (between 40 and 100 t/ha year of dry weight) 10–50 times faster than higher plants [183]. Additionally, most microalgae show high robustness when exposed to different types of abiotic stress, such as extreme temperatures, nutrient depletion, high salinity, or the presence of pollutants. These characteristics make it possible for microalgae to inhabit almost all ecosystems on Earth, demonstrating an enormous range of ecological adaptation capacities [184].

In addition, microalgae are at the base of the marine food chain, being the most abundant primary producers in aquatic ecosystems [185]. They are widely used as a nutritional supplement in aquaculture due to their high nutritional qualities and their potential to produce a wide range of high-value compounds, such as polyunsaturated fatty acids (PUFAs), sterols, vitamins, polyphenols, or carotenoids [186]. Some of these compounds are also synthesized for the pharmaceutical or cosmetic industry and marked as food supplements.

2.4.2. Sustainability Pros and Cons

There is an increasing interest in addressing the integral utilization of all microalgal components, following a biorefinery approach, and in integrating their production with the utilization of residual waste waters or the capture of CO_2 through a circular economy perspective to reduce the costs of microalgal biomass culturing and harvesting [187,188]. These studies are mainly focused on the removal of pollutants and the production of low-value compounds, such as biofuel, bioethanol, or animal feed. An example of this is the cultivation of *Chlorella vulgaris* and *Chlorella sorokiniana* in wastewater from milk processing, dairy wastewater, or wastewater from the wine industry that obtains large amounts of lipids, mainly palmitic, oleic, and linoleic acids [189–191]. *Arthrospira platensis* (Spirulina) was also grown in an anaerobic digestion effluent containing Cu^{2+} and Zn^{2+}, and the obtained biomass was used as feed for animals [192]. Moreover, there are some studies that describe the culture of the microalgae *Dunaliella* sp. and *Haematococcus pluvialis* in digested poultry litter or treated primary wastewater for the production of β-carotene and astaxanthin, respectively. These microalgal species are the first natural source of these highly demanded carotenoids, and their culture in wastewater can improve their feasibility [193,194].

2.4.3. Main Species (and Compounds) in the Context of Agro-Food and Health

Carotenoids are among the high-value compounds produced from microalgae, with several applications and high commercial interest. Some carotenoids, such as β-carotene, astaxanthin, lutein, or fucoxanthin, obtained from different microalgae are widely commercialized for cosmetic and pharmaceutical applications [186]. Although many species of microalgae are able to produce these compounds, only a few of them synthesize enough amounts of carotenoids to make industrial processes economically feasible (Table 4).

Dunaliella salina is one of the main producers of β-carotene and is considered the best commercial source of this naturally produced compound. This green microalga can accumulate up to 13% of its total biomass in β-carotene, making it a much better producer than other species such as *Chromochloris zofingiensis* or *Arthrospira platensis* (Spirulina), which yield up to 1–2% of their dry biomass [145]. The β-carotene synthesized by *Dunaliella salina* has exhibited health-promoting actions in experiments of different nature, such as protective effects against atherosclerosis, protection against UV-induced erythema, and oxidative damage in humans [195].

Another microalga able to produce high amounts of carotenoids in industrial processes is *Haematococcus pluvialis*. This microalga belongs, as *Dunaliella salina*, to the chlorophyte phylum and is the main producer of astaxanthin, a keto-carotenoid with high antioxidant capacity in vitro. *Haematococcus pluvialis* can produce up to 4–5% of its dry biomass as astaxanthin, obtaining higher yields than other strains capable of producing this compound, such as *Chromochloris zofingiensis* or *Chlorococcum* sp. [196]. Natural astaxanthin produced by *Haematococcus pluvialis* has previously been reported to have 10 times the in vitro antioxidative capacity of other carotenoids, such as zeaxanthin, lutein, canthaxanthin, or β-carotene, as well as anti-aging, anti-inflammatory, and anti-atherosclerotic properties [196].

Other green microalgae belonging to the *Chlorella*, *Muriellopsis*, or *Scenedesmus* genera are able to produce considerable amounts of lutein. However, its lutein content (0.5–1.2% of dry biomass) is much lower than the β-carotene or astaxanthin content in *Dunaliella* or *Haematococcus*, respectively [197]. This low productivity has restricted the industrial production of lutein from microalgae, whose main industrial source continues to be the petals of higher plants, such as marigolds [145]. However, microalgae can produce lutein in its free form, while in marigolds, lutein is mainly esterified with fatty acids [145]. One of the most promising species for the production of lutein is *Chlorella vulgaris*. This microalga can accumulate around 0.8% of the total dry weight in lutein and is one of the microalgae species allowed for human consumption [198,199]. However, further studies are necessary for the industrial production of *Chlorella vulgaris* as a producer of this health-promoting carotenoid [195].

Some marine microalgal species are also able to produce carotenoids. Diatoms, including *Phaedoactylum tricornutum*, *Cylindroteca closterium*, *Nitzschia laevis*, and *Odontella aurita*, among others, can produce outstanding amounts of fucoxanthin, up to 2.6% of dry biomass [200].

Table 4. Examples of carotenoid content in marine microalgae.

Species	Carotenoids	Collection Sites	Experimental Conditions	Quantification/ Identification Methodologies	Contents	References
Chlorella vulgaris 211/52	β-carotene, lutein	Grüental campus of the Zurich University of Applied Sciences, Wädenswil, Switzerland	Open thin-layer bioreactor	HPLC	50 μg/g β-carotene, 90 μg/g lutein, DW	[201]
Dunaliella sp. FACHB-558	β-carotene	Institute of Hydrobiology, Chinese Academy of Sciences, China	Two-step cultivation in anaerobically digested poultry litter wastewater	Spectrophotometry	7.26 mg/L, FW	[193]
Isochrysis zhangjiangensis	Fucoxanthin	Chinese Academy of Sciences, China	Photo-autotrophically cultured at 23 °C in a bubble column photobioreactors, supplemented with f/2 medium (nutrient, trace metal, and vitamin solutions)	HPLC	23 μg/g, DW	[202]
Haematococcus plivualis	Astaxanthin	Umeå University, Sweden	First step: inoculum in Bold's basal medium, OD_{750} of 0.1, 20 °C, cool white LED, 1 L/min of air, 7 days. Second step: grown in a multi-cultivator with cool white LED	HPLC	19.1 mg/g, DW	[203]
Desmodesmus sp.	Lutein and zeaxanthin	Isolated from a wastewater treatment system, Kalundborg Kommune, Copenhagen, Denmark	Industrial wastewater in Schott bottles, stirred, aerated (2% CO_2), and with fluorescent lights (200 μmol photon/m^2·s)	HPLC	5.11 mg/g lutein and 0.28 mg/g zeaxanthin, DW	[204]

Table 4. *Cont.*

Species	Carotenoids	Collection Sites	Experimental Conditions	Quantification/ Identification Methodologies	Contents	References
Nannochloropsis salina 40.85	β-carotene and violaxanthin	Algae culture collection (SAG), University of Gottingen, Germany	Industrial wastewater in Schott bottles, stirred, aerated (2% CO_2), and with fluorescent lights (200 μmol photon/m^2·s)	HPLC	2.22 mg/g β-carotene and 1.68 mg/g violaxanthin, DW	[204]
Phaeodactylum tricornutum	Diadinoxanthin and diatoxanthin	-	Industrial wastewater in Schott bottles, stirred, aerated (2% CO_2), and with fluorescent lights (300 μmol photon/m^2·s)	HPLC	2.17 μg/g diadinoxanthin and 1.55 μg/g diatoxanthin, DW	[204]
Chlorella C.S1	β-carotene and lutein	-	Industrial wastewater in flat panel reactors with fluorescent lights (2000 μmol photon/m^2·s)	HPLC	1.4 mg/g β-carotene and 3.22 mg/g lutein, DW	[204]
Nannochloropsis limnetica 18.99	Neoxanthin and violaxanthin	Algal culture collection (SAG), University of Gottingen, Germany	Industrial wastewater in Schott bottles, stirred, aerated (2% CO_2), and with fluorescent lights (200 μmol photon/m^2·s)	HPLC	0.42 mg/g neoxanthin and 1.22 mg/g violaxanthin, DW	[204]
Chlorococcum sp.	Lutein	Umeå University, Sweden	Multi-cultivator with high light/cold stress in BG11	HPLC	15.5 mg/g, DW	[205]
Scenedesmus sp.	Lutein	Umeå University, Sweden	Multi-cultivator with high light/cold stress in BG11	HPLC	10.7 mg/g, DW	[205]

DW, dry weight; HPLC, high-performance liquid chromatography.

2.4.4. Production Approaches and Production Data in Different Continents

The industrial production of microalgae throughout the world is under development. The microalgal global market was estimated at between USD 3.4 and USD 3.9 billion in the period 2018–2020, and it is expected to increase to USD 4.6–5.1 billion by 2027 [206,207]. Although the values of global markets will reach USD 800 million for astaxanthin, USD 620 million for β-carotene, or USD 358 million for lutein by the end of 2026 [145], around 95% of the industrial production of carotenoids is performed by chemical synthesis due to its lower cost.

For astaxanthin, while synthetic production is about USD 1000–2000 per kg; production from *Haematococcus* (around 1% of total production) has a cost between USD 1800 and 7000 per kg, depending on the location [208]. However, naturally produced astaxanthin is thought to have a stronger antioxidant capacity, being over 50 times stronger in singlet oxygen quenching and 20 times stronger in free radical elimination [209]. Natural astaxanthin is mainly obtained from *Haematococcus pluvialis*. This microalga can have two different cell morphologies: a green vegetative form and a red cyst form. Astaxanthin production in *H. pluvialis* is usually performed based on these two different morphologies, with a first stage focusing on developing their green form, which includes the production of biomass, followed by a second red stage where the astaxanthin is produced [145,195]. The production of natural astaxanthin from *H. pluvialis* is mainly concentrated in the USA, Europe, and Asia-Pacific countries, which are the main manufacturers of the companies Algatechnologies Ltd. in Israel, BluebioTech International GmbH in Germany, Algalif in Iceland, Algae Health Sciences and Cyanotech Corporation in the USA, AstaReal AB in Japan, BGG (Beijing Ginko Group) in China, and Parry's Pharmaceuticals in India [208].

In the case of β-carotene, which is the other pigment produced by microalgae on an industrial scale, the cost of its production from *Dunaliella salina* is still higher than the chemically synthesized one. Nevertheless, synthetic β-carotene is composed only of the all-*trans* isomer, whereas natural β-carotene, as found in this microalga, is a mixture of the all-*trans* and 9-*cis* isomers, which may exhibit more beneficial properties [210]. The main strategies to increase the production of β-carotene in *D. salina* are to subject them to salt stress, high light, and N depletion. However, these stress conditions can reduce the production of biomass and, as a consequence, the yield of this carotenoid. Different solutions have been exploited to balance biomass and β-carotene production being the most common of the two-stage processes [211,212]. In these systems, there is a first stage focused on biomass production and a second one that improves the production of β-carotene. The natural production of β-carotene from *D. salina* is mainly found in Asia-Pacific countries, followed by Europe and Australia. The main manufacturers of this compound are Seagrass

Tech Private Limited in India, Hangzhou OuQi Food Co. and Shaanxi Rebecca Bio-Tech Co. in China, Algalimento SL and Monzón Biotech SL in Spain, IBR Ltd. in Israel, and Plankton Australia Pty Ltd. in Australia.

The industrial production of fucoxanthin by the above-mentioned diatoms (Section 2.4.3) could be a more economically feasible approach if combined with the co-production of other high-value compounds. Diatoms are also able to produce large amounts of PUFAs, including eicosapentaenoic acid (EPA) and docosahexaenoic acid (DHA). Thus, the extraction of these PUFAs in combination with fucoxanthin can be an excellent approach that has to be further studied in marine microalgae [200,202].

2.4.5. Current and Prospective Applications

Microalgal carotenoids have a wide range of applications, including in the cosmetic, pharmaceutical, food, and feed industries. Additionally, the development of aquaculture processes including microalgae as a nutritional supplement for fish can stimulate the production of microalgae in high-scale processes, with the reduction in cultivation and harvesting costs being one of the main objectives in microalgae biotechnology. Although the cost of producing synthetic carotenoids is lower than that of natural ones, the continuous increase in demand for natural products in the global market points to carotenoids produced by microalgae as highly demanded products in the following years. This, along with the sustainable production of microalgae, can make them outstanding candidates for the development of "green" biofactories in the near future [213].

2.4.6. Research Needs

Although microalgae have a high potential for carotenoid production, further research is needed relating to the isolation, cultivation, extraction, or processing of algal biomass in order to reduce the cost of their industrial implementation. For example, isolating species from different extreme environments, such as high/low temperatures, high/low pH values, or high salinity, could be an interesting approach to finding strains able to produce higher amounts of carotenoids, such as lutein, fucoxanthin, or zeaxanthin. Additionally, genetic engineering is another strategy to improve the productivity of these compounds and achieve their economically feasible production. Reducing the cultivation costs, both by using alternative nutrient sources and improving the cultivation system, could also help boost productivity. This low productivity is one of the main problems in the industrial production of microalgae, which makes its competition with synthetic systems difficult. On the other hand, although there is much research about the use of emerging technologies (pressurized liquid extraction, pulsed electric fields, moderate electric fields, high-voltage electric discharges, high-pressure homogenization, microwave-assisted extraction, subcritical fluid extraction, and supercritical fluid extraction, among others) for the extraction of microalgal carotenoids [214,215], their industrial scaling-up remains a bottleneck. These concerns are the main challenge for future research in microalgal biotechnology to produce a more efficient and sustainable industry for the natural production of carotenoids from microalgae.

2.5. Yeast

2.5.1. Ecological Importance

Yeasts are single-celled eukaryotic microorganisms that are members of the fungi kingdom. Since they are organotrophic and heterotrophic microorganisms, they require organic compounds as energy and carbon sources, and their growth is always associated with the presence of organic matter. Yeasts are globally distributed. They have been found in diverse habitats, including aquatic environments such as marine water, freshwater, glacier meltwater, groundwater, and the deep sea [216,217]. The concentration and diversity of yeasts in different aquatic habitats vary depending on the amount and type of organic matter and other environmental parameters. It has been observed that yeast populations decrease as the distance from the land increases. In coastal waters, thousands of yeast cells can be found per liter, while deep-sea regions can contain lower levels of yeast (10 cells/L, or even fewer) [20]. Despite the low numbers, the diversity is not low. For example, when studying the diversity of culturable yeasts in five locations in coastal waters on King George Island (maritime Antarctica), Garmendia et al. [218] identified eleven genera. The analysis by massive sequencing of the region ITS2 showed great diversity since 31 yeast genera were found. Kutty and Philip [20] reviewed studies on the diversity of marine yeasts around the world and found that the most frequently reported genera were *Candida, Cryptococcus, Debaryomyces*, and *Rhodotorula*. Of them, ascomycetous genera are predominant in shallow waters, and basidiomycetous yeasts predominate in the deep sea.

In both terrestrial and aquatic environments, yeasts play an important role as organic matter decomposers, facilitating the mineralization and recycling of nutrients. They also actively participate in nitrogen, sulfur, and phosphorous cycles [219]. Their ability to produce hydrolytic enzymes such as proteases, glycosidases, esterases, and lipases allows them to break down many organic polymers, producing lower molecular weight molecules that can serve as carbon or energy sources for themselves and for other organisms. In this way, their ecological role is very important in oligotrophic habitats. They also contribute to the bioremediation of contaminated areas due to their ability to decompose organic pollutants such as alkanes, phenolic compounds, or dyes [220,221]. Most yeasts undergo aerobic respiration, but some of them are also capable of obtaining energy by fermentation or anaerobic respiration using nitrate or nitrite as electron acceptors that produce NO or N_2O, which are released into the atmosphere. In these cases, yeast would act as denitrifying microorganisms, causing the depletion of fixed nitrogen forms.

Some aquatic yeasts have evolved molecular and metabolic adaptations to survive or even grow under harsh conditions, such as low or high temperatures, high solar radiation, low nutrient availability, high salt concentrations, or even low or high pH [222]. In particular, in polar regions or at high altitudes, yeasts living in clear waters without natural shade are exposed to high UV radiation, which can influence diversity and population size [223]. UV radiation produces detrimental effects on living cells, causing direct or indirect damage to DNA, proteins, and lipids and accumulating ROS. To cope with such effects, some yeasts synthesize UV protectants and antioxidants such as carotenoids and accumulate them intracellularly [224]. The antioxidant properties of carotenoids may help extend the survival times of yeast in their natural habitats. Differences in growth between strains differing in carotenoid content showed a protective effect of those compounds [224].

2.5.2. Sustainability Pros and Cons

Many aquatic yeasts can accumulate intracellularly high levels of carotenoids [225]; thus, they can be considered a natural and renewable source for carotenoid production.

As with other marine organisms, yeast, as a source of carotenoids, can be cultivated on agro-industrial waste or by-products in processes supporting the concept of a circular economy. As an example, the production of provitamin A carotenoids by *Rhodotorula glutinis* using goat cheese whey as a substrate has been recently described [226]. In addition, lycopene can also be obtained from yeasts by applying cheap agro-industrial waste such as flour extracts, whey, rice, and glycerol, among others [227]. In another interesting study, it was observed that the use of waste glycerol from the biodiesel production process in the yeast culture significantly increased the productivity and concentration of torularhodin compared to pure glycerol [228].

2.5.3. Main Species (and Compounds) in the Context of Agro-Food and Health

A great diversity of carotenoids from different yeast species has been reported (Table 5). In most cases, various carotenoids can be obtained from a single yeast species. For example, as shown in the carotenoid database (http://carotenoiddb.jp, accessed on 30 May 2023), *Rhodotorula aurantiaca*, a psychrophilic red yeast isolated from Antarctica, can produce ten different carotenoids. Some yeast carotenoids can be found in other organisms. This is the case for β-carotene, which can be extracted from vegetables and fruits, or astaxanthin, which is also present in some microalgae, bacteria, crustaceans, or fish such as salmon and trout [229]. However, other yeast carotenoids, such as torulene and torularhodin, are specific fungal metabolites.

Yeasts that accumulate remarkable levels of carotenoids are basidiomycetous yeasts belonging to the Cystofilobasidiales and Sporidiobolales orders. Among Cystofilobasidiales, only one species of *Phaffia rodozyme* (teleomorph X. *dendrorhous*) has been reported to be carotenogenic [230]. This species, which has been isolated from some tree exudates, leaves, fungal stromata, and freshwater [231], accumulates astaxanthin, a carotenoid with many biotechnological applications [230]. Among Sporidiobolales, yeasts that accumulate the highest levels of carotenoids belong to the genera *Rhodotorula* and *Sporobolomyces* and to their teleomorphs, *Rhodosporidium* and *Sporidiobolus* [37]. The carotenoid profile produced by these yeasts is mainly represented by γ-carotene, β-carotene, torulene, and torularhodin [232], but other compounds have also been described. Carotenogenic yeasts from Sporidiobolales have been isolated from diverse aquatic habitats. Ueno et al. [225] isolated 40 yeast strains from different aquatic environments in Japan. These yeasts accumulated carotenoids in amounts greater than 200 μg/g of their dry cell mass. They were identified as *Rhodotorula* spp. and *Rhodosporidium* spp. and produced different amounts of intracellular β-carotene, torularhodin, γ-carotene, torulene, and β-zeacarotene, the first three compounds being the most abundant. Carotenogenic isolates from different species within the *Rhodotorula* genus have been isolated from the coastal waters of King George Island (maritime Antarctica) [218] and from oligotrophic lakes in Patagonia, Argentina [233]. Pigmented *Rhodotorula* spp. strains that accumulated intracellular carotenoids were also isolated from seawater near Chile [234]. One of the strains accumulated carotenoids, representing 19% of its dry weight, when cultured in glycerol as the only carbon source. The carotenoid profile was not very common and included β-carotene, adonirrubin, 3-hydroxyechinenone, and canthaxanthin. Strains from Sporidiobolales identified as *R. mucilaginosa*, *R. toruloides*, *R. diobovata*, and *S. roseus* recognized as species that accumulate significant amounts of carotenoids [235] were also found in oligotrophic hypersaline coastal waters of the Arabian Gulf [236].

Carotenoids are also produced in fewer amounts by basidiomycetous yeasts of the Sporidiobolales order. Strains belonging to different species from the genera *Bullera*, *Cystobasidium*, *Cystofilobasidium*, and *Dioszegia*, many of which have been isolated from aquatic environments, produce intracellular carotenoids [237]. Most of the compounds found in these yeasts are carotenes. However, two uncommon xanthophylls, 16-hydroxytorulene and torularhodinaldehyde, which were previously known only from chemical synthesis, have been found in certain *Cystofilobasidium* species [238]. Moreover, some isolates belonging to the genus *Dioszegia* produce almost exclusively plectaniaxanthin, a xanthophyll that is a dihydroxylated derivative of torulene [239]. These carotenoids represent a minor fraction of the carotenoids produced by yeast species but are illustrative of their chemical diversity.

Undoubtedly, yeasts from aquatic environments throughout the world are a source of common and rare carotenoids that should be further explored.

Many biological properties have been demonstrated for the main yeast carotenoids, which include carotenes such as β-carotene, γ-carotene, and torulene, and xanthophylls, including astaxanthin, canthaxanthin, and torularhodin [240]. Some of them, such as γ-carotene, β-carotene, torulene, and torularhodin, are vitamin A precursors [241]. Antibacterial and anti-fungal activity has been demonstrated for torularhodin, a specific fungal carotenoid [242,243]. Moreover, due to its important antimicrobial activity, it was successfully used to protect implanted medical devices against microbial contamination [244,245]. In the case of torulene and torularhodin, Du et al. [246] postulated that their action in the inhibition of prostate cancer in nude mice was associated with the apoptosis of tumor cells.

Table 5. Examples of carotenoid content in marine yeast.

Species	Carotenoids	Collection Sites	Experimental Conditions	Quantification/ Identification Methodologies	Contents	References
Sporidiobolus salmonicolor	β-carotene, 2,3 dihydroxy-γ-carotene, 4-ketotorulene, and torulene	Union Glacier, Antarctic	30 mL YM broth in 300-mL baffled flasks, 150 rpm, 20 °C, 5 days	HPLC	2,3-dihydroxy-γ-carotene was the main carotenoid	[247]
Sporidiobolus metaroseus	β-carotene, β-cryptoxanthin, 4-ketotorulene, and spirilloxanthin	Union Glacier, Antarctic	30 mL YM broth in 300-mL baffled flasks, 150 rpm, 20 °C, 5 days	HPLC	β-carotene and 4-ketotorulene were the main carotenoids	[247]
Rhodotorula mucilaginosa	Astaxanthin, β-carotene, and lycopene	Chiloe, 30 km southeast of Puerto Montt, Chile	4 g yeast/L seaweed (25% *v*/*v*), 150 rpm, 25 °C, 6 days	HPLC	1.84 ± 0.03 mg/L total carotenoids. Carotenoids proportion: 1.8 ± 0.3% astaxanthin, 21.8 ± 1.5% β-carotene, and 38.4 ± 9.4% lycopene	[248]
Rhodosporidium babjevae	β-carotene, γ-carotene, torularhodin, and torulene	Grøtsundet, Northern Norway	10 g/L Difco Marine broth 2216 (10 g/LDifco Bacto-peptone, 10 g/L glucose, 15 g/L NaCl, and 15 g/L agar), 6 °C, 140 h	HPLC	Torularhodin > torulene > β-carotene > γ-carotene	[249]
Rhodotorula mucilaginosa	β-carotene, torularhodin, and torulene	Escondido lake, North-western Patagonia, Argentina	20 mL of minimal medium salt broth (MMS), 25 °C, 250 rpm, 24 h	TLC (for pigment profile), spectrophotometry, and HPLC	205 ± 15 μg/g total carotenoids, DW. Carotenoid proportions: 10.8% β-carotene, 83.4% torularhodin, and 5.7% torulene	[250]
Rhodotorula dairenensis	β-carotene, γ-carotene, torularhodin, and torulene	Freshwater in the middle of the Sagamigawa River, Japan	20 mL of YPD liquid medium (2% glucose, 2% peptone, and 1% yeast extract), 25 °C, 120 rpm, 36 h	HPLC	267 μg/g total carotenoids, DW	[225]

DW, dry weight; HPLC, high-performance liquid chromatography; TLC, thin-layer chromatography.

2.5.4. Production Approaches and Production Data in Different Continents

The production of carotenoids by yeast cultures can be considered a sustainable process. Although the use of yeast for carotenoid production is not economically profitable yet, efforts to improve yields and reduce production costs have merged to meet global demands for the use of natural products.

The biosynthesis of carotenoids by yeast is influenced by many factors that can modify yields and affect operating costs. The yeast strains, culture medium composition (type and concentration of nutrient sources), and growth conditions (temperature, pH, oxygen, and light) influence the amount and profile of carotenoids accumulated within the yeast cell [251,252] (Table 5).

Carotenoid biosynthesis in yeast begins in the late logarithmic phase and continues in the stationary phase; thus, the optimal harvest time to obtain the maximum content of carotenoids should be determined precisely in each case. The type and concentration of the main nutrients, such as carbon and nitrogen sources, are also important factors that significantly regulate the carotenogenesis process. Many studies on carotenoid synthesis in yeast cultures have been conducted in synthetic media with pure carbon sources, such as glucose, sucrose, xylose, cellobiose, glycerol, or a combination of some of them [253]. To lower production costs and in the search for circular economy strategies, many by-products, wastes, and raw materials from agro-industries, such as raw glycerol, brewery effluents, molasses, grape must, and milk whey, have been proposed as alternative carbon sources for carotenoid production [226,254–256].

An important fact for obtaining high levels of intracellular carotenoids is to consider that carbon sources cannot be the limiting factor in yeast growth. However, high initial concentrations of sugars are detrimental to carotenoid accumulation in batch cultures of yeasts that exhibit the Crabtree effect, such as *Phaffia rodozyma* [257]. In these cases, at high concentrations of carbon sources, fermentation occurs even under aerobic conditions. In consequence, pyruvate from glycolysis is partially transformed in ethanol and not totally derived to acetyl-CoA, which is a precursor for the mevalonate pathway, the first step in the carotenoid biosynthesis pathway [256]. The best strategy to improve biomass and carotenoid production under such conditions is the implementation of fed-batch cultures [258].

The C/N ratio is an important parameter to optimize carotenoid accumulation in yeast. Much work related to this topic has been published; however, there are no general rules to follow. Although it is fairly agreed that high C/N ratios improve carotenoid synthesis [257], there are some cases in which the results are different [259]. Differences in carotenoid profiles have also been observed with different C/N ratios [29].

In the case of *R. glutinis*, the effect of the C/S and C/P ratios on the accumulation of carotenoids has also been studied [252,260]. In both cases, an increase in the sulfur and phosphorus source concentrations improved the production of carotenoids.

Light also influences the production of carotenoids by yeast. To avoid possible damage caused by light exposure, carotenogenesis is induced when yeasts grow in the presence of light. The effect of light depends on the yeast strain and is related to an increase in the activity of certain enzymes associated with the carotenoid biosynthetic pathway [261]. Light induction of carotenoid biosynthesis has been reported for different yeast species such as *R. glutinis* [262], *Rhodosporidium toruloides* [263], and *P. rodozyma* [264].

Another factor that influences the production of carotenoids by yeasts is temperature. Studies on the effect of temperature on carotenoid content in *Rhodosporidiobolus colostri* [265], *Rhodotorula babjevae* [249], *R. mucilaginosa* [266], and *R. glutinis* [267] coincide in that at higher growth temperatures an increase in torularhodin is obtained in detriment of the β-carotene content. The effect on torulene varied with the specific yeast strain. According to Zhao and Li [267], the enhancement in torularhodin production at higher temperatures might be due to its high antioxidant activity and the protection it would exert by scavenging ROS induced by high-temperature stress.

Carotenogenic yeasts are mainly aerobic microorganisms. The oxygen supply is an essential factor to enhance their growth in culture conditions, and it influences the amount and profile of carotenoids accumulated inside the cells [268]. Oxygen levels influence the relative concentration of carotenes and xanthophylls since the latter are formed through the oxidation of the former [29]. The synthesis of astaxanthin by *P. rodozyma* is greatly stimulated by oxygen supply. In the presence of low oxygen levels, yellowish biomass is formed due to the accumulation of β-carotene [258].

The yeast *P. rodozyma* accumulates intracellular astaxanthin but in much lower amounts than the microalgae *Haematococcus pluvialis*, so its commercial production is not as extensive. At present, there is at least a commercial product called Astaferm based on astaxanthin obtained from *P. rodozyma*. It is produced for human consumption in various presentations by the company Nextferm (Israel) and is also sold in the USA. Information about these products can be found on the company website (https://www.nextferm.com/, accessed on 30 May 2023).

2.5.5. Current and Prospective Applications

Currently, the carotenoids found in yeasts have many commercial applications. However, their production at the industrial level, in most cases, does not involve their extraction from yeasts. The main reasons for that are the low yields of the processes and the high costs associated with microbial production and product purification. Astaxanthin, a xanthophyll produced by *P. rodozyma*, is a clear example of it. Astaxanthin is a carotenoid extensively used in aquaculture as a feed additive for imparting reddish coloration to the flesh of salmon, trout, shrimp, and ornamental fish [229]. It is also used in poultry to enhance the color of egg yolks and the skin and meat tissue of broilers, resulting in better acceptance in the market [269]. Due to its various health benefits, astaxanthin has been incorporated into food and beverages, nutraceutical preparations, and cosmetics [229]. Its industrial production is mainly based on chemical synthesis, except for products devoted to human direct consumption, for which natural astaxanthin is preferred [36].

Other yeast carotenoids, such as torulene and torularhodin, are not currently produced or used in industry; however, they have many potential applications. They have a reddish color and could be used as pigments in animal feeding in aquaculture or poultry [270]. Due to their proven activities as antioxidants and vitamin A precursors, they could also have applications as food and cosmetic additives. Moreover, both carotenoids have been associated with tumor prevention, so they could also be postulated as nutraceutical products [240]. These carotenoids are synthesized only by yeast and fungi, and their effect on humans and animals is still unknown.

2.5.6. Research Needs

The carotenoids present in yeasts have the potential to be used as food and cosmetic ingredients, as well as pigments in aquaculture and poultry. However, much research is still needed so that the production of carotenoids in yeast is competitive in the market. Some yeast carotenoids, such as torulene and torularhodin, have not been used in animals yet due to fewer studies on their safety and effects on health [240]. Further research is also needed to determine the properties of other carotenoids produced by aquatic yeasts that have not been explored yet. On the other hand, the extraction of yeast carotenoids has been scarcely studied.

3. Roles of Carotenoids in Marine Organisms

Carotenoids are biosynthesized by all photosynthetic organisms (plants, algae, and cyanobacteria), as well as some fungi, non-photosynthetic archaea, bacteria, and arthropods. The vast majority of animals obtain them through their diet [35].

3.1. Photosynthesis

Carotenoids are essential in photosynthesis, intervening in actions such as light harvesting, energy transfer, photoprotection, or the assembly of pigment-protein complexes in the photosynthetic apparatus. Some microalgae can biosynthesize large amounts of carotenoids in response to stress conditions. They are deposited in lipid globules that can be found outside the photosynthetic apparatus (for instance, astaxanthin produced by *Haematococcus*) or within it (as appears to be the case for the massive β-carotene accumulation observed in some *Dunaliella* species). Carotenoids are also thought to somehow intervene in photoreception and phototaxis in some flagellate microalgae, which have a photosensitive eyespot capable of sensing variations in light intensity. The photoreceptors are usually membrane-bound rhodopsins (consisting of an apoprotein bound to retinal, a carotenoid derivative) located close to layers of photoprotective/reflective carotenoid-rich globules. Photoreception and phototaxis are important in deciding where to go for optimal light or to prevent predation [271].

There is a diverse body of evidence that carotenoids can protect aquatic organisms from various stressors. As can be inferred from the information provided below, some of the possible mechanisms could be protection against oxidation and modulation of membrane properties.

3.2. Secondary Antenna in Retinal Proteins

The important role that carotenoids can play as secondary antennas of type I rhodopsins has recently been demonstrated [272]. Many marine and freshwater bacteria, as well as some eukaryotes, are capable of obtaining energy using sunlight, thanks to type I rhodopsins. These proteins, evolutionary-divergent but structurally related to the rhodopsin II of animals, have retinal as their main chromophore. The retinal of the rhodopsin proteins and the chlorophylls of the photosystems are the only two types of pigments with the ability to convert solar energy into chemical energy that, to date, are known. Since the discovery of the first type I rhodopsin in the 1970s, bacterioruberin, which pumps H^+ thanks to the sunlight, generating a proton-motive force used by ATP synthase for the production of ATP, many other similar light-driven pumps have been discovered [273,274].

In 2005, it was announced for the first time that a carotenoid, the keto-carotenoid salinixanthin, could act as a secondary antenna for a retinal rhodopsin. The study showed that salinixanthin, found in the halophile bacteria *Salinibacter ruber*, is able to harvest light energy and transfer it to retinal [275]. The recent discovery that the hydroxylated carotenoids, lutein and zeaxanthin, can also act as secondary antennas demonstrates that the ability of salinixanthin is not an exceptional case and uncovers a new essential role of carotenoids as auxiliary antennas in retinal rhodopsins. Furthermore, the study estimates, on the basis of the sequence and structural features of many known rhodopsins, that carotenoids could be the second chromophore in about half the microbial retinal rhodopsin proteins, having an essential role in the microbial photoautotroph metabolism in the oceans and continental waters [272].

3.3. Oxidative Stress

Halobacillus halophilus is a moderately halophilic bacterium that can tolerate up to 3 M sodium chloride. It accumulates C30 carotenoids structurally related to staphyloxanthin that impart a yellow hue. By using diphenylamine (an inhibitor of the biosynthesis of colored carotenoids) and the oxidant duroquinone, it was shown that the C30-colored carotenoids were essential for the survival of the bacteria under oxidative stress conditions [276].

Cyanobacteria that fix nitrogen (diazotrophs) are especially susceptible to reactive oxygen species due to the sensitivity of nitrogenase to oxygen. The total antioxidative potential of various diazotrophic cyanobacteria (*Cyanothece*, *Anabaena*, and especially *Trichodesmium* spp., a bloom-forming cyanobacteria that contributes to a great extent to global nitrogen fixation) and non-diazotrophic cyanobacteria (*Prochlorothrix*, *Synechococcus*) has been evaluated by means of the ferric reducing/antioxidant power (FRAP) assay. To gain insight into the different cell components that contribute to total antioxidant activity, non-polar and polar organic extracts as well as crude protein extracts were tested. *Trichodesmium* sp. IMS101 showed the highest activity among the samples tested. The relative contribution to the total antioxidant activity ranged from 54% for the non-polar organic extract to only 13% for the protein extract, a result that contrasted with the average relative contribution of each extract of the other cyanobacterial species tested (86% for the protein extract, 9% for the non-polar organic extract, and 5% for the polar organic extract). As a result of subsequent studies (bioassay-guided fractionation and HPLC profiling of purified fractions), the authors pinpointed β-carotene and retinyl palmitate as possible molecules contributing to the high in vitro antioxidant activity of the non-polar organic extract obtained from *Trichodesmium* sp. IMS101 [277].

3.4. Salt Stress

Halophilic archaea or haloarchaea are extremophiles that require high salt concentrations for optimal growth. They can be found in saline environments, including salt lakes, solar salterns, and salted foods. The predominant carotenoids present in them are C50 carotenoids, such as bacterioruberin and its derivatives, monoanhydrobacterioruberin and bisanhydrobacterioruberin [278].

Carotenoid-rich microbial communities are well known to occur in solar salterns where there are environments with different salt concentrations, ranging from seawater to NaCl saturation. In crystallizer ponds, the green microalga *Dunaliella* appears as the sole primary producer. It occurs together with dense communities of heterotrophic halophilic archaea that thrive due to the carbon photosynthetically fixed by the microalga. The red hue of the crystallizer ponds increases light absorption, raises the temperature, and enhances salt production. The coloration is due to the massive accumulation of β-carotene by *Dunaliella* and the carotenoid and retinal protein-based pigments of the Archaea (for example, *Haloquadratum walsbyi* and other *Halobacteriaceae* or *Salinibacter*) [279].

Bacterioruberin and other archaeal C50 carotenoids are believed to increase membrane rigidity and provide protection against UV light. The presence of carotenoids in the membrane could help microbes cope with hypersaline conditions by acting as a water barrier and allowing the passage of ions and oxygen molecules through the cell membrane. Bacterioruberin has also been shown to be a component of specific transmembrane proteins and modulate membrane dynamics and physics [19].

3.5. Low Temperatures

The cryosphere has been colonized by a variety of microbes collectively known as psychrophiles. Temperatures close to 0 °C lead to increased viscosity of solvents and solubility of gases, reduced solubility of nutrients and solutes, decreased diffusion and amplified desiccation, osmotic stress, and ice development. Other characteristics of these environments are high salinity, low water activity and availability of nutrients, oxidative stress, freeze–thaw cycles, extremes of light (high at high altitudes and low in frosted lakes, permafrost, and deeper ice sheets), and exposure to UV radiation. Among the strategies that cold-adapted microbes have developed to cope with these harsh conditions is the production of pigments, which are very often carotenoids [280].

The biosynthesis of carotenoids (haloxanthin, monoanhydrobacterioruberin, bacterioruberin, bacterioruberin monoglycoside, and bacterioruberin diglycoside) of the Antarctic psychrotrophic bacterium *Micrococcus roseus* was shown to increase at 5 °C compared to 25 °C; additionally, more polar carotenoids (the two glycosides) were produced at 5 °C [281].

The carotenoid content of the Antarctic psychrotolerant bacterium *Sphingobacterium antarcticus* and the mesophilic bacterium *Sphingobacterium multivorum* has been studied, revealing that the main carotenoids were zeaxanthin, β-cryptoxanthin, and β-carotene. Interestingly, the relative amounts of polar pigments were higher in cells grown at 5 °C than in cells grown at 25 °C, and the synthesis of polar carotenoids was higher in the psychrotolerant strain [282].

Arthrobacter is a ubiquitous bacterial genus. Some *Arthrobacter* species produce the rare C50 carotenoid bacterioruberin, which is found mainly in haloarchaea. Interestingly, strains of *A. agilis* and *A. bussei* exhibited increased bacterioruberin biosynthesis when the culture temperature was reduced from 30 to 10 °C. The increased levels of bacterioruberin were correlated with increased membrane fluidity at low-temperature growth and with improved cell resistance to freeze–thaw stress [283]. In another study, heterotrophic Antarctic bacteria were subjected to freeze–thaw cycles and simulated solar radiation, and a higher survivability was observed in carotenoid-pigmented bacteria [77].

3.6. High Temperatures

Bacteria of the genus *Thermus* can also tolerate high temperatures. *Thermus thermophilus* is known to accumulate uncommon carotenoid (specifically zeaxanthin) diglucoside esters and monoglucoside esters, generically termed thermozeaxanthins. In thermophilic bacteria, carotenoids span the membrane and are believed to stabilize it by modulating its fluidity, which is essential for its survival in such harsh conditions. This hypothesis is supported by experiments in which thermozeaxanthins were incorporated into egg phosphatidylcholine liposomes, resulting in their stabilization in the temperature range of 30–80 °C. On the other hand, polar carotenoids in artificial membranes can reduce the rate of oxygen diffusion and protect membrane lipids from oxidation [284].

Blastochloris tepida is a recently described thermophilic purple bacterium. In an interesting experiment, the light-harvesting one reaction center core complexes of this species were compared with those of its mesophilic counterpart, *B. viridis*. Both a higher total carotenoid content and a different carotenoid profile (with higher levels of carotenoids with more than nine conjugated double bonds) were detected in *B. tepida*. Additionally, it was observed that thermostability decreased when *B. tepida* was treated with diphenylalanine, an inhibitor of the biosynthesis of colored carotenoids. Altogether, the results indicated that *B. tepida* carotenoids are important for high-temperature tolerance [285].

3.7. Roles in Marine Animals

The carotenoids of marine algae and microbes are incorporated into herbivorous marine animals (e.g., sponges, sea anemones, bivalves, microcrustaceans, and tunicates) and can be modified through metabolism; then these herbivores serve as foods for carnivorous animals, such as snails, crustaceans, starfish, and fish. Thus, in marine animals, it is common to find metabolites of β-carotene, fucoxanthin, peridinin, diatoxanthin, alloxanthin, and astaxanthin [286]. Concerning the role of carotenoids in marine animals, those with an unsubstituted β-ring (for instance α-carotene, β-carotene, or β-cryptoxanthin) can be converted into vitamin A. Unlike in humans, carotenoids with substituted β-rings (canthaxanthin, lutein, zeaxanthin, and astaxanthin) have been reported to be precursors of vitamin A forms in some fish [286]. In addition to their presumed antioxidant and immunomodulatory activities, carotenoids are usually found in integuments, where they can be important for photoprotection, camouflage, and signaling for breeding. Besides, carotenoids are common in marine animal gonads, and there is evidence that they are important for reproduction (for instance, in ovary development, fertilization, hatching, or larval growth and survival) [286].

4. Carotenoids: Versatile Compounds with Health-Promoting for Foods, Cosmetics and Other Products

The importance of carotenoids for food security is beyond any doubt, as they are essential for photosynthesis, the engine of life on Earth, and the primary driver of food production. Besides carotenoids and/or apocarotenoids (i.e., compounds obtained by the carotenoid cleavage, such as many volatiles, the phytohormones strigolactones, and abscisic acid), they are important for pollination, seed dispersal, plant resilience to diverse stresses, etc. Their roles as colorants and, for some of them, as vitamin A precursors have been long known. Since the last three decades, there has been an expanding interest in their possible health-promoting biological actions, which could also be attributed, at least to some extent, to apocarotenoids [33,35]. The possible mechanisms are several: direct antioxidant actions (quenching, scavenging), pro-oxidant actions, enhancement of gap junctional intercellular communication, modulation of signaling pathways or immune function, and absorption of visible light (or UV in the case of the colorless carotenoids phytoene and phytofluene), which may interact. As a result, different effects such as antioxidative, prooxidative, anticarcinogenic, or anti-inflammatory can take place. Thus, carotenoids are thought to contribute to reducing the risk of cancer, cardiovascular diseases, metabolic, bone, skin, or eye disorders. Positive effects on cognitive performance and during pregnancy and early life are also being increasingly reported [33,35,287–289]. They can also provide aesthetic benefits as, together with melanin and hemoglobin derivatives, they are contributors to skin color and may have other cosmetic benefits. Therefore, carotenoids can be used in products for human consumption (foods, supplements, nutraceuticals, nutricosmetics, etc.) as colorants, vitamin A precursors, antioxidants, and health-promoting compounds [33,34,290]. Research on carotenoids has been propelled in recent years by cohesive and functional research networks, such as IBERCAROT in Ibero-America (http://www.cyted.org/?q=es/detalle_proyecto&un=829, accessed on 30 May 2023), EUROCAROTEN in Europe (https://www.cost.eu/actions/CA15136/#tabs|Name:overview, accessed on 30 May 2023) or the Spanish Carotenoid Network (CaRed).

5. Potential Health-Promoting Actions of Carotenoids from Aquatic Organisms

In human physiology, ROS are used as important signaling molecules, but their production has to be tightly controlled. Unbalanced ROS generation, resulting in oxidative stress and closely related to inflammatory signaling, contributes to many chronic diseases [291,292]. Although, as has already been commented, carotenoids, including those from marine organisms, can be involved in different actions contributing to health promotion, much attention is paid to their possible antioxidant and anti-inflammatory effects. Astaxanthin, torulene, and torularhodin, which are carotenoids commonly found in marine organisms, have been shown to exhibit higher antioxidant capacity than β-carotene under certain conditions [29,229]. In addition, anticancer activity has been demonstrated for these carotenoids [246,293–295]. Among the carotenoids of marine organisms, there are some, such as astaxanthin, that have been shown to exhibit both antioxidant and anti-inflammatory activity, and thus they are considered potential health-promoting agents against atherosclerotic cardiovascular disease [296]. Astaxanthin also exhibits antimicrobial activity. Its antifungal and antibiofilm activity against *Candida albicans* and *Candida glabrata* has recently been reported [297]. Studies in this respect involving "marine carotenoids" are summarized in Table 6. Only a few examples are included in this table, but extensive works can be found in the literature on some marine microorganisms, such as microalgae and seaweed [298,299].

Table 6. Some examples of carotenoids in marine organisms with potential antioxidant and anti-inflammatory health-promoting activities.

Organisms	Species	Carotenoids	Biological Activities	Methodologies	References
Archaea	*Halobacterium halobium*	Bacterioruberin [1]	Antioxidant	Exposing hepatoma cell lines with carotenoid extract to arachidonic acid or H_2O_2. The protective effect against oxidative stress was also assessed by the MTT assay	[48]
	Haloarcula japonica	Bacterioruberin [1]	Antioxidant	DPPH assay	[49]
	Haloterrigena turkmenica	Bacterioruberin [1]	Antioxidant	DPPH and FRP assays	[51]
	Haloferax volcanii, Halogranum rubrum, Haloplanus inordinatus, Halogeometricum limi, and Haloplanus vescus	Bacterioruberin [1]	Antioxidant	DPPH assay and the evaluation of the inhibition of H_2O_2-induced hemolysis of mouse erythrocytes	[57]
	Haloarcula hispanica and *Halobacterium salinarum*	Bacterioruberin [1]	Antioxidant/ anti-inflammatory	Antioxidant: DPPH, ABTS, NO, FRAP, CCA, and ICA assays. Anti-inflammatory: against COX-2	[59]
	Halorubrum sp.	Bacterioruberin [1]	Antioxidant	DPPH and ABTS assays	[52]
	Haloarcula sp. and *Halorubrum tebenquichense*	Bacterioruberin [1]	Antioxidant	DPPH, ABTS, and FRAP assays	[53]
	Haloferax mediterranei	Bacterioruberin [1]	Antioxidant	DPPH, ABTS, and FRAP assays	[54]
Bacteria	*Arthrobacter* sp. G20	Crocin [1]	Antioxidant	DPPH assay	[300]
	strain 04OKA-13-27	(3*R*)-saproxanthin [2]	Antioxidant	Against free-radical-induced lipid peroxidation in a rat brain homogenate	[301]
	strain YM6-073	(3*R*,2′*S*)-myxo, (3*R*,3′*R*)-zeaxanthin [2]	Antioxidant	Against free-radical-induced lipid peroxidation in a rat brain homogenate	[301]
	strain 04OKA-17-12	(3*R*,3′*R*)-zeaxanthin [2]	Antioxidant	Against free-radical-induced lipid peroxidation in a rat brain homogenate	[301]
	Rubritalea squalenifaciens	Diapolycopenedioic acid [2] xylosyl esters	Antioxidant	$^1O_2$ suppression model	[302]
	Exiguobacterium acetylicum S01	Diapolycopenedioic-acid-diglucosyl ester [2]	Antioxidant/ anti-inflammatory	Antioxidant: DPPH assay. Anti-inflammatory: inhibition of NO production and TNF-α protein levels in LPS-induced oxidative stress in PBMC	[303]
	Exiguobacterium acetylicum S01	Keto-myxocoxanthinglucoside ester [2]	Antioxidant/ anti-inflammatory	Antioxidant: DPPH assay. Anti-inflammatory: inhibition of NO production and TNF-α protein levels in LPS-induced oxidative stress in PBMC	[303]
	Micrococcus yunnanensis	Sarcinaxanthin [2]	Antioxidant	$^1O_2$ suppression model	[304]
	Micrococcus yunnanensis	Sarcinaxanthin monoglucoside [2]	Antioxidant	$^1O_2$ suppression model	[304]
	Micrococcus yunnanensis	Sarcinaxanthin diglucoside [2]	Antioxidant	$^1O_2$ suppression model	[304]
	Kocuria sp. *RAM1*	Bisanhydrobacterioruberin derivative, trisanhydrobacterioruberin, and 3,4,3′,4′-tetrahydrospirilloxanthin [1]	Antioxidant/ anti-inflammatory	Antioxidant: DPPH assay. Anti-inflammatory: hypotonic solution-induced hemolysis	[305]
	Halobacillus halophilus (mutant)	Hydroxy-3,4-dehydro-apo-8′-lycopene [2]	Antioxidant	$^1O_2$ suppression model	[306]
	Halobacillus halophilus (mutant)	Methyl hydroxy-3,4-dehydro-apo-8′-lycopenoate [2]	Antioxidant	$^1O_2$ suppression model	[306]
	Planococcus maritimus	Methyl glucosyl-3,4-dehydro-apo-8′-lycopenoate [2]	Antioxidant	$^1O_2$ suppression model	[307]
	Planococcus sp. *ANT_H30*	Unidentified [1]	Antioxidant	DPPH assay	[308]
	Rhodococcus sp. *ANT_H53B*	Dihydroxyneurosporene, hydroxyechinenone, and 4 unidentified [1]	Antioxidant	DPPH assay	[308]
	Planococcus sp. *Eg-Natrun*	Astaxanthin and β-carotene [1]	Antioxidant	Fenton reaction	[87]
	Erythrobacter citreus LAMA 915	Zeaxanthin, caloxanthin, nostoxanthin, adonixanthin, canthaxanthin, and erythroxanthin sulfate [1]	Antioxidant	DPPH assay	[85]
Cyanobacteria	*Trichodesmiumemi* IMS101	Zeaxanthin, all-*trans*- and 9-*cis*-β-carotene [1]	Antioxidant	FRAP method	[277]
	Aphanothece microscopica Nageli	13-*cis*-Antheroxanthin, 15-*cis*- and all-*trans*-lutein, all-*trans*-zeaxanthin, all-*trans*-cantaxanthin, all-*trans*-myxoxanthophyll, β-carotene-5,6-epoxide, all-*trans*-β-cryptoxanthin, all-*trans*-crocoxanthin, all-*trans*- and 9-*cis*-echineone, and all-*trans*- 9-*cis*-, and 13-*cis*-β-carotene [1]	Antioxidant	Fluorescence decay resulting from the ROO· induced oxidation of the C_{11}-BODIPY$^{581/591}$ probe	[309]
	Alkalinema aff. pantanalense	Zeaxanthin, lutein derivatives, echinenone derivative, and unknown carotenoids [1]	Antioxidant	$^1O_2$ suppression model	[310]

Table 6. *Cont.*

Organisms	Species	Carotenoids	Biological Activities	Methodologies	References
	Cuspidothrix issatschenkoi	Canthaxanthin, all-*trans*- and 13-*cis*-β-carotene, α-carotene derivative, and unknown carotenoids [1]	Antioxidant	$^1O_2$ suppression model	[310]
	Leptolyngbya-like sp.	β-carotene oxygenated derivatives, lutein derivative, lutein, zeaxanthin, echinenone, all-*trans* and 13-*cis*-β-carotene, α-carotene derivative, and unknown carotenoids [1]	Anti-inflammatory	Inhibition of NO production in macrophage cells	[310]
Macroalga (brown)	*Hijikia fusiformis*	Fucoxanthin [1]	Antioxidant	DPPH assay	[311]
	Myagropsis myagroide	Fucoxanthin [1]	Anti-inflammatory	Inhibition of NO in LPS-induced macrophage cells	[312]
	Sargassum muticum	-	Antioxidant	Evaluation of the total antioxidant capacity after supplementation in humans	[174]
	Sargassum hemiphyllum	Fucoxanthin and fucoidan (polysaccharide) [1]	Anti-inflammatory	Evaluation of hepatic inflammation through modulation of leptin/adiponectin axis after supplementation in humans with NAFLD	[175]
Macroalga (green)	*Halimeda opuntia*	Unknown carotenoids [1]	Antioxidant	DPPH assay	[313]
Macroalga (red)	*Eucheuma denticulatum*	Lutein and zeaxanthin [1]	Antioxidant	ORAC assay	[314]
Microalga	*Dunaliella salina*	Lutein, zeaxanthin, α-carotene, all-*trans*- and 9-*cis*-β-carotene, and one unknown compound [1]	Antioxidant/anti-inflammatory	Antioxidant: reducing capacity, chelating activity, DPPH and $^1O_2$ suppression model. Anti-inflammatory: against COX-2 on human oral squamous carcinoma cells	[315]
	Haematococcus pluvialis	Astaxanthin [3]	Antioxidant	Evaluation of oxidative damage in rats caused by high fructose consumption after supplementation of astaxanthin	[316]
	Phaeodactylum tricornutum	Fucoxanthin [1]	Anti-inflammatory	Inhibition of NF-κB and NLRP3 inflammasome activation induced by the combination of LPS and ATP in bone marrow-derived immune cells and astrocytes	[317]
	Haematococcus pluvialis	3*S*,3′*S*-astaxanthin and 3*S*,3′*S*-astaxanthin esters [1]	Antioxidant	$^1O_2$ suppression model	[209]
	Brown microalga	Fucoxanthin [1]	Anti-inflammatory	Inhibition of COX-2 and iNOS expression in in macrophage cells incubated with LPS	[318]
Yeast	*Rhodosporidium paludigenum*	Carotenoids	Antioxidant	Evaluation of MDA levels in the muscle, and the activities of serum T-AOC, CAT, SOD and GPx, and hepatopancreases SOD and GPx after supplementation of shrimps with live yeast	[319]
	Rhodotorula sp.	Torularhodin	Antioxidant	Chemiluminescence and photochemiluminescence (Trolox) methods	[243] [4]

[1] Evaluation of an algae extract; [2] evaluation of the isolated carotenoid; [3] evaluation of the standard carotenoid. [4] In these works, the antioxidant activity of torularhodin extract produced from *Rhodotorula rubra*, which was not reported as aquatic yeast, was evaluated. However, torularhodin is produced by species of marine *Rhodotorula* [249]. ABTS, 2,2′-azino-bis(3-ethylbenzothiazoline-6-sulfonic acid; CAT, catalase; CCA, copper chelating assay; COX-2, cyclooxygenase-2; DPPH, 2,2-diphenyl-1-picrylhydrazyl; FRAP, ferric reducing antioxidant power; FRP, ferric reducing power; GPx, glutathione peroxidase; ICA, iron chelating assay, iNOS, inducible nitric oxide synthase; LPS, lipopolysaccharide; MDA, malondialdehyde; MTT, 3-(4,5-dimethylthiazol-2-yl)-2,5-diphenyl-2H-tetrazolium bromide; NAFLD, non-alcoholic fatty liver disease; NLRP3, nod-like receptor family pyrin domain containing 3; NO, nitric oxide; ORAC, oxygen radical absorbance capacity; PBMC, peripheral blood mononuclear cell; SOD, superoxide dismutase; T-AOC, total antioxidant competence; TNF-α, tumor necrosis factor-α; and Trolox, 6-hydroxy-2,5,7,8-tetramethylchroman-2-carboxylicacid.

6. Advantages and Disadvantages of Using Marine Organisms as a Source of Carotenoids over Chemical Synthesis

In general, microorganisms, including those of marine origin, have unique benefits for carotenoid production, such as their short life cycle, adaptability to various seasons and climates, the ability to generate a diverse range of carotenoids with varying colors and biological properties, and easier scalability of production [320]. Some of these characteristics can be attributed to other maritime organisms, such as seaweeds. In fact, the microalgae *Dunaliella salina* and *Haematococcus pluvialis* have been long used for the commercial production of β-carotene and astaxanthin, respectively. Besides, yeasts of the Rhodotorula genus and bacteria of the *Flavobacterium* genus, just to mention a couple of examples, elicit increased interest in the commercial production of carotenoids, highlighting the current and future competitiveness of fermentative processes for the carotenoid market [321].

One of the main drawbacks of using carotenoids from natural sources directly from the matrices (without prior isolation/extraction) is their frequently lower bioavailability relative to formulated products. This is because the matrix must be broken down during digestion before the carotenoid component can be released and utilized. This common low bioavailability of carotenoids present in natural sources can result in low bioactivity. For example, in a study by Edgar et al. [322] in which the resistance of rainbow trouts to a viral pathogen after oral ingestion of synthetic carotenoids (β-carotene, astaxanthin, and canthaxanthin) or natural sources of carotenoids (such as *Dunaliella salina* and *Phaffia rhodozyma*) was examined, it was observed that mortality was markedly reduced in the fish fed with astaxanthin, while it was only slightly reduced (with no statistical difference) in those fed with *D. salina*. Nonetheless, in the last few years, modern processing techniques have been used to overcome these limitations, leading to a rise in commercially available natural sources [322].

On the other hand, an advantage of using the complete matrix as a source of carotenoids is that it can contain other bioactive compounds. These compounds may also be present in carotenoid extracts obtained from natural sources, which is not the case with synthetic carotenoids. As an example, β-carotene from *Dunaliella* contains numerous carotenoids and essential nutrients that are not present in synthetic β-carotene [323].

In the production of carotenoids, it is important to consider the isomers of the carotenoids that are obtained, since different isomers of a carotenoid may have different bioavailability and biological activity. In this regard, it should be noted that, in some cases, chemical synthesis would have to be optimized in order to be able to isolate only those isomers with greater bioavailability or biological activity, which could be obtained from natural sources. This optimization could increase the cost of production. Astaxanthin represents a clear example of the advantages of obtaining a carotenoid from natural sources. Natural astaxanthin, for example, from *H. pluvialis* and *Paracoccus*, occurs in the trans form (3*S*, 3*S*), whereas the synthetic production of astaxanthin generally obtains an isomeric mixture of (3*S*,3′*S*), (3*R*,3′*S*), and (3*R*,3′*R*) in a ratio of 1:2:1. The separation of the active isomer from the synthetic mixture could increase production costs. In several studies, the in vitro or biological activity of the isomeric mixture of astaxanthin synthetically produced has been shown to be lower than that of the carotenoid obtained from natural sources. Lastly, natural astaxanthin has been reported to have higher stability and better bioavailability compared to some of the stereoisomers obtained synthetically [36,324]. In relation to fucoxanthin, some authors indicate that, although this carotenoid can be synthesized chemically, its extraction from brown seaweed is a more accessible, safe, and economical method. However, the content in brown seaweed varies greatly by species, geographical location, season, temperature, salinity, light intensity, and interactions among these factors [325].

Another aspect to consider is the presence of carotenoid esters naturally present in some marine organisms. Generally, carotenoid esters have a higher stability than the respective free carotenoids. As an example, in microalgal extracts were found ca. 70% astaxanthin-monoesters, 25% astaxanthin-diesters and 5% free astaxanthin [326]. In relation to chemical synthesis, the esterification of carotenoids could lead to an increase in the price of production.

It is also important to consider that both synthetic and natural carotenoids must pass strict analyses before they are allowed for human consumption. In this regard, both chemical synthesis and isolation from natural sources could generate undesirable compounds that should be eliminated from the final product. In some cases, for human consumption, it has been preferred to opt for natural sources instead of chemical synthesis. This is the case with astaxanthin. The natural form of this carotenoid was introduced as a human nutraceutical supplement in the late 1990s, following approval by the Food and Drug Administration (FDA) as a new dietary ingredient, while synthetic astaxanthin has not been officially registered for direct human use in any country by regulatory authorities and has been used predominantly for animal feed [209,320].

Another advantage of using marine organisms as sources of carotenoids is related to consumer acceptance. Consumers generally feel that synthetic compounds are less safe than those obtained from natural sources. For example, in a recent study, it was observed how chemicals of natural origin are considered to be healthier and safer than synthetic ones and how consumers frequently associated the word "synthetic" with "unnatural", "health hazard", and "environmental hazard" [327]. As a consequence, some consumers are willing to pay high prices for products of natural origin [320]. However, as mentioned earlier, one approach to enhancing the productivity of marine organisms as a carotenoid source is through genetic engineering. In this regard, it is important to note that a considerable portion of consumers who oppose the consumption of synthetically derived carotenoids are likely to hold similar reservations about consuming carotenoids obtained from genetically modified organisms.

The increased demand for natural carotenoids as food additives has spurred advancements in technological innovations for the manufacturing of microbial carotenoids. Numerous research studies have emphasized sustainable and biocompatible methods for manufacturing carotenoids from natural sources, promoting the use of safer carotenoids obtained through alternative means rather than chemical synthesis. The retro-sustainable biotechnological pathway can be accomplished through the implementation of a circular economy approach. This approach not only enables carotenoid recovery but also the possibility of obtaining additional value-added substances such as lipids and proteins associated with the recycling of raw materials for bioproduction and solvents used in extraction procedures. On the other hand, it is frequently thought that the production of synthetic carotenoids does not follow the principles of the circular economy and is destructive to the ecosystem. Thus, synthesis using nonrenewable sources reveals a negative impact on the environment and consumers [320].

However, there are important hurdles to overcome in order to increase the use of marine organisms to extract carotenoids instead of obtaining carotenoids from chemical synthesis. The main barriers are the acceptability by some consumers of new sources, the low bioavailability of carotenoids from natural sources (which can result in low bioactivity), and, of course, the safety of the product as food ingredients. Several research studies are needed, in particular about the metabolization of carotenoids from natural sources by humans and their health benefits, although there are many others. Further, the extraction of carotenoids from natural origin cannot yet compete in price with the chemical synthesis of carotenoids and could still present some negative environmental impacts. Thus, strategies and technologies (e.g., genetic engineering, CRISPR, fine-tuned cultivation conditions, sustainable and scalable extraction methodologies, and green solvents for biorefineries) may therefore be required to increase the production and use of carotenoid-rich natural products.

7. Concluding Remarks

Altogether, the organisms investigated in this review can accumulate both common dietary carotenoids (β-carotene, lutein, zeaxanthin, astaxanthin, canthaxanthin, or fucoxanthin in some countries) and unusual ones (e.g., bacterioruberin, salinixanthin, myxoxanthophyll, and β-zeacarotene), some of which are unique to some taxa. The fact that some studies provide evidence that unusual "marine carotenoids" exhibit important properties such as distinctive colors, high in vitro antioxidant capacity, or even health-promoting biological actions in studies of diverse nature is encouraging.

Some advantages of using marine organisms as a source of carotenoids over chemical synthesis are that, in some cases, chemical synthesis produces isomers with lower activity, bioavailability, and stability. In addition, marine organisms have a great capacity to synthesize a wide variety of carotenoids, which can be obtained in a renewable manner without depleting natural resources. Thus, marine organisms have the potential to be key sustainable sources of these compounds, which could help Europe achieve its Green Deal and Recovery Plan. However, to consolidate the industrial production of carotenoids from marine organisms, more research is needed to increase productivity and reduce costs. In summary, the need to promote the blue economy to help produce sustainable and health-promoting foods offers an exciting opportunity to tap into aquatic ecosystems and valorize carotenoid-rich organisms.

Author Contributions: P.M.-B. and A.J.M.-M. conceptualization and design of the idea, literature research, writing—original draft preparation, prepared figures and tables, supervision, writing—review and editing; P.G.-V., M.L.G., A.L.-V., R.L., J.M., C.R., V.S., S.V. and E.V. literature research, writing—original draft preparation, prepared figures and tables, writing—review and editing. All authors have read and agreed to the published version of the manuscript.

Funding: C.R. and J.M. contributions were financed by The Research Council of Norway (326803 SUSKELPFOOD; 319577 SAFER-IMTA, 294946 SBP-N) and Møreforsking AS. P.M.-B. and A.J.M.-M. acknowledge Grant PID2019-110438RB-C21 (NEWCARFOODS), funded by MCIN/AEI/10.13039/501100011033. P.G.-V., A.L.-V. and R.L. acknowledge FITORIBOPEP-P20_00728. PROY. I + D + I JUNTA ANDALUCIA 2020, AGL2019-110438RB-C22-AEI/FEDER, Fertinagro Biotech Foundation (Cátedra Fertinagro Biotech de la Universidad de Huelva). P.M.-B. was supported by a posdoc fellowship from Consejería de Transformación Económica, Industria, Conocimiento y Universidades de la Junta de Andalucía (PAIDI 2020). P.G.-V. and A.L.-V. wants to thank also Next Generation European Funds and the Ministry of Universities of Spain for funding the Recualificación del Profesorado Universitario system. V.S. and E.V. acknowledge CSIC I + D 2018 250 UdelaR, ANII POS NAC 2014 1 102321, and Ph.D. grant CAP UdelaR 2017–2019.

Acknowledgments: This article is based upon work from COST Actions SEAWHEAT and EUROCAROTEN (CA20106, https://seawheatcost.haifa.ac.il/ (accessed on 30 May 2023) & CA15136 www.eurocaroten.eu (accessed on 30 May 2023)), supported by COST (European Cooperation in Science and Technology). A.J.M.-M. and P.M.-B. acknowledge MICROAGRO network (Red 121RT0110, https://www.cyted.org/?q=es/detalle_proyecto&un=1013 (accessed on 30 May 2023)) funded by CYTED (http://www.cyted.org/ (accessed on 30 May 2023)) and are member of the Carotenoid Network: from microbia and plants to food and health (CaRed) (BIO2015-71703-REDT), funded by the Spanish Ministry of Economy and Competitiveness, and CaRed: Red española de carotenoides (BIO2017-90877-REDT), funded by the Spanish Ministry of Economy, Industry and Competitiveness. M.L.G. and S.V. acknowledge MICROAGRO network (Red 121RT0110, https://www.cyted.org/?q=es/detalle_proyecto&un=1013 (accessed on 30 May 2023)) funded by CYTED (http://www.cyted.org/ (accessed on 30 May 2023)).

Conflicts of Interest: A.J.M.-M. carries out consultancy work for diverse companies.

References

1. Willett, W.; Rockström, J.; Loken, B.; Springmann, M.; Lang, T.; Vermeulen, S.; Garnett, T.; Tilman, D.; DeClerck, F.; Wood, A.; et al. Food in the Anthropocene: The EAT—Lancet Commission on Healthy Diets from Sustainable Food Systems. *Lancet* **2019**, *393*, 447–492. [CrossRef]
2. European Commission. *Blue Economy Report 2022*; European Union: Brussels, Belgium, 2022.
3. Meléndez-Martínez, A.J.; Mandić, A.I.; Bantis, F.; Böhm, V.; Borge, G.I.A.; Brnčić, M.; Bysted, A.; Cano, M.P.; Dias, M.G.; Elgersma, A.; et al. A Comprehensive Review on Carotenoids in Foods and Feeds: Status Quo, Applications, Patents, and Research Needs. *Crit. Rev. Food Sci. Nutr.* **2021**, *62*, 1999–2049. [CrossRef]
4. Sosa-Hernández, J.; Romero-Castillo, K.; Parra-Arroyo, L.; Aguilar-Aguila-Isaías, M.; García-Reyes, I.; Ahmed, I.; Parra-Saldivar, R.; Bilal, M.; Iqbal, H. Mexican Microalgae Biodiversity and State-Of-The-Art Extraction Strategies to Meet Sustainable Circular Economy Challenges: High-Value Compounds and Their Applied Perspectives. *Mar. Drugs* **2019**, *17*, 174. [CrossRef]
5. Cotas, J.; Gomes, L.; Pacheco, D.; Pereira, L. Ecosystem Services Provided by Seaweeds. *Hydrobiology* **2023**, *2*, 75–96. [CrossRef]
6. Caporgno, M.P.; Mathys, A. Trends in Microalgae Incorporation Into Innovative Food Products with Potential Health Benefits. *Front. Nutr.* **2018**, *5*, 1–10. [CrossRef]
7. Krause-Jensen, D.; Duarte, C.M. Substantial Role of Macroalgae in Marine Carbon Sequestration. *Nat. Geosci.* **2016**, *9*, 737–742. [CrossRef]
8. Enzing, C.; Ploeg, M.; Barbosa, M.; Sijtsma, L. *Microalgae-Based Products for the Food and Feed Sector: An Outlook for Europe*; Vigani, M., Parisi, C., Rodriguez Cerezo, E., Eds.; Publications Office of the European Union: Luxembourg, 2014; ISBN 9789279340376.
9. Ferdouse, F.; Løvstad Holdt, S.; Smith, R.; Murúa, P.; Yang, L. The Global Status of Seaweed Production. Trade and Utilization, Globefish Research Program. *FAO* **2018**, *124*, 1–120.

10. European Market Observatory for Fisheries and Aquaculture (EUMOFA). *Blue Bioeconomy: Situation Report and Perspectives*; EUMOFA: Luxembourg, 2018.
11. Nova, P.; Martins, A.P.; Teixeira, C.; Abreu, H.; Silva, J.G.; Silva, A.M.; Freitas, A.C.; Gomes, A.M. Foods with Microalgae and Seaweeds Fostering Consumers Health: A Review on Scientific and Market Innovations. *J. Appl. Phycol.* **2020**, *32*, 1789–1802. [CrossRef]
12. Venkata Mohan, S.; Nikhil, G.N.; Chiranjeevi, P.; Nagendranatha Reddy, C.; Rohit, M.V.; Kumar, A.N.; Sarkar, O. Waste Biorefinery Models towards Sustainable Circular Bioeconomy: Critical Review and Future Perspectives. *Bioresour. Technol.* **2016**, *215*, 2–12. [CrossRef] [PubMed]
13. Venil, C.K.; Aruldass, C.A.; Dufossé, L.; Zakaria, Z.A.; Ahmad, W.A. Current Perspective on Bacterial Pigments: Emerging Sustainable Compounds with Coloring and Biological Properties for the Industry—An Incisive Evaluation. *RSC Adv.* **2014**, *4*, 39523. [CrossRef]
14. Bourdichon, F.; Casaregola, S.; Farrokh, C.; Frisvad, J.C.; Gerds, M.L.; Hammes, W.P.; Harnett, J.; Huys, G.; Laulund, S.; Ouwehand, A.; et al. Food Fermentations: Microorganisms with Technological Beneficial Use. *Int. J. Food Microbiol.* **2012**, *154*, 87–97. [CrossRef]
15. Gerhauser, C. Impact of Dietary Gut Microbial Metabolites on the Epigenome. *Philos. Trans. R. Soc. B Biol. Sci.* **2018**, *373*, 20170359. [CrossRef] [PubMed]
16. Suez, J.; Zmora, N.; Segal, E.; Elinav, E. The Pros, Cons, and Many Unknowns of Probiotics. *Nat. Med.* **2019**, *25*, 716–729. [CrossRef] [PubMed]
17. Hadrich, D. Microbiome Research Is Becoming the Key to Better Understanding Health and Nutrition. *Front. Genet.* **2018**, *9*, 212. [CrossRef]
18. Nevejan, N.; De Schryver, P.; Wille, M.; Dierckens, K.; Baruah, K.; Van Stappen, G. Bacteria as Food in Aquaculture: Do They Make a Difference? *Rev. Aquac.* **2018**, *10*, 180–212. [CrossRef]
19. Grivard, A.; Goubet, I.; de Souza Duarte Filho, L.M.; Thiéry, V.; Chevalier, S.; de Oliveira-Junior, R.G.; El Aouad, N.; Guedes da Silva Almeida, J.R.; Sitarek, P.; Quintans-Junior, L.J.; et al. Archaea Carotenoids: Natural Pigments with Unexplored Innovative Potential. *Mar. Drugs* **2022**, *20*, 524. [CrossRef] [PubMed]
20. Kutty, S.N.; Philip, R. Marine Yeasts-a Review. *Yeast* **2008**, *25*, 465–483. [CrossRef] [PubMed]
21. Chi, Z.; Liu, G.-L.; Lu, Y.; Jiang, H.; Chi, Z.-M. Bio-Products Produced by Marine Yeasts and Their Potential Applications. *Bioresour. Technol.* **2016**, *202*, 244–252. [CrossRef]
22. Carlozzi, P.; Touloupakis, E.; Di Lorenzo, T.; Giovannelli, A.; Seggiani, M.; Cinelli, P.; Lazzeri, A. Whey and Molasses as Inexpensive Raw Materials for Parallel Production of Biohydrogen and Polyesters via a Two-Stage Bioprocess: New Routes towards a Circular Bioeconomy. *J. Biotechnol.* **2019**, *303*, 37–45. [CrossRef]
23. Gupta, V.; Goel, R. Managing Dissolved Methane Gas in Anaerobic Effluents Using Microbial Resource Management-Based Strategies. *Bioresour. Technol.* **2019**, *289*, 121601. [CrossRef]
24. Sarsaiya, S.; Jain, A.; Kumar Awasthi, S.; Duan, Y.; Kumar Awasthi, M.; Shi, J. Microbial Dynamics for Lignocellulosic Waste Bioconversion and Its Importance with Modern Circular Economy, Challenges and Future Perspectives. *Bioresour. Technol.* **2019**, *291*, 121905. [CrossRef]
25. Tao, Y.; Ersahin, M.E.; Ghasimi, D.S.M.; Ozgun, H.; Wang, H.; Zhang, X.; Guo, M.; Yang, Y.; Stuckey, D.C.; van Lier, J.B. Biogas Productivity of Anaerobic Digestion Process Is Governed by a Core Bacterial Microbiota. *Chem. Eng. J.* **2020**, *380*, 122425. [CrossRef]
26. López, A.R.; Rodríguez, S.B.; Vallejo, R.A.; García, P.G.; Macías-Sánchez, M.D.; Díaz, M.G.; Librán, R.G.; Acero, F.J.F. Sustainable Cultivation of *Nannochloropsis gaditana* Microalgae in Outdoor Raceways Using Flue Gases for a Complete 2-Year Cycle: A Circular Economy Challenge. *J. Appl. Phycol.* **2019**, *31*, 1515–1523. [CrossRef]
27. Stiles, W.A.V.; Styles, D.; Chapman, S.P.; Esteves, S.; Bywater, A.; Melville, L.; Silkina, A.; Lupatsch, I.; Fuentes Grünewald, C.; Lovitt, R.; et al. Using Microalgae in the Circular Economy to Valorise Anaerobic Digestate: Challenges and Opportunities. *Bioresour. Technol.* **2018**, *267*, 732–742. [CrossRef]
28. Rodríguez, P.D.; Arce Bastias, F.; Arena, A.P. Modeling and Environmental Evaluation of a System Linking a Fishmeal Facility with a Microalgae Plant within a Circular Economy Context. *Sustain. Prod. Consum.* **2019**, *20*, 356–364. [CrossRef]
29. Mata-Gómez, L.C.; Montañez, J.C.; Méndez-Zavala, A.; Aguilar, C.N. Biotechnological Production of Carotenoids by Yeasts: An Overview. *Microb. Cell Fact.* **2014**, *13*, 1–11. [CrossRef] [PubMed]
30. Ding, S.; Liu, Y. Adsorption of CO_2 from Flue Gas by Novel Seaweed-Based KOH-Activated Porous Biochars. *Fuel* **2020**, *260*, 116382. [CrossRef]
31. Israel, A.; Gavrieli, J.; Glazer, A.; Friedlander, M. Utilization of Flue Gas from a Power Plant for Tank Cultivation of the Red Seaweed *Gracilaria cornea*. *Aquaculture* **2005**, *249*, 311–316. [CrossRef]
32. Batista, A.P.; Raymundo, A.; Sousa, I.; Empis, J. Rheological Characterization of Coloured Oil-in-Water Food Emulsions with Lutein and Phycocyanin Added to the Oil and Aqueous Phases. *Food Hydrocoll.* **2006**, *20*, 44–52. [CrossRef]
33. Meléndez-Martínez, A.J. An Overview of Carotenoids, Apocarotenoids and Vitamin A in Agro-Food, Nutrition, Health and Disease. *Mol. Nutr. Food Res.* **2019**, *63*, 1801045. [CrossRef]
34. Meléndez-Martínez, A.J.; Stinco, C.M.; Mapelli-Brahm, P. Skin Carotenoids in Public Health and Nutricosmetics: The Emerging Roles and Applications of the UV Radiation-Absorbing Colourless Carotenoids Phytoene and Phytofluene. *Nutrients* **2019**, *11*, 1093. [CrossRef] [PubMed]
35. Rodriguez-Concepcion, M.; Avalos, J.; Bonet, M.L.; Boronat, A.; Gomez-Gomez, L.; Hornero-Mendez, D.; Carmen Limon, M.; Meléndez-Martínez, A.J.; Olmedilla-Alonso, B.; Palou, A.; et al. A Global Perspective on Carotenoids: Metabolism, Biotechnology, and Benefits for Nutrition and Health. *Prog. Lipid Res.* **2018**, *70*, 62–93. [CrossRef] [PubMed]

36. Stachowiak, B.; Szulc, P. Astaxanthin for the Food Industry. *Molecules* **2021**, *26*, 2666. [CrossRef] [PubMed]
37. Rapoport, A.; Guzhova, I.; Bernetti, L.; Buzzini, P.; Kieliszek, M.; Kot, A.M. Carotenoids and Some Other Pigments from Fungi and Yeasts. *Metabolites* **2021**, *11*, 92. [CrossRef]
38. Woese, C.R.; Kandler, O.; Wheelis, M.L. Towards a Natural System of Organisms: Proposal for the Domains Archaea, Bacteria, and Eucarya. *Proc. Natl. Acad. Sci. USA* **1990**, *87*, 4576–4579. [CrossRef] [PubMed]
39. DeLong, E.F. Exploring Marine Planktonic Archaea: Then and Now. *Front. Microbiol.* **2021**, *11*, 616086. [CrossRef]
40. Heal, K.R.; Qin, W.; Ribalet, F.; Bertagnolli, A.D.; Coyote-Maestas, W.; Hmelo, L.R.; Moffett, J.W.; Devol, A.H.; Armbrust, E.V.; Stahl, D.A.; et al. Two Distinct Pools of B 12 Analogs Reveal Community Interdependencies in the Ocean. *Proc. Natl. Acad. Sci. USA* **2017**, *114*, 364–369. [CrossRef]
41. Santoro, A.E.; Richter, R.A.; Dupont, C.L. Planktonic Marine Archaea. *Ann. Rev. Mar. Sci.* **2019**, *11*, 131–158. [CrossRef]
42. Bomberg, M.; Montonen, L.; Münster, U.; Jurgens, G. Diversity and Function of Archaea in Freshwater Habitats. *Curr. Trends Microbiol.* **2008**, *4*, 61–89.
43. Juottonen, H.; Fontaine, L.; Wurzbacher, C.; Drakare, S.; Peura, S.; Eiler, A. Archaea in Boreal Swedish Lakes Are Diverse, Dominated by Woesearchaeota and Follow Deterministic Community Assembly. *Environ. Microbiol.* **2020**, *22*, 3158–3171. [CrossRef]
44. Chaban, B.; Ng, S.Y.M.; Jarrell, K.F. Archaeal Habitats—From the Extreme to the Ordinary. *Can. J. Microbiol.* **2006**, *52*, 73–116. [CrossRef]
45. Miralles-Robledillo, J.M.; Bernabeu, E.; Giani, M.; Martínez-Serna, E.; Martínez-Espinosa, R.M.; Pire, C. Distribution of Denitrification among Haloarchaea: A Comprehensive Study. *Microorganisms* **2021**, *9*, 1669. [CrossRef] [PubMed]
46. Pfeifer, K.; Ergal, İ.; Koller, M.; Basen, M.; Schuster, B.; Rittmann, S.K.-M.R. Archaea Biotechnology. *Biotechnol. Adv.* **2021**, *47*, 107668. [CrossRef] [PubMed]
47. Halorubin. Available online: https://cosmetics.specialchem.com/product/i-adeka-halorubin (accessed on 30 May 2023).
48. Abbes, M.; Baati, H.; Guermazi, S.; Messina, C.; Santulli, A.; Gharsallah, N.; Ammar, E. Biological Properties of Carotenoids Extracted from *Halobacterium halobium* Isolated from a Tunisian Solar Saltern. *BMC Complement. Altern. Med.* **2013**, *13*, 255. [CrossRef]
49. Yatsunami, R.; Ando, A.; Yang, Y.; Takaichi, S.; Kohno, M.; Matsumura, Y.; Ikeda, H.; Fukui, T.; Nakasone, K.; Fujita, N.; et al. Identification of Carotenoids from the Extremely Halophilic Archaeon *Haloarcula japonica*. *Front. Microbiol.* **2014**, *5*, 1–5. [CrossRef] [PubMed]
50. de la Vega, M.; Sayago, A.; Ariza, J.; Barneto, A.G.; León, R. Characterization of a Bacterioruberin-Producing Haloarchaea Isolated from the Marshlands of the Odiel River in the Southwest of Spain. *Biotechnol. Prog.* **2016**, *32*, 592–600. [CrossRef]
51. Squillaci, G.; Parrella, R.; Carbone, V.; Minasi, P.; La Cara, F.; Morana, A. Carotenoids from the Extreme Halophilic Archaeon *Haloterrigena turkmenica*: Identification and Antioxidant Activity. *Extremophiles* **2017**, *21*, 933–945. [CrossRef]
52. Sahli, K.; Gomri, M.A.; Esclapez, J.; Gómez-Villegas, P.; Ghennai, O.; Bonete, M.-J.; León, R.; Kharroub, K. Bioprospecting and Characterization of Pigmented Halophilic Archaeal Strains from Algerian Hypersaline Environments with Analysis of Carotenoids Produced by *Halorubrum* sp. BS2. *J. Basic Microbiol.* **2020**, *60*, 624–638. [CrossRef]
53. Lizama, C.; Romero-Parra, J.; Andrade, D.; Riveros, F.; Bórquez, J.; Ahmed, S.; Venegas-Salas, L.; Cabalín, C.; Simirgiotis, M.J. Analysis of Carotenoids in *Haloarchaea* Species from Atacama Saline Lakes by High Resolution UHPLC-Q-Orbitrap-Mass Spectrometry: Antioxidant Potential and Biological Effect on Cell Viability. *Antioxidants* **2021**, *10*, 1230. [CrossRef]
54. Giani, M.; Gervasi, L.; Loizzo, M.R.; Martínez-Espinosa, R.M. Carbon Source Influences Antioxidant, Antiglycemic, and Antilipidemic Activities of Haloferax Mediterranei Carotenoid Extracts. *Mar. Drugs* **2022**, *20*, 659. [CrossRef]
55. Yang, Y.; Yatsunami, R.; Ando, A.; Miyoko, N.; Fukui, T.; Takaichi, S.; Nakamura, S. Complete Biosynthetic Pathway of the C 50 Carotenoid Bacterioruberin from Lycopene in the Extremely Halophilic Archaeon *Haloarcula japonica*. *J. Bacteriol.* **2015**, *197*, 1614–1623. [CrossRef]
56. Alvares, J.J.; Furtado, I.J. Characterization of Multicomponent Antioxidants from *Haloferax alexandrinus* GUSF-1 (KF796625). *3 Biotech* **2021**, *11*, 1–12. [CrossRef]
57. Hou, J.; Cui, H.-L. In Vitro Antioxidant, Antihemolytic, and Anticancer Activity of the Carotenoids from Halophilic Archaea. *Curr. Microbiol.* **2018**, *75*, 266–271. [CrossRef] [PubMed]
58. Hegazy, G.E.; Abu-Serie, M.M.; Abo-Elela, G.M.; Ghozlan, H.; Sabry, S.A.; Soliman, N.A.; Abdel-Fattah, Y.R. In Vitro Dual (Anticancer and Antiviral) Activity of the Carotenoids Produced by Haloalkaliphilic Archaeon *Natrialba* sp. M6. *Sci. Rep.* **2020**, *10*, 5986. [CrossRef] [PubMed]
59. Gómez-Villegas, P.; Vigara, J.; Vila, M.; Varela, J.; Barreira, L.; Léon, R. Antioxidant, Antimicrobial, and Bioactive Potential of Two New Haloarchaeal Strains Isolated from Odiel Salterns (Southwest Spain). *Biology* **2020**, *9*, 298. [CrossRef]
60. Zalazar, L.; Pagola, P.; Miró, M.V.; Churio, M.S.; Cerletti, M.; Martínez, C.; Iniesta-Cuerda, M.; Soler, A.J.; Cesari, A.; De Castro, R. Bacterioruberin Extracts from a Genetically Modified Hyperpigmented *Haloferax volcanii* Strain: Antioxidant Activity and Bioactive Properties on Sperm Cells. *J. Appl. Microbiol.* **2019**, *126*, 796–810. [CrossRef] [PubMed]
61. D'Souza, S.E.; Altekar, W.; D'Souza, S.F. Adaptive Response of Haloferax Mediterranei to Low Concentrations of NaCl (<20%) in the Growth Medium. *Arch. Microbiol.* **1997**, *168*, 68–71. [CrossRef] [PubMed]
62. Fang, C.-J.; Ku, K.-L.; Lee, M.-H.; Su, N.-W. Influence of Nutritive Factors on C50 Carotenoids Production by Haloferax Mediterranei ATCC 33500 with Two-Stage Cultivation. *Bioresour. Technol.* **2010**, *101*, 6487–6493. [CrossRef]
63. Calegari-Santos, R.; Diogo, R.A.; Fontana, J.D.; Bonfim, T.M.B. Carotenoid Production by Halophilic Archaea under Different Culture Conditions. *Curr. Microbiol.* **2016**, *72*, 641–651. [CrossRef]
64. Tatiana, A.; Yasynska, O.; Rivas, P.C.; Lilia, E.; Jose, M. Journal of Drug Delivery Science and Technology Improved Stability and Biological Activity of Bacterioruberin in Nanovesicles. *J. Drug Deliv. Sci. Technol.* **2022**, *77*, 103896. [CrossRef]

65. Chaari, M.; Theochari, I.; Papadimitriou, V.; Xenakis, A.; Ammar, E. Encapsulation of Carotenoids Extracted from Halophilic Archaea in Oil-in-Water (O/W) Micro- and Nano-Emulsions. *Colloids Surf. B Biointerfaces* **2018**, *161*, 219–227. [CrossRef]
66. Pagels, F.; Vasconcelos, V.; Guedes, A.C. Carotenoids from Cyanobacteria: Biotechnological Potential and Optimization Strategies. *Biomolecules* **2021**, *11*, 735. [CrossRef]
67. George, D.M.; Vincent, A.S.; Mackey, H.R. An Overview of Anoxygenic Phototrophic Bacteria and Their Applications in Environmental Biotechnology for Sustainable Resource Recovery. *Biotechnol. Rep.* **2020**, *28*, e00563. [CrossRef]
68. Sandmann, G. Diversity and Origin of Carotenoid Biosynthesis: Its History of Coevolution towards Plant Photosynthesis. *N. Phytol.* **2021**, *232*, 479–493. [CrossRef]
69. Béjà, O.; Suzuki, M.T. Photoheterotrophic Marine Prokaryotes. In *Microbial Ecology of the Oceans*; John Wiley & Sons, Inc.: Hoboken, NJ, USA, 2008; pp. 131–157.
70. Imhoff, J. True Marine and Halophilic Anoxygenic Phototrophic Bacteria. *Arch. Microbiol.* **2001**, *176*, 243–254. [CrossRef] [PubMed]
71. Hubas, C.; Jesus, B.; Passarelli, C.; Jeanthon, C. Tools Providing New Insight into Coastal Anoxygenic Purple Bacterial Mats: Review and Perspectives. *Res. Microbiol.* **2011**, *162*, 858–868. [CrossRef] [PubMed]
72. Hess, W.R.; Garczarek, L.; Pfreundt, U.; Partensky, F. Phototrophic Microorganisms: The Basis of the Marine Food Web. In *The Marine Microbiome*; Springer International Publishing: Berlin, Germany, 2016; pp. 57–97.
73. Gómez-Consarnau, L.; Akram, N.; Lindell, K.; Pedersen, A.; Neutze, R.; Milton, D.L.; González, J.M.; Pinhassi, J. Proteorhodopsin Phototrophy Promotes Survival of Marine Bacteria during Starvation. *PLoS Biol.* **2010**, *8*, e1000358. [CrossRef]
74. Béjà, O.; Spudich, E.N.; Spudich, J.L.; Leclerc, M.; DeLong, E.F. Proteorhodopsin Phototrophy in the Ocean. *Nature* **2001**, *411*, 786–789. [CrossRef]
75. Armstrong, G.A. Genetics of Eubacterial Carotenoids Biosynthesis: A Colorful Tale. *Annu. Rev. Microbiol.* **1997**, *51*, 629–659. [CrossRef]
76. Stafsnes, M.H.; Josefsen, K.D.; Kildahl-Andersen, G.; Valla, S.; Ellingsen, T.E.; Bruheim, P. Isolation and Characterization of Marine Pigmented Bacteria from Norwegian Coastal Waters and Screening for Carotenoids with UVA-Blue Light Absorbing Properties. *J. Microbiol.* **2010**, *48*, 16–23. [CrossRef] [PubMed]
77. Dieser, M.; Greenwood, M.; Foreman, C.M. Carotenoid Pigmentation in Antarctic Heterotrophic Bacteria as a Strategy to Withstand Environmental Stresses. *Arct. Antarct. Alp. Res.* **2010**, *42*, 396–405. [CrossRef]
78. Ram, S.; Mitra, M.; Shah, F.; Tirkey, S.R.; Mishra, S. Bacteria as an Alternate Biofactory for Carotenoid Production: A Review of Its Applications, Opportunities and Challenges. *J. Funct. Foods* **2020**, *67*, 103867. [CrossRef]
79. Kristjansdottir, T.; Ron, E.Y.C.; Molins-Delgado, D.; Fridjonsson, O.H.; Turner, C.; Bjornsdottir, S.H.; Gudmundsson, S.; van Niel, E.W.J.; Karlsson, E.N.; Hreggvidsson, G.O. Engineering the Carotenoid Biosynthetic Pathway in *Rhodothermus marinus* for Lycopene Production. *Metab. Eng. Commun.* **2020**, *11*, e00140. [CrossRef]
80. Park, S.; Cho, S.W.; Lee, Y.; Choi, M.; Yang, J.; Lee, H.; Seo, S.W. Engineering *Vibrio* sp. SP1 for the Production of Carotenoids Directly from Brown Macroalgae. *Comput. Struct. Biotechnol. J.* **2021**, *19*, 1531–1540. [CrossRef]
81. Shindo, K.; Misawa, N. New and Rare Carotenoids Isolated from Marine Bacteria and Their Antioxidant Activities. *Mar. Drugs* **2014**, *12*, 1690–1698. [CrossRef] [PubMed]
82. Yoon, J.; Lee, E.-Y.; Nam, S.-J. Erythrobacter *Rubeus* sp. Nov., a Carotenoid-Producing Alphaproteobacterium Isolated from Coastal Seawater. *Arch. Microbiol.* **2022**, *204*, 125. [CrossRef]
83. Vila, E.; Hornero-Méndez, D.; Lareo, C.; Saravia, V. Biotechnological Production of Zeaxanthin by an Antarctic *Flavobacterium*: Evaluation of Culture Conditions. *J. Biotechnol.* **2020**, *319*, 54–60. [CrossRef]
84. Liu, H.; Zhang, C.; Zhang, X.; Tan, K.; Zhang, H.; Cheng, D.; Ye, T.; Li, S.; Ma, H.; Zheng, H. A Novel Carotenoids-Producing Marine Bacterium from Noble Scallop *Chlamys nobilis* and Antioxidant Activities of Its Carotenoid Compositions. *Food Chem.* **2020**, *320*, 126629. [CrossRef]
85. Niero, H.; da Silva, M.A.C.; de Felicio, R.; Trivella, D.B.B.; Lima, A.O.d.S. Carotenoids Produced by the Deep-Sea Bacterium *Erythrobacter citreus* LAMA 915: Detection and Proposal of Their Biosynthetic Pathway. *Folia Microbiol.* **2021**, *66*, 441–456. [CrossRef]
86. Setiyono, E.; Heriyanto; Pringgenies, D.; Shioi, Y.; Kanesaki, Y.; Awai, K.; Brotosudarmo, T.H.P. Sulfur-Containing Carotenoids from A Marine Coral Symbiont *Erythrobacter flavus* Strain KJ5. *Mar. Drugs* **2019**, *17*, 349. [CrossRef]
87. Sayed, A.; Elbalasy, I.; Mohamed, M.S. Novel β-Carotene and Astaxanthin-Producing Marine *Planococcus* sp.: Insights into Carotenogenesis Regulation and Genetic Aspects. *Appl. Biochem. Biotechnol.* **2022**, *195*, 217–235. [CrossRef] [PubMed]
88. Asker, D. Isolation and Characterization of a Novel, Highly Selective Astaxanthin-Producing Marine Bacterium. *J. Agric. Food Chem.* **2017**, *65*, 9101–9109. [CrossRef]
89. Ide, T.; Hoya, M.; Tanaka, T.; Harayama, S. Enhanced Production of Astaxanthin in *Paracoccus* sp. Strain N-81106 by Using Random Mutagenesis and Genetic Engineering. *Biochem. Eng. J.* **2012**, *65*, 37–43. [CrossRef]
90. Sowmya, R.; Sachindra, N.M. Biochemical and Molecular Characterization of Carotenogenic Flavobacterial Isolates from Marine Waters. *Pol. J. Microbiol.* **2016**, *65*, 77–88. [CrossRef] [PubMed]
91. Khomlaem, C.; Aloui, H.; Oh, W.-G.; Kim, B.S. High Cell Density Culture of *Paracoccus* sp. LL1 in Membrane Bioreactor for Enhanced Co-Production of Polyhydroxyalkanoates and Astaxanthin. *Int. J. Biol. Macromol.* **2021**, *192*, 289–297. [CrossRef]
92. Choi, S.S.; Seo, Y.B.; Nam, S.-W.; Kim, G.-D. Enhanced Production of Astaxanthin by Co-Culture of *Paracoccus haeundaensis* and Lactic Acid Bacteria. *Front. Mar. Sci.* **2021**, *7*, 597553. [CrossRef]
93. Joshi, C.; Singhal, R.S. Modelling and Optimization of Zeaxanthin Production by *Paracoccus zeaxanthinifaciens* ATCC 21588 Using Hybrid Genetic Algorithm Techniques. *Biocatal. Agric. Biotechnol.* **2016**, *8*, 228–235. [CrossRef]

94. Joshi, C.; Singhal, R.S. Zeaxanthin Production by Paracoccus Zeaxanthinifaciens ATCC 21588 in a Lab-Scale Bubble Column Reactor: Artificial Intelligence Modelling for Determination of Optimal Operational Parameters and Energy Requirements. *Korean J. Chem. Eng.* **2018**, *35*, 195–203. [CrossRef]
95. Takaichi, S.; Mochimaru, M.; Maoka, T.; Katoh, H. Myxol and 4-Ketomyxol 2′-Fucosides, Not Rhamnosides, from *Anabaena* sp. PCC 7120 and Nostoc Punctiforme PCC 73102, and Proposal for the Biosynthetic Pathway of Carotenoids. *Plant Cell Physiol.* **2005**, *46*, 497–504. [CrossRef]
96. Takaichi, S.; Mochimaru, M.; Maoka, T. Presence of Free Myxol and 4-Hydroxymyxol and Absence of Myxol Glycosides in *Anabaena variabilis* ATCC 29413, and Proposal of a Biosynthetic Pathway of Carotenoids. *Plant Cell Physiol.* **2006**, *47*, 211–216. [CrossRef]
97. Takaichi, S.; Maoka, T.; Masamoto, K. Myxoxanthophyll in *Synechocystis* sp. PCC 6803 Is Myxol 2′-Dimethyl-Fucoside, (3R,2′S)-Myxol 2′-(2,4-Di-O-Methyl-α-l-Fucoside), Not Rhamnoside. *Plant Cell Physiol.* **2001**, *42*, 756–762. [CrossRef] [PubMed]
98. Britton, G.; Brown, D.J.; Goodwin, T.W.; Leuenberger, F.J.; Schocher, A.J. The Carotenoids of *Flavobacterium* Strain R1560. *Arch. Microbiol.* **1977**, *113*, 33–37. [CrossRef] [PubMed]
99. Sowmya, R.; Sachindra, N.M. Carotenoid Production by *Formosa* sp. KMW, a Marine Bacteria of *Flavobacteriaceae* Family: Influence of Culture Conditions and Nutrient Composition. *Biocatal. Agric. Biotechnol.* **2015**, *4*, 559–567. [CrossRef]
100. Vila, E.; Hornero-Méndez, D.; Azziz, G.; Lareo, C.; Saravia, V. Carotenoids from Heterotrophic Bacteria Isolated from Fildes Peninsula, King George Island, Antarctica. *Biotechnol. Rep.* **2019**, *21*, e00306. [CrossRef] [PubMed]
101. Lee, J.H.; Kim, Y.S.; Choi, T.-J.; Lee, W.J.; Kim, Y.T. Paracoccus *Haeundaensis* sp. Nov., a Gram-Negative, Halophilic, Astaxanthin-Producing Bacterium. *Int. J. Syst. Evol. Microbiol.* **2004**, *54*, 1699–1702. [CrossRef]
102. Berry, A.; Janssens, D.; Hümbelin, M.; Jore, J.P.M.; Hoste, B.; Cleenwerck, I.; Vancanneyt, M.; Bretzel, W.; Mayer, A.F.; Lopez-Ulibarri, R.; et al. *Paracoccus zeaxanthinifaciens* sp. Nov., a Zeaxanthin-Producing Bacterium. *Int. J. Syst. Evol. Microbiol.* **2003**, *53*, 231–238. [CrossRef]
103. Takaichi, S.; Mochimaru, M. Carotenoids and Carotenogenesis in Cyanobacteria: Unique Ketocarotenoids and Carotenoid Glycosides. *Cell. Mol. Life Sci.* **2007**, *64*, 2607–2619. [CrossRef]
104. Sarnaik, A.; Nambissan, V.; Pandit, R.; Lali, A. Recombinant *Synechococcus elongatus* PCC 7942 for Improved Zeaxanthin Production under Natural Light Conditions. *Algal Res.* **2018**, *36*, 139–151. [CrossRef]
105. Shimada, N.; Okuda, Y.; Maeda, K.; Umeno, D.; Takaichi, S.; Ikeuchi, M. Astaxanthin Production in a Model Cyanobacterium *Synechocystis* sp. PCC 6803. *J. Gen. Appl. Microbiol.* **2020**, *66*, 116–120. [CrossRef]
106. Sandmann, G.; Albrecht, M.; Schnurr, G.; Knörzer, O.; Böger, P. The Biotechnological Potential and Design of Novel Carotenoids by Gene Combination in *Escherichia coli*. *Trends Biotechnol.* **1999**, *17*, 233–237. [CrossRef]
107. Schmidt-Dannert, C. Engineering Novel Carotenoids in Microorganisms. *Curr. Opin. Biotechnol.* **2000**, *11*, 255–261. [CrossRef]
108. Henke, N.; Heider, S.; Peters-Wendisch, P.; Wendisch, V. Production of the Marine Carotenoid Astaxanthin by Metabolically Engineered *Corynebacterium glutamicum*. *Mar. Drugs* **2016**, *14*, 124. [CrossRef] [PubMed]
109. Buijs, Y.; Bech, P.K.; Vazquez-Albacete, D.; Bentzon-Tilia, M.; Sonnenschein, E.C.; Gram, L.; Zhang, S.-D. Marine Proteobacteria as a Source of Natural Products: Advances in Molecular Tools and Strategies. *Nat. Prod. Rep.* **2019**, *36*, 1333–1350. [CrossRef] [PubMed]
110. Takemura, M.; Takagi, C.; Aikawa, M.; Araki, K.; Choi, S.-K.; Itaya, M.; Shindo, K.; Misawa, N. Heterologous Production of Novel and Rare C30-Carotenoids Using *Planococcus* Carotenoid Biosynthesis Genes. *Microb. Cell Fact.* **2021**, *20*, 194. [CrossRef]
111. Bech, P.K.; Lysdal, K.L.; Gram, L.; Bentzon-Tilia, M.; Strube, M.L. Marine Sediments Hold an Untapped Potential for Novel Taxonomic and Bioactive Bacterial Diversity. *mSystems* **2020**, *5*, e00782-20. [CrossRef]
112. Joint, I.; Mühling, M.; Querellou, J. Culturing Marine Bacteria—An Essential Prerequisite for Biodiscovery. *Microb. Biotechnol.* **2010**, *3*, 564–575. [CrossRef]
113. Overmann, J.; Lepleux, C. Marine Bacteria and Archaea: Diversity, Adaptations, and Culturability. In *The Marine Microbiome*; Springer International Publishing: Berlin, Germany, 2016; pp. 21–55.
114. Matallana-Surget, S.; Joux, F.; Lebaron, P.; Cavicchioli, R. Isolement et Caractérisation de Bactéries Marines Oligotrophes. *J. Soc. Biol.* **2007**, *201*, 41–50. [CrossRef] [PubMed]
115. Rygaard, A.M.; Thøgersen, M.S.; Nielsen, K.F.; Gram, L.; Bentzon-Tilia, M. Effects of Gelling Agent and Extracellular Signaling Molecules on the Culturability of Marine Bacteria. *Appl. Environ. Microbiol.* **2017**, *83*, e00243-17. [CrossRef]
116. Stewart, E.J. Growing Unculturable Bacteria. *J. Bacteriol.* **2012**, *194*, 4151–4160. [CrossRef]
117. Kato, S.; Yamagishi, A.; Daimon, S.; Kawasaki, K.; Tamaki, H.; Kitagawa, W.; Abe, A.; Tanaka, M.; Sone, T.; Asano, K.; et al. Isolation of Previously Uncultured Slow-Growing Bacteria by Using a Simple Modification in the Preparation of Agar Media. *Appl. Environ. Microbiol.* **2018**, *84*, e00807-18. [CrossRef]
118. Ben-Dov, E.; Kramarsky-Winter, E.; Kushmaro, A. An in Situ Method for Cultivating Microorganisms Using a Double Encapsulation Technique. *FEMS Microbiol. Ecol.* **2009**, *68*, 363–371. [CrossRef] [PubMed]
119. Manca, F.; Mulà, C.; Gustafsson, C.; Mauri, A.; Roslin, T.; Thomas, D.N.; Benedetti-Cecchi, L.; Norkko, A.; Strona, G. Unveiling the Complexity and Ecological Function of Aquatic Macrophyte—Animal Networks in Coastal Ecosystems. *Biol. Rev.* **2022**, *97*, 1306–1324. [CrossRef] [PubMed]
120. Steneck, R.S.; Graham, M.H.; Bourque, B.J.; Corbett, D.; Erlandson, J.M.; Estes, J.A.; Tegner, M.J. Kelp Forest Ecosystems: Biodiversity, Stability, Resilience and Future. *Environ. Conserv.* **2002**, *29*, 436–459. [CrossRef]
121. Singh, C.L.; Huggett, M.J.; Lavery, P.S.; Säwström, C.; Hyndes, G.A. Kelp-Associated Microbes Facilitate Spatial Subsidy in a Detrital-Based Food Web in a Shoreline Ecosystem. *Front. Mar. Sci.* **2021**, *8*, 678222. [CrossRef]

122. Lotze, H.K.; Milewski, I.; Fast, J.; Kay, L.; Worm, B. Ecosystem-Based Management of Seaweed Harvesting. *Bot. Mar.* **2019**, *62*, 395–409. [CrossRef]
123. Tano, S.A.; Eggertsen, M.; Wikström, S.A.; Berkström, C.; Buriyo, A.S.; Halling, C. Tropical Seaweed Beds as Important Habitats for Juvenile Fish. *Mar. Freshw. Res.* **2017**, *68*, 1921. [CrossRef]
124. Yong, W.T.L.; Thien, V.Y.; Rupert, R.; Rodrigues, K.F. Seaweed: A Potential Climate Change Solution. *Renew. Sustain. Energy Rev.* **2022**, *159*, 112222. [CrossRef]
125. Costanza, R.; de Groot, R.; Sutton, P.; van der Ploeg, S.; Anderson, S.J.; Kubiszewski, I.; Farber, S.; Turner, R.K. Changes in the Global Value of Ecosystem Services. *Glob. Environ. Chang.* **2014**, *26*, 152–158. [CrossRef]
126. McGlathery, K.; Sundbäck, K.; Anderson, I. Eutrophication in Shallow Coastal Bays and Lagoons: The Role of Plants in the Coastal Filter. *Mar. Ecol. Prog. Ser.* **2007**, *348*, 1–18. [CrossRef]
127. Watanabe, K.; Yoshida, G.; Hori, M.; Umezawa, Y.; Moki, H.; Kuwae, T. Macroalgal Metabolism and Lateral Carbon Flows Can Create Significant Carbon Sinks. *Biogeosciences* **2020**, *17*, 2425–2440. [CrossRef]
128. Kim, S.-K.; Pangestuti, R. Biological Activities and Potential Health Benefits of Fucoxanthin Derived from Marine Brown Algae. *Adv. Food Nutr. Res.* **2011**, *64*, 111–128. [CrossRef]
129. Larkum, A.W.D.; Kühl, M. Chlorophyll d: The Puzzle Resolved. *Trends Plant Sci.* **2005**, *10*, 355–357. [CrossRef] [PubMed]
130. Clark, M.; Macdiarmid, J.; Jones, A.D.; Ranganathan, J.; Herrero, M.; Fanzo, J. The Role of Healthy Diets in Environmentally Sustainable Food Systems. *Food Nutr. Bull.* **2020**, *41*, 31S–58S. [CrossRef]
131. WHO; FAO. *Sustainable Healthy Diets: Guiding Principles*; WHO: Geneva, Switzerland, 2019; ISBN 9789241516648.
132. Eismann, A.I.; Perpetuo Reis, R.; Ferreira da Silva, A.; Negrão Cavalcanti, D. *Ulva* spp. Carotenoids: Responses to Environmental Conditions. *Algal Res.* **2020**, *48*, 101916. [CrossRef]
133. Xie, X.; Lu, X.; Wang, L.; He, L.; Wang, G. High Light Intensity Increases the Concentrations of β-Carotene and Zeaxanthin in Marine Red Macroalgae. *Algal Res.* **2020**, *47*, 101852. [CrossRef]
134. Quitério, E.; Grosso, C.; Ferraz, R.; Delerue-Matos, C.; Soares, C. A Critical Comparison of the Advanced Extraction Techniques Applied to Obtain Health-Promoting Compounds from Seaweeds. *Mar. Drugs* **2022**, *20*, 677. [CrossRef]
135. Hasselström, L.; Thomas, J.-B.E. A Critical Review of the Life Cycle Climate Impact in Seaweed Value Chains to Support Carbon Accounting and Blue Carbon Financing. *Clean. Environ. Syst.* **2022**, *6*, 100093. [CrossRef]
136. Koesling, M.; Kvadsheim, N.P.; Halfdanarson, J.; Emblemsvåg, J.; Rebours, C. Environmental Impacts of Protein-Production from Farmed Seaweed: Comparison of Possible Scenarios in Norway. *J. Clean. Prod.* **2021**, *307*, 127301. [CrossRef]
137. Hentati, F.; Tounsi, L.; Djomdi, D.; Pierre, G.; Delattre, C.; Ursu, A.V.; Fendri, I.; Abdelkafi, S.; Michaud, P. Bioactive Polysaccharides from Seaweeds. *Molecules* **2020**, *25*, 3152. [CrossRef]
138. El-Beltagi, H.S.; Mohamed, A.A.; Mohamed, H.I.; Ramadan, K.M.A.; Barqawi, A.A.; Mansour, A.T. Phytochemical and Potential Properties of Seaweeds and Their Recent Applications: A Review. *Mar. Drugs* **2022**, *20*, 342. [CrossRef] [PubMed]
139. Din, N.A.S.; Mohd Alayudin, 'A.S.; Sofian-Seng, N.-S.; Rahman, H.A.; Mohd Razali, N.S.; Lim, S.J.; Wan Mustapha, W.A. Brown Algae as Functional Food Source of Fucoxanthin: A Review. *Foods* **2022**, *11*, 2235. [CrossRef] [PubMed]
140. Mikami, K.; Hosokawa, M. Biosynthetic Pathway and Health Benefits of Fucoxanthin, an Algae-Specific Xanthophyll in Brown Seaweeds. *Int. J. Mol. Sci.* **2013**, *14*, 13763–13781. [CrossRef] [PubMed]
141. Zarekarizi, A.; Hoffmann, L.; Burritt, D. Approaches for the Sustainable Production of Fucoxanthin, a Xanthophyll with Potential Health Benefits. *J. Appl. Phycol.* **2019**, *31*, 281–299. [CrossRef]
142. Shen, Y.; Li, J.; Gu, R.; Yue, L.; Wang, H.; Zhan, X.; Xing, B. Carotenoid and Superoxide Dismutase Are the Most Effective Antioxidants Participating in ROS Scavenging in Phenanthrene Accumulated Wheat Leaf. *Chemosphere* **2018**, *197*, 513–525. [CrossRef] [PubMed]
143. Van den Hoek, C.; Mann, D.G.; Jahns, H.M. *Algae. An Introduction to Phycology*; Guiry, M.D., Ed.; Issue 2; Cambridge University Press: Cambridge, UK, 1995; ISBN 0521304199.
144. Matsuno, T. Aquatic Animal Carotenoids. *Fish. Sci.* **2001**, *67*, 771–783. [CrossRef]
145. Ren, Y.; Sun, H.; Deng, J.; Huang, J.; Chen, F. Carotenoid Production from Microalgae: Biosynthesis, Salinity Responses and Novel Biotechnologies. *Mar. Drugs* **2021**, *19*, 713. [CrossRef]
146. Fukushima, Y.; Taguchi, C.; Kishimoto, Y.; Kondo, K. Japanese Carotenoid Database with α- and β-Carotene, β-Cryptoxanthin, Lutein, Zeaxanthin, Lycopene, and Fucoxanthin and Intake in Adult Women. *Int. J. Vitam. Nutr. Res.* **2023**, *93*, 42–53. [CrossRef]
147. Nomura, M.; Kamogawa, H.; Susanto, E.; Kawagoe, C.; Yasui, H.; Saga, N.; Hosokawa, M.; Miyashita, K. Seasonal Variations of Total Lipids, Fatty Acid Composition, and Fucoxanthin Contents of Sargassum Horneri (Turner) and Cystoseira Hakodatensis (Yendo) from the Northern Seashore of Japan. *J. Appl. Phycol.* **2013**, *25*, 1159–1169. [CrossRef]
148. Terasaki, M.; Hirose, A.; Narayan, B.; Baba, Y.; Kawagoe, C.; Yasui, H.; Saga, N.; Hosokawa, M.; Miyashita, K. Evaluation of Recoverable Functional Lipid Components of Several Brown Seaweeds (Phaeophyta) from Japan with Special Reference to Fucoxanthin and Fucosterol Content. *J. Phycol.* **2009**, *45*, 974–980. [CrossRef]
149. Bhat, I.; Haripriya, G.; Jogi, N.; Mamatha, B.S. Carotenoid Composition of Locally Found Seaweeds of Dakshina Kannada District in India. *Algal Res.* **2021**, *53*, 102154. [CrossRef]
150. Koizumi, J.; Takatani, N.; Kobayashi, N.; Mikami, K.; Miyashita, K.; Yamano, Y.; Wada, A.; Maoka, T.; Hosokawa, M. Carotenoid Profiling of a Red Seaweed *Pyropia yezoensis*: Insights into Biosynthetic Pathways in the Order Bangiales. *Mar. Drugs* **2018**, *16*, 426. [CrossRef] [PubMed]

151. Duarte, C.M.; Holmer, M.; Olsen, Y.; Soto, D.; Marbà, N.; Guiu, J.; Black, K.; Karakassis, I. Will the Oceans Help Feed Humanity? *Bioscience* **2009**, *59*, 967–976. [CrossRef]
152. Duarte, C.M.; Bruhn, A.; Krause-Jensen, D. A Seaweed Aquaculture Imperative to Meet Global Sustainability Targets. *Nat. Sustain.* **2022**, *5*, 185–193. [CrossRef]
153. Rebours, C.; Marinho-Soriano, E.; Zertuche-González, J.A.; Hayashi, L.; Vásquez, J.A.; Kradolfer, P.; Soriano, G.; Ugarte, R.; Abreu, M.H.; Bay-Larsen, I.; et al. Seaweeds: An Opportunity for Wealth and Sustainable Livelihood for Coastal Communities. *J. Appl. Phycol.* **2014**, *26*, 1939–1951. [CrossRef] [PubMed]
154. Barbier, M.; Araújo, R.; Rebours, C.; Jacquemin, B.; Holdt, S.L.; Charrier, B. Development and Objectives of the PHYCOMORPH European Guidelines for the Sustainable Aquaculture of Seaweeds (PEGASUS). *Bot. Mar.* **2020**, *63*, 5–16. [CrossRef]
155. Smale, D.A.; Pessarrodona, A.; King, N.; Burrows, M.T.; Yunnie, A.; Vance, T.; Moore, P. Environmental Factors Influencing Primary Productivity of the Forest-Forming Kelp Laminaria Hyperborea in the Northeast Atlantic. *Sci. Rep.* **2020**, *10*, 12161. [CrossRef]
156. White, L.; Loisel, S.; Sevin, L.; Davoult, D. In Situ Estimates of Kelp Forest Productivity in Macro-tidal Environments. *Limnol. Oceanogr.* **2021**, *66*, 4227–4239. [CrossRef]
157. Bae, M.; Kim, M.-B.; Park, Y.-K.; Lee, J.-Y. Health Benefits of Fucoxanthin in the Prevention of Chronic Diseases. *Biochim. Biophys. Acta–Mol. Cell Biol. Lipids* **2020**, *1865*, 158618. [CrossRef]
158. Hosokawa, M. Health-Promoting Functions of the Marine Carotenoid Fucoxanthin. *Adv. Exp. Med. Biol.* **2021**, *1261*, 273–284. [CrossRef]
159. Liu, M.; Li, W.; Chen, Y.; Wan, X.; Wang, J. Fucoxanthin: A Promising Compound for Human Inflammation-Related Diseases. *Life Sci.* **2020**, *255*, 117850. [CrossRef]
160. Olsthoorn, S.E.M.; Wang, X.; Tillema, B.; Vanmierlo, T.; Kraan, S.; Leenen, P.J.M.; Mulder, M.T. Brown Seaweed Food Supplementation: Effects on Allergy and Inflammation and Its Consequences. *Nutrients* **2021**, *13*, 2613. [CrossRef] [PubMed]
161. Kong, Z.-L.; Sudirman, S.; Hsu, Y.-C.; Su, C.-Y.; Kuo, H.-P. Fucoxanthin-Rich Brown Algae Extract Improves Male Reproductive Function on Streptozotocin-Nicotinamide-Induced Diabetic Rat Model. *Int. J. Mol. Sci.* **2019**, *20*, 4485. [CrossRef]
162. Kong, Z.-L.; Kao, N.-J.; Hu, J.-Y.; Wu, C.-S. Fucoxanthin-Rich Brown Algae Extract Decreases Inflammation and Attenuates Colitis-Associated Colon Cancer in Mice. *J. Food Nutr. Res.* **2016**, *4*, 137–147. [CrossRef]
163. Wang, P.-T.; Sudirman, S.; Hsieh, M.-C.; Hu, J.-Y.; Kong, Z.-L. Oral Supplementation of Fucoxanthin-Rich Brown Algae Extract Ameliorates Cisplatin-Induced Testicular Damage in Hamsters. *Biomed. Pharmacother.* **2020**, *125*, 109992. [CrossRef] [PubMed]
164. Yang, Y.-P.; Tong, Q.-Y.; Zheng, S.-H.; Zhou, M.-D.; Zeng, Y.-M.; Zhou, T.-T. Anti-Inflammatory Effect of Fucoxanthin on Dextran Sulfate Sodium-Induced Colitis in Mice. *Nat. Prod. Res.* **2020**, *34*, 1791–1795. [CrossRef] [PubMed]
165. Sugiura, Y.; Kinoshita, Y.; Usui, M.; Tanaka, R.; Matsushita, T.; Miyata, M. The Suppressive Effect of a Marine Carotenoid, Fucoxanthin, on Mouse Ear Swelling through Regulation of Activities and MRNA Expression of Inflammation-Associated Enzymes. *Food Sci. Technol. Res.* **2016**, *22*, 227–234. [CrossRef]
166. Tan, C.; Hou, Y. First Evidence for the Anti-Inflammatory Activity of Fucoxanthin in High-Fat-Diet-Induced Obesity in Mice and the Antioxidant Functions in PC12 Cells. *Inflammation* **2014**, *37*, 443–450. [CrossRef]
167. Jiang, X.; Wang, G.; Lin, Q.; Tang, Z.; Yan, Q.; Yu, X. Fucoxanthin Prevents Lipopolysaccharide-Induced Depressive-like Behavior in Mice via AMPK- NF-KB Pathway. *Metab. Brain Dis.* **2019**, *34*, 431–442. [CrossRef]
168. Miyashita, K.; Hosokawa, M. Fucoxanthin in the Management of Obesity and Its Related Disorders. *J. Funct. Foods* **2017**, *36*, 195–202. [CrossRef]
169. Hitoe, S.; Shimoda, H. Seaweed Fucoxanthin Supplementation Improves Obesity Parameters in Mild Obese Japanese Subjects. *Funct. Foods Health Dis.* **2017**, *7*, 246. [CrossRef]
170. Hu, L.; Chen, W.; Tian, F.; Yuan, C.; Wang, H.; Yue, H. Neuroprotective Role of Fucoxanthin against Cerebral Ischemic/Reperfusion Injury through Activation of Nrf2/HO-1 Signaling. *Biomed. Pharmacother.* **2018**, *106*, 1484–1489. [CrossRef]
171. Sun, G.; Xin, T.; Zhang, R.; Liu, C.; Pang, Q. Fucoxanthin Attenuates Behavior Deficits and Neuroinflammatory Response in 1-Methyl-4-Phenyl-1,2,3,6-Tetrahydropyridine-Induced Parkinson's Disease in Mice. *Pharmacogn. Mag.* **2020**, *16*, 51. [CrossRef]
172. Aziz, E.; Batool, R.; Khan, M.U.; Rauf, A.; Akhtar, W.; Heydari, M.; Rehman, S.; Shahzad, T.; Malik, A.; Mosavat, S.H.; et al. An Overview on Red Algae Bioactive Compounds and Their Pharmaceutical Applications. *J. Complement. Integr. Med.* **2020**, *17*. [CrossRef]
173. Ali, I.; Manzoor, Z.; Koo, J.-E.; Kim, J.-E.; Byeon, S.-H.; Yoo, E.-S.; Kang, H.-K.; Hyun, J.-W.; Lee, N.-H.; Koh, Y.-S. 3-Hydroxy-4,7-Megastigmadien-9-One, Isolated from Ulva Pertusa, Attenuates TLR9-Mediated Inflammatory Response by down-Regulating Mitogen-Activated Protein Kinase and NF-KB Pathways. *Pharm. Biol.* **2017**, *55*, 435–440. [CrossRef]
174. Young Park, S.; Su Seo, I.; Joo Lee, S.; Pyung Lee, S. Study on the Health Benefits of Brown Algae (*Sargassum muticum*) in Volunteers. *J. Food Nutr. Res.* **2015**, *3*, 126–130. [CrossRef]
175. Shih, P.-H.; Shiue, S.-J.; Chen, C.-N.; Cheng, S.-W.; Lin, H.-Y.; Wu, L.-W.; Wu, M.-S. Fucoidan and Fucoxanthin Attenuate Hepatic Steatosis and Inflammation of NAFLD through Modulation of Leptin/Adiponectin Axis. *Mar. Drugs* **2021**, *19*, 148. [CrossRef] [PubMed]
176. Asai, A.; Sugawara, T.; Ono, H.; Nagao, A. Biotransformation of Fucoxanthinol into Amarouciaxanthin A in Mice and HepG2 Cells: Formation and Cytotoxicity of Fucoxanthin Metabolites. *Drug Metab. Dispos.* **2004**, *32*, 205–211. [CrossRef]
177. Hashimoto, T.; Ozaki, Y.; Taminato, M.; Das, S.K.; Mizuno, M.; Yoshimura, K.; Maoka, T.; Kanazawa, K. The Distribution and Accumulation of Fucoxanthin and Its Metabolites after Oral Administration in Mice. *Br. J. Nutr.* **2009**, *102*, 242–248. [CrossRef]
178. Hashimoto, T.; Ozaki, Y.; Mizuno, M.; Yoshida, M.; Nishitani, Y.; Azuma, T.; Komoto, A.; Maoka, T.; Tanino, Y.; Kanazawa, K. Pharmacokinetics of Fucoxanthinol in Human Plasma after the Oral Administration of Kombu Extract. *Br. J. Nutr.* **2012**, *107*, 1566–1569. [CrossRef] [PubMed]

179. Mikami, N. A Protocol for Human Serum Fucoxanthinol Quantitation Using LC-MS/MS System. *J. Nutr. Med. Diet Care* **2016**, *2*, 019. [CrossRef]
180. Novotny, J.A.; Kurilich, A.C.; Britz, S.J.; Clevidence, B.A. Plasma Appearance of Labeled β-Carotene, Lutein, and Retinol in Humans after Consumption of Isotopically Labeled Kale. *J. Lipid Res.* **2005**, *46*, 1896–1903. [CrossRef]
181. Mohibbullah, M.; Haque, M.N.; Sohag, A.A.M.; Hossain, M.T.; Zahan, M.S.; Uddin, M.J.; Hannan, M.A.; Moon, I.S.; Choi, J.-S. A Systematic Review on Marine Algae-Derived Fucoxanthin: An Update of Pharmacological Insights. *Mar. Drugs* **2022**, *20*, 279. [CrossRef] [PubMed]
182. Irisson, J.-O.; Ayata, S.-D.; Lindsay, D.J.; Karp-Boss, L.; Stemmann, L. Machine Learning for the Study of Plankton and Marine Snow from Images. *Ann. Rev. Mar. Sci.* **2022**, *14*, 277–301. [CrossRef]
183. Fernández, F.G.A.; Reis, A.; Wijffels, R.H.; Barbosa, M.; Verdelho, V.; Llamas, B. The Role of Microalgae in the Bioeconomy. *N. Biotechnol.* **2021**, *61*, 99–107. [CrossRef] [PubMed]
184. Fabris, M.; Abbriano, R.M.; Pernice, M.; Sutherland, D.L.; Commault, A.S.; Hall, C.C.; Labeeuw, L.; McCauley, J.I.; Kuzhiuparambil, U.; Ray, P.; et al. Emerging Technologies in Algal Biotechnology: Toward the Establishment of a Sustainable, Algae-Based Bioeconomy. *Front. Plant Sci.* **2020**, *11*, 279. [CrossRef] [PubMed]
185. Adarme-Vega, T.C.; Lim, D.K.Y.; Timmins, M.; Vernen, F.; Li, Y.; Schenk, P.M. Microalgal Biofactories: A Promising Approach towards Sustainable Omega-3 Fatty Acid Production. *Microb. Cell Fact.* **2012**, *11*, 96. [CrossRef]
186. Russell, C.; Rodriguez, C.; Yaseen, M. High-Value Biochemical Products & Applications of Freshwater Eukaryotic Microalgae. *Sci. Total Environ.* **2022**, *809*, 151111. [CrossRef]
187. López-Sánchez, A.; Silva-Gálvez, A.L.; Aguilar-Juárez, Ó.; Senés-Guerrero, C.; Orozco-Nunnelly, D.A.; Carrillo-Nieves, D.; Gradilla-Hernández, M.S. Microalgae-Based Livestock Wastewater Treatment (MbWT) as a Circular Bioeconomy Approach: Enhancement of Biomass Productivity, Pollutant Removal and High-Value Compound Production. *J. Environ. Manag.* **2022**, *308*, 114612. [CrossRef]
188. Srimongkol, P.; Sangtanoo, P.; Songserm, P.; Watsuntorn, W.; Karnchanatat, A. Microalgae-Based Wastewater Treatment for Developing Economic and Environmental Sustainability: Current Status and Future Prospects. *Front. Bioeng. Biotechnol.* **2022**, *10*, 904046. [CrossRef]
189. Verma, R.; Suthar, S.; Chand, N.; Mutiyar, P.K. Phycoremediation of Milk Processing Wastewater and Lipid-Rich Biomass Production Using *Chlorella vulgaris* under Continuous Batch System. *Sci. Total Environ.* **2022**, *833*, 155110. [CrossRef]
190. Iliopoulou, A.; Zkeri, E.; Panara, A.; Dasenaki, M.; Fountoulakis, M.S.; Thomaidis, N.S.; Stasinakis, A.S. Treatment of Different Dairy Wastewater with *Chlorella sorokiniana*: Removal of Pollutants and Biomass Characterization. *J. Chem. Technol. Biotechnol.* **2022**, *97*, 3193–3201. [CrossRef]
191. León-Vaz, A.; León, R.; Díaz-Santos, E.; Vigara, J.; Raposo, S. Using Agro-Industrial Wastes for Mixotrophic Growth and Lipids Production by the Green Microalga *Chlorella sorokiniana*. *N. Biotechnol.* **2019**, *51*, 31–38. [CrossRef] [PubMed]
192. Zhou, T.; Li, X.; Zhang, Q.; Dong, S.; Liu, H.; Liu, Y.; Chaves, A.V.; Ralph, P.J.; Ruan, R.; Wang, Q. Ecotoxicological Response of Spirulina Platensis to Coexisted Copper and Zinc in Anaerobic Digestion Effluent. *Sci. Total Environ.* **2022**, *837*, 155874. [CrossRef] [PubMed]
193. Han, T.; Lu, H.; Zhao, Y.; Xu, H.; Zhang, Y.; Li, B. Two-Step Strategy for Obtaining *Dunaliella* sp. Biomass and β-Carotene from Anaerobically Digested Poultry Litter Wastewater. *Int. Biodeterior. Biodegradation* **2019**, *143*, 104714. [CrossRef]
194. Shah, M.M.R. Astaxanthin Production by Microalgae *Haematococcus pluvialis* through Wastewater Treatment: Waste to Resource. In *Application of Microalgae in Wastewater Treatment*; Springer International Publishing: Berlin, Germany, 2019; pp. 17–39.
195. Sathasivam, R.; Radhakrishnan, R.; Hashem, A.; Abd_Allah, E.F. Microalgae Metabolites: A Rich Source for Food and Medicine. *Saudi J. Biol. Sci.* **2019**, *26*, 709–722. [CrossRef] [PubMed]
196. Patel, A.K.; Tambat, V.S.; Chen, C.-W.; Chauhan, A.S.; Kumar, P.; Vadrale, A.P.; Huang, C.-Y.; Dong, C.-D.; Singhania, R.R. Recent Advancements in Astaxanthin Production from Microalgae: A Review. *Bioresour. Technol.* **2022**, *364*, 128030. [CrossRef]
197. Muhammad, G.; Butler, T.O.; Chen, B.; Lv, Y.; Xiong, W.; Zhao, X.; Solovchenko, A.E.; Zhao, A.; Mofijur, M.; Xu, J.; et al. Sustainable Production of Lutein-an Underexplored Commercially Relevant Pigment from Microalgae. *Biomass Convers. Biorefinery* **2022**, 1–22. [CrossRef]
198. Gong, M.; Bassi, A. Investigation of Chlorella Vulgaris UTEX 265 Cultivation under Light and Low Temperature Stressed Conditions for Lutein Production in Flasks and the Coiled Tree Photo-Bioreactor (CTPBR). *Appl. Biochem. Biotechnol.* **2017**, *183*, 652–671. [CrossRef]
199. Safafar, H.; Uldall Nørregaard, P.; Ljubic, A.; Møller, P.; Løvstad Holdt, S.; Jacobsen, C. Enhancement of Protein and Pigment Content in Two *Chlorella* Species Cultivated on Industrial Process Water. *J. Mar. Sci. Eng.* **2016**, *4*, 84. [CrossRef]
200. Sun, H.; Yang, S.; Zhao, W.; Kong, Q.; Zhu, C.; Fu, X.; Zhang, F.; Liu, Z.; Zhan, Y.; Mou, H.; et al. Fucoxanthin from Marine Microalgae: A Promising Bioactive Compound for Industrial Production and Food Application. *Crit. Rev. Food Sci. Nutr.* **2022**, 1–17. [CrossRef]
201. Damergi, E.; Schwitzguébel, J.-P.; Refardt, D.; Sharma, S.; Holliger, C.; Ludwig, C. Extraction of Carotenoids from *Chlorella vulgaris* Using Green Solvents and Syngas Production from Residual Biomass. *Algal Res.* **2017**, *25*, 488–495. [CrossRef]
202. Li, Y.; Sun, H.; Wu, T.; Fu, Y.; He, Y.; Mao, X.; Chen, F. Storage Carbon Metabolism of Isochrysis Zhangjiangensis under Different Light Intensities and Its Application for Co-Production of Fucoxanthin and Stearidonic Acid. *Bioresour. Technol.* **2019**, *282*, 94–102. [CrossRef] [PubMed]

203. Martínez, J.M.; Gojkovic, Z.; Ferro, L.; Maza, M.; Álvarez, I.; Raso, J.; Funk, C. Use of Pulsed Electric Field Permeabilization to Extract Astaxanthin from the Nordic Microalga *Haematococcus pluvialis*. *Bioresour. Technol.* **2019**, *289*, 121694. [CrossRef] [PubMed]
204. Safafar, H.; Van Wagenen, J.; Møller, P.; Jacobsen, C. Carotenoids, Phenolic Compounds and Tocopherols Contribute to the Antioxidative Properties of Some *Microalgae* Species Grown on Industrial Wastewater. *Mar. Drugs* **2015**, *13*, 7339–7356. [CrossRef] [PubMed]
205. León-Vaz, A.; León, R.; Vigara, J.; Funk, C. Exploring Nordic Microalgae as a Potential Novel Source of Antioxidant and Bioactive Compounds. *N. Biotechnol.* **2023**, *73*, 1–8. [CrossRef]
206. Yusof, Z.; Khong, N.M.H.; Choo, W.S.; Foo, S.C. Opportunities for the Marine Carotenoid Value Chain from the Perspective of Fucoxanthin Degradation. *Food Chem.* **2022**, *383*, 132394. [CrossRef]
207. Loke Show, P. Global Market and Economic Analysis of Microalgae Technology: Status and Perspectives. *Bioresour. Technol.* **2022**, *357*, 127329. [CrossRef]
208. Jannel, S.; Caro, Y.; Bermudes, M.; Petit, T. Novel Insights into the Biotechnological Production of *Haematococcus pluvialis*-Derived Astaxanthin: Advances and Key Challenges to Allow Its Industrial Use as Novel Food Ingredient. *J. Mar. Sci. Eng.* **2020**, *8*, 789. [CrossRef]
209. Capelli, B.; Bagchi, D.; Cysewski, G.R. Synthetic Astaxanthin Is Significantly Inferior to Algal-Based Astaxanthin as an Antioxidant and May Not Be Suitable as a Human Nutraceutical Supplement. *Nutrafoods* **2013**, *12*, 145–152. [CrossRef]
210. Singh, R.V.; Sambyal, K. An Overview of β-Carotene Production: Current Status and Future Prospects. *Food Biosci.* **2022**, *47*, 101717. [CrossRef]
211. Mohebi Najafabadi, M.; Naeimpoor, F. Boosting β-Carotene and Storage Materials Productivities by Two-Stage Mixed and Monochromatic Exposure Stresses on *Dunaliella salina*. *Int. J. Phytoremediation* **2022**, *25*, 609–620. [CrossRef] [PubMed]
212. Prieto, A.; Pedro Cañavate, J.; García-González, M. Assessment of Carotenoid Production by *Dunaliella salina* in Different Culture Systems and Operation Regimes. *J. Biotechnol.* **2011**, *151*, 180–185. [CrossRef] [PubMed]
213. Calijuri, M.L.; Silva, T.A.; Magalhães, I.B.; de Paula Pereira, A.S.A.; Marangon, B.B.; de Assis, L.R.; Lorentz, J.F. Bioproducts from Microalgae Biomass: Technology, Sustainability, Challenges and Opportunities. *Chemosphere* **2022**, *305*, 135508. [CrossRef]
214. Poojary, M.; Barba, F.; Aliakbarian, B.; Donsì, F.; Pataro, G.; Dias, D.; Juliano, P. Innovative Alternative Technologies to Extract Carotenoids from Microalgae and Seaweeds. *Mar. Drugs* **2016**, *14*, 214. [CrossRef] [PubMed]
215. Wang, M.; Morón-Ortiz, Á.; Zhou, J.; Benítez-González, A.; Mapelli-Brahm, P.; Meléndez-Martínez, A.J.; Barba, F.J. Effects of Pressurized Liquid Extraction with Dimethyl Sulfoxide on the Recovery of Carotenoids and Other Dietary Valuable Compounds from the Microalgae *Spirulina, chlorella* and *Phaeodactylum tricornutum*. *Food Chem.* **2022**, *405*, 134885. [CrossRef]
216. Monapathi, M.E.; Bezuidenhout, C.C.; James Rhode, O.H. Aquatic Yeasts: Diversity, Characteristics and Potential Health Implications. *J. Water Health* **2020**, *18*, 91–105. [CrossRef] [PubMed]
217. Nagahama, T. Yeast Biodiversity in Freshwater, Marine and Deep-Sea Environments BT–Biodiversity and Ecophysiology of Yeasts. In *Biodiversity and Ecophysiology of Yeasts*; Péter, G., Rosa, C., Eds.; Springer: Berlin/Heidelberg, Germany, 2006; pp. 241–262. ISBN 978-3-540-30985-7.
218. Garmendia, G.; Alvarez, A.; Villarreal, R.; Martínez-Silveira, A.; Wisniewski, M.; Vero, S. Fungal Diversity in the Coastal Waters of King George Island (Maritime Antarctica). *World J. Microbiol. Biotechnol.* **2021**, *37*, 142. [CrossRef]
219. Vero, S.; Garmendia, G.; Martínez-Silveira, A.; Cavello, I.; Wisniewski, M. Yeast Activities Involved in Carbon and Nitrogen Cycles in Antarctica. In *The Ecological Role of Micro-Organisms in the Antarctic Environment*; Castro-Sowinski, S., Ed.; Springer Polar Sciences: Berlin/Heidelberg, Germany, 2019; pp. 45–64. ISBN 978-3-030-02785-8.
220. Jafari, N.; Soudi, M.R.; Kasra-Kermanshahi, R. Biodegradation Perspectives of Azo Dyes by Yeasts. *Microbiology* **2014**, *83*, 484–497. [CrossRef]
221. Ruscasso, F.; Cavello, I.; Butler, M.; Loveira, E.L.; Curutchet, G.; Cavalitto, S. Biodegradation and Detoxification of Reactive Orange 16 by Candida Sake 41E. *Bioresour. Technol. Rep.* **2021**, *15*, 100726. [CrossRef]
222. Segal-Kischinevzky, C.; Romero-Aguilar, L.; Alcaraz, L.D.; López-Ortiz, G.; Martínez-Castillo, B.; Torres-Ramírez, N.; Sandoval, G.; González, J. Yeasts Inhabiting Extreme Environments and Their Biotechnological Applications. *Microorganisms* **2022**, *10*, 794. [CrossRef]
223. Libkind, D.; Moliné, M.; Sampaio, J.P.; Van Broock, M. Yeasts from High-Altitude Lakes: Influence of UV Radiation. *FEMS Microbiol. Ecol.* **2009**, *69*, 353–362. [CrossRef] [PubMed]
224. Moliné, M.; Flores, M.R.; Libkind, D.; del Carmen Carmen Diéguez, M.; Farías, M.E.; van Broock, M. Photoprotection by Carotenoid Pigments in the Yeast *Rhodotorula mucilaginosa*: The Role of Torularhodin. *Photochem. Photobiol. Sci.* **2010**, *9*, 1145–1151. [CrossRef]
225. Ueno, R.; Hamada-Sato, N.; Ishida, M.; Urano, N. Potential of Carotenoids in Aquatic Yeasts as a Phylogenetically Reliable Marker and Natural Colorant for Aquaculture. *Biosci. Biotechnol. Biochem.* **2011**, *75*, 1654–1661. [CrossRef]
226. Mata-Gómez, L.C.; Mapelli-Brahm, P.; Meléndez-Martínez, A.J.; Méndez-Zavala, A.; Morales-Oyervides, L.; Montañez, J. Microbial Carotenoid Synthesis Optimization in Goat Cheese Whey Using the Robust Taguchi Method: A Sustainable Approach to Help Tackle Vitamin A Deficiency. *Foods* **2023**, *12*, 658. [CrossRef]
227. Hernández-Almanza, A.; Montañez, J.; Martínez, G.; Aguilar-Jiménez, A.; Contreras-Esquivel, J.C.; Aguilar, C.N. Lycopene: Progress in Microbial Production. *Trends Food Sci. Technol.* **2016**, *56*, 142–148. [CrossRef]
228. Cardoso, L.A.C.; Jäckel, S.; Karp, S.G.; Framboisier, X.; Chevalot, I.; Marc, I. Improvement of *Sporobolomyces ruberrimus* Carotenoids Production by the Use of Raw Glycerol. *Bioresour. Technol.* **2016**, *200*, 374–379. [CrossRef]
229. Ambati, R.; Phang, S.-M.; Ravi, S.; Aswathanarayana, R. Astaxanthin: Sources, Extraction, Stability, Biological Activities and Its Commercial Applications—A Review. *Mar. Drugs* **2014**, *12*, 128–152. [CrossRef] [PubMed]

230. Kurtzman, C.P.; Mateo, R.Q.; Kolecka, A.; Theelen, B.; Robert, V.; Boekhout, T. Advances in Yeast Systematics and Phylogeny and Their Use as Predictors of Biotechnologically Important Metabolic Pathways. *FEMS Yeast Res.* **2015**, *15*, fov050. [CrossRef] [PubMed]
231. Libkind, D.; Moliné, M.; Colabella, F. Isolation and Selection of New Astaxanthin-Producing Strains of *Phaffia rhodozyma*. In *Methods in Molecular Biology*; Springer: Berlin/Heidelberg, Germany, 2018; pp. 297–310.
232. Buzzini, P.; Innocenti, M.; Turchetti, B.; Libkind, D.; van Broock, M.; Mulinacci, N. Carotenoid Profiles of Yeasts Belonging to the Genera *Rhodotorula, Rhodosporidium, Sporobolomyces*, and *Sporidiobolus*. *Can. J. Microbiol.* **2007**, *53*, 1024–1031. [CrossRef]
233. Brandão, L.R.; Libkind, D.; Vaz, A.B.M.; Espírito Santo, L.C.; Moliné, M.; de García, V.; van Broock, M.; Rosa, C.A. Yeasts from an Oligotrophic Lake in Patagonia (Argentina): Diversity, Distribution and Synthesis of Photoprotective Compounds and Extracellular Enzymes. *FEMS Microbiol. Ecol.* **2011**, *76*, 1–13. [CrossRef]
234. Pino-Maureira, N.L.; González-Saldía, R.R.; Capdeville, A.; Srain, B. Rhodotorula Strains Isolated from Seawater That Can Biotransform Raw Glycerol into Docosahexaenoic Acid (DHA) and Carotenoids for Animal Nutrition. *Appl. Sci.* **2021**, *11*, 2824. [CrossRef]
235. Mannazzu, I.; Landolfo, S.; da Silva, T.L.; Buzzini, P. Red Yeasts and Carotenoid Production: Outlining a Future for Non-Conventional Yeasts of Biotechnological Interest. *World J. Microbiol. Biotechnol.* **2015**, *31*, 1665–1673. [CrossRef] [PubMed]
236. Fotedar, R.; Chatting, M.; Kolecka, A.; Zeyara, A.; Al Malki, A.; Kaul, R.; Bukhari, S.J.; Moaiti, M.A.; Febbo, E.J.; Boekhout, T.; et al. Communities of Culturable Yeasts and Yeast-like Fungi in Oligotrophic Hypersaline Coastal Waters of the Arabian Gulf Surrounding Qatar. *Antonie Van Leeuwenhoek* **2022**, *115*, 609–633. [CrossRef] [PubMed]
237. Moliné, M.; Libkind, D.; de Garcia, V.; Giraudo, M.R. Production of Pigments and Photo-Protective Compounds by Cold-Adapted Yeasts. In *Cold-Adapted Yeasts*; Springer: Berlin/Heidelberg, Germany, 2014; pp. 193–224.
238. Herz, S.; Weber, R.W.S.; Anke, H.; Mucci, A.; Davoli, P. Intermediates in the Oxidative Pathway from Torulene to Torularhodin in the Red Yeasts *Cystofilobasidium infirmominiatum* and *C. Capitatum* (Heterobasidiomycetes, Fungi). *Phytochemistry* **2007**, *68*, 2503–2511. [CrossRef] [PubMed]
239. Madhour, A.; Anke, H.; Mucci, A.; Davoli, P.; Weber, R.W.S. Biosynthesis of the Xanthophyll Plectaniaxanthin as a Stress Response in the Red Yeast *Dioszegia* (Tremellales, Heterobasidiomycetes, Fungi). *Phytochemistry* **2005**, *66*, 2617–2626. [CrossRef]
240. Kot, A.M.; Błażejak, S.; Gientka, I.; Kieliszek, M.; Bryś, J. Torulene and Torularhodin: "New" Fungal Carotenoids for Industry? *Microb. Cell Fact.* **2018**, *17*, 49. [CrossRef]
241. Moliné, M.; Libkind, D.; van Broock, M. Production of Torularhodin, Torulene, and β-Carotene by *Rhodotorula* Yeasts. In *Methods Molecular Biology*; Humana Press: Totowa, NJ, USA, 2012; pp. 275–283.
242. Manimala, M.R.A.; Murugesan, R. In Vitro Antioxidant and Antimicrobial Activity of Carotenoid Pigment Extracted from *Sporobolomyces* sp. Isolated from Natural Source. *J. Appl. Nat. Sci.* **2014**, *6*, 649–653. [CrossRef]
243. Ungureanu, C.; Ferdes, M. Evaluation of Antioxidant and Antimicrobial Activities of Torularhodin. *Adv. Sci. Lett.* **2012**, *18*, 50–53. [CrossRef]
244. Ungureanu, C.; Popescu, S.; Purcel, G.; Tofan, V.; Popescu, M.; Sălăgeanu, A.; Pîrvu, C. Improved Antibacterial Behavior of Titanium Surface with Torularhodin–Polypyrrole Film. *Mater. Sci. Eng. C* **2014**, *42*, 726–733. [CrossRef]
245. Ungureanu, C.; Dumitriu, C.; Popescu, S.; Enculescu, M.; Tofan, V.; Popescu, M.; Pirvu, C. Enhancing Antimicrobial Activity of TiO2/Ti by Torularhodin Bioinspired Surface Modification. *Bioelectrochemistry* **2016**, *107*, 14–24. [CrossRef]
246. Du, C.; Li, Y.; Guo, Y.; Han, M.; Zhang, W.; Qian, H. The Suppression of Torulene and Torularhodin Treatment on the Growth of PC-3 Xenograft Prostate Tumors. *Biochem. Biophys. Res. Commun.* **2016**, *469*, 1146–1152. [CrossRef]
247. Barahona, S.; Yuivar, Y.; Socias, G.; Alcaíno, J.; Cifuentes, V.; Baeza, M. Identification and Characterization of Yeasts Isolated from Sedimentary Rocks of Union Glacier at the Antarctica. *Extremophiles* **2016**, *20*, 479–491. [CrossRef] [PubMed]
248. Leyton, A.; Flores, L.; Mäki-Arvela, P.; Lienqueo, M.E.; Shene, C. *Macrocystis pyrifera* Source of Nutrients for the Production of Carotenoids by a Marine Yeast *Rhodotorula mucilaginosa*. *J. Appl. Microbiol.* **2019**, *127*, 1069–1079. [CrossRef] [PubMed]
249. Sperstad, S.; Lutnæs, B.F.; Stormo, S.K.; Liaaen-Jensen, S.; Landfald, B. Torularhodin and Torulene Are the Major Contributors to the Carotenoid Pool of Marine *Rhodosporidium babjevae* (Golubev). *J. Ind. Microbiol. Biotechnol.* **2006**, *33*, 269–273. [CrossRef]
250. Libkind, D.; Van Broock, M. Biomass and Carotenoid Pigment Production by Patagonian Native Yeasts. *World J. Microbiol. Biotechnol.* **2006**, *22*, 687–692. [CrossRef]
251. Allahkarami, S.; Sepahi, A.A.; Hosseini, H.; Razavi, M.R. Isolation and Identification of Carotenoid-Producing *Rhodotorula* sp. from Pinaceae Forest Ecosystems and Optimization of in Vitro Carotenoid Production. *Biotechnol. Rep.* **2021**, *32*, e00687. [CrossRef] [PubMed]
252. Elfeky, N.; Elmahmoudy, M.; Bao, Y. Manipulation of Culture Conditions: Tool for Correlating/Improving Lipid and Carotenoid Production by *Rhodotorula glutinis*. *Processes* **2020**, *8*, 140. [CrossRef]
253. Aksu, Z.; Eren, A.T. Production of Carotenoids by the Isolated Yeast of *Rhodotorula glutinis*. *Biochem. Eng. J.* **2007**, *35*, 107–113. [CrossRef]
254. Cipolatti, E.P.; Remedi, R.D.; dos Santos Sá, C.; Rodrigues, A.B.; Gonçalves Ramos, J.M.; Veiga Burkert, C.A.; Furlong, E.B.; Fernandes de Medeiros Burkert, J. Use of Agroindustrial Byproducts as Substrate for Production of Carotenoids with Antioxidant Potential by Wild Yeasts. *Biocatal. Agric. Biotechnol.* **2019**, *20*, 101208. [CrossRef]
255. de Carvalho, J.C.; Goyzueta-Mamani, L.D.; Molina-Aulestia, D.T.; Magalhães Júnior, A.I.; Iwamoto, H.; Ambati, R.; Ravishankar, G.A.; Soccol, C.R. Microbial Astaxanthin Production from Agro-Industrial Wastes-Raw Materials, Processes, and Quality. *Fermentation* **2022**, *8*, 484. [CrossRef]
256. Igreja, W.S.; de Andrade Maia, F.; Lopes, A.S.; Chisté, R.C. Biotechnological Production of Carotenoids Using Low Cost-Substrates Is Influenced by Cultivation Parameters: A Review. *Int. J. Mol. Sci.* **2021**, *22*, 8819. [CrossRef] [PubMed]

257. Braunwald, T.; Schwemmlein, L.; Graeff-Hönninger, S.; French, W.T.; Hernandez, R.; Holmes, W.E.; Claupein, W. Effect of Different C/N Ratios on Carotenoid and Lipid Production by *Rhodotorula glutinis*. *Appl. Microbiol. Biotechnol.* **2013**, *97*, 6581–6588. [CrossRef] [PubMed]
258. Ramirez, J.; Obledo, N.; Arellano, M.; Herrera, E. *Astaxanthin Production by Phaffia Rhodozyma in a Fedbatch Culture Using a Low Cost Medium Feeding*; Springer: Guadalajara, México, 2006; Volume 4.
259. Tkáčová, J.; Čaplová, J.; Klempová, T.; Čertík, M. Correlation between Lipid and Carotenoid Synthesis in Torularhodin-Producing *Rhodotorula glutinis*. *Ann. Microbiol.* **2017**, *67*, 541–551. [CrossRef]
260. Elfeky, N.; Elmahmoudy, M.; Zhang, Y.; Guo, J.; Bao, Y. Lipid and Carotenoid Production by *Rhodotorula glutinis* with a Combined Cultivation Mode of Nitrogen, Sulfur, and Aluminium Stress. *Appl. Sci.* **2019**, *9*, 2444. [CrossRef]
261. Frengova, G.I.; Beshkova, D.M. Carotenoids from *Rhodotorula* and *Phaffia*: Yeasts of Biotechnological Importance. *J. Ind. Microbiol. Biotechnol.* **2009**, *36*, 163–180. [CrossRef] [PubMed]
262. Zhang, Z.; Zhang, X.; Tan, T. Lipid and Carotenoid Production by *Rhodotorula glutinis* under Irradiation/High-Temperature and Dark/Low-Temperature Cultivation. *Bioresour. Technol.* **2014**, *157*, 149–153. [CrossRef] [PubMed]
263. Pham, K.D.; Shida, Y.; Miyata, A.; Takamizawa, T.; Suzuki, Y.; Ara, S.; Yamazaki, H.; Masaki, K.; Mori, K.; Aburatani, S.; et al. Effect of Light on Carotenoid and Lipid Production in the Oleaginous Yeast *Rhodosporidium toruloides*. *Biosci. Biotechnol. Biochem.* **2020**, *84*, 1501–1512. [CrossRef]
264. Stachowiak, B. Effect of Illumination Intensities on Astaxanthin Synthesis by *Xanthophyllomyces dendrorhous* and Its Mutants. *Food Sci. Biotechnol.* **2013**, *22*, 1033–1038. [CrossRef]
265. Li, C.; Xu, Y.; Li, Z.; Cheng, P.; Yu, G. Transcriptomic and Metabolomic Analysis Reveals the Potential Mechanisms Underlying the Improvement of β-Carotene and Torulene Production in *Rhodosporidiobolus colostri* under Low Temperature Treatment. *Food Res. Int.* **2022**, *156*, 111158. [CrossRef]
266. Cheng, Y.-T.; Yang, C.-F. Using Strain *Rhodotorula mucilaginosa* to Produce Carotenoids Using Food Wastes. *J. Taiwan Inst. Chem. Eng.* **2016**, *61*, 270–275. [CrossRef]
267. Zhao, D.; Li, C. Multi-Omics Profiling Reveals Potential Mechanisms of Culture Temperature Modulating Biosynthesis of Carotenoids, Lipids, and Exopolysaccharides in Oleaginous Red Yeast *Rhodotorula glutinis* ZHK. *LWT* **2022**, *171*, 114103. [CrossRef]
268. Davoli, P.; Mierau, V.; Weber, R.W.S. Carotenoids and Fatty Acids in Red Yeasts *Sporobolomyces roseus* and *Rhodotorula glutinis*. *Appl. Biochem. Microbiol.* **2004**, *40*, 392–397. [CrossRef]
269. Akiba, Y.; Sato, K.; Takahashi, K.; Matsushita, K.; Komiyama, H.; Tsunekawa, H.; Nagao, H. Meat Color Modification in Broiler Chickens by Feeding Yeast *Phaffia rhodozyma* Containing High Concentrations of Astaxanthin. *J. Appl. Poult. Res.* **2001**, *10*, 154–161. [CrossRef]
270. Zoz, L.; Carvalho, J.C.; Soccol, V.T.; Casagrande, T.C.; Cardoso, L. Torularhodin and Torulene: Bioproduction, Properties and Prospective Applications in Food and Cosmetics—A Review. *Braz. Arch. Biol. Technol.* **2015**, *58*, 278–288. [CrossRef]
271. Varela, J.C.; Pereira, H.; Vila, M.; León, R. Production of Carotenoids by Microalgae: Achievements and Challenges. *Photosynth. Res.* **2015**, *125*, 423–436. [CrossRef]
272. Chazan, A.; Das, I.; Fujiwara, T.; Murakoshi, S.; Rozenberg, A.; Molina-Márquez, A.; Sano, F.K.; Tanaka, T.; Gómez-Villegas, P.; Larom, S.; et al. Phototrophy by Antenna-Containing Rhodopsin Pumps in Aquatic Environments. *Nature* **2023**, *615*, 535–540. [CrossRef]
273. Béjà, O.; Aravind, L.; Koonin, E.V.; Suzuki, M.T.; Hadd, A.; Nguyen, L.P.; Jovanovich, S.B.; Gates, C.M.; Feldman, R.A.; Spudich, J.L.; et al. Bacterial Rhodopsin: Evidence for a New Type of Phototrophy in the Sea. *Science* **2000**, *289*, 1902–1906. [CrossRef]
274. Pinhassi, J.; DeLong, E.F.; Béjà, O.; González, J.M.; Pedrós-Alió, C. Marine Bacterial and Archaeal Ion-Pumping Rhodopsins: Genetic Diversity, Physiology, and Ecology. *Microbiol. Mol. Biol. Rev.* **2016**, *80*, 929–954. [CrossRef] [PubMed]
275. Balashov, S.P.; Imasheva, E.S.; Boichenko, V.A.; Antón, J.; Wang, J.M.; Lanyi, J.K. Xanthorhodopsin: A Proton Pump with a Light-Harvesting Carotenoid Antenna. *Science* **2005**, *309*, 2061–2064. [CrossRef] [PubMed]
276. Köcher, S.; Breitenbach, J.; Müller, V.; Sandmann, G. Structure, Function and Biosynthesis of Carotenoids in the Moderately Halophilic Bacterium *Halobacillus halophilus*. *Arch. Microbiol.* **2009**, *191*, 95–104. [CrossRef] [PubMed]
277. Kelman, D.; Ben-Amotz, A.; Berman-Frank, I. Carotenoids Provide the Major Antioxidant Defence in the Globally Significant N 2 -Fixing Marine *Cyanobacterium trichodesmium*. *Environ. Microbiol.* **2009**, *11*, 1897–1908. [CrossRef] [PubMed]
278. Giani, M.; Garbayo, I.; Vílchez, C.; Martínez-Espinosa, R.M. Haloarchaeal Carotenoids: Healthy Novel Compounds from Extreme Environments. *Mar. Drugs* **2019**, *17*, 524. [CrossRef] [PubMed]
279. Oren, A. Saltern Evaporation Ponds as Model Systems for the Study of Primary Production Processes under Hypersaline Conditions. *Aquat. Microb. Ecol.* **2009**, *56*, 193–204. [CrossRef]
280. Sajjad, W.; Din, G.; Rafiq, M.; Iqbal, A.; Khan, S.; Zada, S.; Ali, B.; Kang, S. Pigment Production by Cold-Adapted Bacteria and Fungi: Colorful Tale of Cryosphere with Wide Range Applications. *Extremophiles* **2020**, *24*, 447–473. [CrossRef]
281. Chattopadhyay, M.K.; Jagannadham, M.V.; Vairamani, M.; Shivaji, S. Carotenoid Pigments of an Antarctic Psychrotrophic Bacterium *Micrococcus roseus*: Temperature Dependent Biosynthesis, Structure, and Interaction with Synthetic Membranes. *Biochem. Biophys. Res. Commun.* **1997**, *239*, 85–90. [CrossRef]
282. Jagannadham, M.V.; Chattopadhyay, M.K.; Subbalakshmi, C.; Vairamani, M.; Narayanan, K.; Mohan Rao, C.; Shivaji, S. Carotenoids of an Antarctic Psychrotolerant Bacterium, *Sphingobacterium antarcticus*, and a Mesophilic Bacterium, *Sphingobacterium multivorum*. *Arch. Microbiol.* **2000**, *173*, 418–424. [CrossRef]
283. Flegler, A.; Lipski, A. The C50 Carotenoid Bacterioruberin Regulates Membrane Fluidity in Pink-Pigmented *Arthrobacter* Species. *Arch. Microbiol.* **2022**, *204*, 70. [CrossRef]

284. Tian, B.; Hua, Y. Carotenoid Biosynthesis in Extremophilic Deinococcus-Thermus Bacteria. *Trends Microbiol.* **2010**, *18*, 512–520. [CrossRef]
285. Seto, R.; Takaichi, S.; Kurihara, T.; Kishi, R.; Honda, M.; Takenaka, S.; Tsukatani, Y.; Madigan, M.T.; Wang-Otomo, Z.-Y.; Kimura, Y. Lycopene-Family Carotenoids Confer Thermostability on Photocomplexes from a New Thermophilic Purple Bacterium. *Biochemistry* **2020**, *59*, 2351–2358. [CrossRef]
286. Maoka, T. Carotenoids in Marine Animals. *Mar. Drugs* **2011**, *9*, 278–293. [CrossRef]
287. Eggersdorfer, M.; Wyss, A. Carotenoids in Human Nutrition and Health. *Arch. Biochem. Biophys.* **2018**, *652*, 18–26. [CrossRef] [PubMed]
288. Johnson, E.J. Role of Lutein and Zeaxanthin in Visual and Cognitive Function throughout the Lifespan. *Nutr. Rev.* **2014**, *72*, 605–612. [CrossRef] [PubMed]
289. Visioli, F.; Artaria, C. Astaxanthin in Cardiovascular Health and Disease: Mechanisms of Action, Therapeutic Merits, and Knowledge Gaps. *Food Funct.* **2017**, *8*, 39–63. [CrossRef] [PubMed]
290. Meléndez-Martínez, A.J.; Böhm, V.; Borge, G.I.; Cano, M.P.; Fikselová, M.; Gruskiene, R.; Lavelli, V.; Loizzo, M.R.; Mandić, A.; Mapelli-Brahm, P.; et al. Carotenoids: Considerations for Their Use in Functional Foods, Nutraceuticals, Nutricosmetics, Supplements, Botanicals and Novel Foods in the Context of Sustainability, Circular Economy and Climate Change. *Annu. Rev. Food Sci. Technol.* **2021**, *12*, 433–460. [CrossRef]
291. Checa, J.; Aran, J.M. Reactive Oxygen Species: Drivers of Physiological and Pathological Processes. *J. Inflamm. Res.* **2020**, *13*, 1057–1073. [CrossRef]
292. Ranneh, Y.; Ali, F.; Akim, A.M.; Hamid, H.A.; Khazaai, H.; Fadel, A. Crosstalk between Reactive Oxygen Species and Pro-Inflammatory Markers in Developing Various Chronic Diseases: A Review. *Appl. Biol. Chem.* **2017**, *60*, 327–338. [CrossRef]
293. Du, C.; Guo, Y.; Cheng, Y.; Han, M.; Zhang, W.; Qian, H. Anti-Cancer Effects of Torulene, Isolated from *Sporidiobolus pararoseus*, on Human Prostate Cancer LNCaP and PC-3 Cells via a Mitochondrial Signal Pathway and the down-Regulation of AR Expression. *RSC Adv.* **2017**, *7*, 2466–2474. [CrossRef]
294. Nakao, R.; Nelson, O.L.; Park, J.S.; Mathison, B.D.; Thompson, P.A.; Chew, B.P. Effect of Dietary Astaxanthin at Different Stages of Mammary Tumor Initiation in BALB/c Mice. *Anticancer Res.* **2010**, *30*, 2171–2175.
295. Prabhu, P.N.; Ashokkumar, P.; Sudhandiran, G. Antioxidative and Antiproliferative Effects of Astaxanthin during the Initiation Stages of 1,2-Dimethyl Hydrazine-Induced Experimental Colon Carcinogenesis. *Fundam. Clin. Pharmacol.* **2009**, *23*, 225–234. [CrossRef]
296. Fassett, R.G.; Coombes, J.S. Astaxanthin: A Potential Therapeutic Agent in Cardiovascular Disease. *Mar. Drugs* **2011**, *9*, 447–465. [CrossRef]
297. Sartaj, K.; Gupta, P.; Tripathi, S.; Poluri, K.M.; Prasad, R. Insights into the Extraction, Characterization and Antifungal Activity of Astaxanthin Derived from Yeast de-Oiled Biomass. *Environ. Technol. Innov.* **2022**, *27*, 102437. [CrossRef]
298. Lakshminarayana, R.; Vijay, K.; Ambedkar, R.; Ranga Rao, A.; Ravishankar, G.A. Biological Activities and Health Benefits of Seaweed Carotenoids with Special Reference to Fucoxanthin. In *Sustainable Global Resources of Seaweeds Volume 2: Food, Pharmaceutical and Health Applications*; Ranga Rao, A., Ravishankar, G.A., Eds.; Springer International Publishing: Berlin, Germany, 2022; pp. 539–558. ISBN 978-3-030-92174-3.
299. Sathasivam, R.; Ki, J.-S. A Review of the Biological Activities of Microalgal Carotenoids and Their Potential Use in Healthcare and Cosmetic Industries. *Mar. Drugs* **2018**, *16*, 26. [CrossRef]
300. Afra, S.; Makhdoumi, A.; Matin, M.M.; Feizy, J. A Novel Red Pigment from Marine *Arthrobacter* sp. G20 with Specific Anticancer Activity. *J. Appl. Microbiol.* **2017**, *123*, 1228–1236. [CrossRef] [PubMed]
301. Shindo, K.; Kikuta, K.; Suzuki, A.; Katsuta, A.; Kasai, H.; Yasumoto-Hirose, M.; Matsuo, Y.; Misawa, N.; Takaichi, S. Rare Carotenoids, (3R)-Saproxanthin and (3R,2′S)-Myxol, Isolated from Novel Marine Bacteria (*Flavobacteriaceae*) and Their Antioxidative Activities. *Appl. Microbiol. Biotechnol.* **2007**, *74*, 1350–1357. [CrossRef] [PubMed]
302. Shindo, K.; Asagi, E.; Sano, A.; Hotta, E.; Minemura, N.; Mikami, K.; Tamesada, E.; Misawa, N.; Maoka, T. Diapolycopenedioic Acid Xylosyl Esters A, B, and C, Novel Antioxidative Glyco-C30-Carotenoic Acids Produced by a New Marine Bacterium *Rubritalea squalenifaciens*. *J. Antibiot.* **2008**, *61*, 185–191. [CrossRef] [PubMed]
303. Jinendiran, S.; Dahms, H.-U.; Dileep Kumar, B.S.; Kumar Ponnusamy, V.; Sivakumar, N. Diapolycopenedioic-Acid-Diglucosyl Ester and Keto-Myxocoxanthin Glucoside Ester: Novel Carotenoids Derived from *Exiguobacterium acetylicum* S01 and Evaluation of Their Anticancer and Anti-Inflammatory Activities. *Bioorg. Chem.* **2020**, *103*, 104149. [CrossRef]
304. Osawa, A.; Ishii, Y.; Sasamura, N.; Morita, M.; Kasai, H.; Maoka, T.; Shindo, K. Characterization and Antioxidative Activities of Rare C50 Carotenoids-Sarcinaxanthin, Sarcinaxanthin Monoglucoside, and Sarcinaxanthin Diglucoside-Obtained from *Micrococcus yunnanensis*. *J. Oleo Sci.* **2010**, *59*, 653–659. [CrossRef]
305. Metwally, R.A.; El-Sersy, N.A.; El Sikaily, A.; Sabry, S.A.; Ghozlan, H.A. Optimization and Multiple in Vitro Activity Potentials of Carotenoids from Marine *Kocuria* sp. RAM1. *Sci. Rep.* **2022**, *12*, 18203. [CrossRef]
306. Osawa, A.; Ishii, Y.; Sasamura, N.; Morita, M.; Köcher, S.; Müller, V.; Sandmann, G.; Shindo, K. Hydroxy-3,4-Dehydro-Apo-8′-Lycopene and Methyl Hydroxy-3,4-Dehydro-Apo-8′-Lycopenoate, Novel C30 Carotenoids Produced by a Mutant of Marine Bacterium *Halobacillus halophilus*. *J. Antibiot.* **2010**, *63*, 291–295. [CrossRef]
307. Shindo, K.; Endo, M.; Miyake, Y.; Wakasugi, K.; Morritt, D.; Bramley, P.M.; Fraser, P.D.; Kasai, H.; Misawa, N. Methyl Glucosyl-3,4-Dehydro-Apo-8′-Lycopenoate, a Novel Antioxidative Glyco-C30-Carotenoic Acid Produced by a Marine Bacterium *Planococcus maritimus*. *J. Antibiot.* **2008**, *61*, 729–735. [CrossRef]
308. Styczynski, M.; Rogowska, A.; Gieczewska, K.; Garstka, M.; Szakiel, A.; Dziewit, L. Genome-Based Insights into the Production of Carotenoids by Antarctic Bacteria, *Planococcus* sp. ANT_H30 and *Rhodococcus* sp. ANT_H53B. *Molecules* **2020**, *25*, 4357. [CrossRef] [PubMed]

309. Patias, L.D.; Fernandes, A.S.; Petry, F.C.; Mercadante, A.Z.; Jacob-Lopes, E.; Zepka, L.Q. Carotenoid Profile of Three Microalgae/Cyanobacteria Species with Peroxyl Radical Scavenger Capacity. *Food Res. Int.* **2017**, *100*, 260–266. [CrossRef]
310. Lopes, G.; Clarinha, D.; Vasconcelos, V. Carotenoids from Cyanobacteria: A Biotechnological Approach for the Topical Treatment of Psoriasis. *Microorganisms* **2020**, *8*, 302. [CrossRef] [PubMed]
311. Yan, X.; Chuda, Y.; Suzuki, M.; Nagata, T. Fucoxanthin as the Major Antioxidant in *Hijikia fusiformis*, a Common Edible Seaweed. *Biosci. Biotechnol. Biochem.* **1999**, *63*, 605–607. [CrossRef] [PubMed]
312. Heo, S.-J.; Yoon, W.-J.; Kim, K.-N.; Ahn, G.-N.; Kang, S.-M.; Kang, D.-H.; Affan, A.; Oh, C.; Jung, W.-K.; Jeon, Y.-J. Evaluation of Anti-Inflammatory Effect of Fucoxanthin Isolated from Brown Algae in Lipopolysaccharide-Stimulated RAW 264.7 Macrophages. *Food Chem. Toxicol.* **2010**, *48*, 2045–2051. [CrossRef]
313. Nazarudin, M.F.; Yasin, I.S.M.; Mazli, N.A.I.N.; Saadi, A.R.; Azizee, M.H.S.; Nooraini, M.A.; Saad, N.; Ferdous, U.T.; Fakhrulddin, I.M. Preliminary Screening of Antioxidant and Cytotoxic Potential of Green Seaweed, *Halimeda opuntia* (Linnaeus) Lamouroux. *Saudi J. Biol. Sci.* **2022**, *29*, 2698–2705. [CrossRef]
314. Balasubramaniam, V.; June Chelyn, L.; Vimala, S.; Mohd Fairulnizal, M.N.; Brownlee, I.A.; Amin, I. Carotenoid Composition and Antioxidant Potential of *Eucheuma denticulatum*, *Sargassum polycystum* and *Caulerpa lentillifera*. *Heliyon* **2020**, *6*, e04654. [CrossRef]
315. Chiu, H.-F.; Liao, J.; Lu, Y.-Y.; Han, Y.-C.; Shen, Y.-C.; Venkatakrishnan, K.; Golovinskaia, O.; Wang, C. Anti-Proliferative, Anti-in Fl Ammatory and pro-Apoptotic e Ff Ects of *Dunaliella salina* on Human KB Oral Carcinoma Cells. *J. Food Biochem.* **2017**, *41*, e12349. [CrossRef]
316. Dokumacioglu, E.; Iskender, H.; Yenice, G.; Kapakin, K.; Sevim, C.; Hayirli, A.; Saral, S.; Comakli, A. Effects of Astaxanthin on Biochemical and Histopathological Parameters Related to Oxidative Stress on Testes of Rats on High Fructose Regime. *Andrologia* **2018**, *50*, e13042. [CrossRef]
317. Lee, A.H.; Shin, H.Y.; Park, J.H.; Koo, S.Y.; Kim, S.M.; Yang, S.-H. Fucoxanthin from Microalgae *Phaeodactylum tricornutum* Inhibits pro- Inflammatory Cytokines by Regulating Both NF-KB and NLRP3 Inflammasome Activation. *Sci. Rep.* **2021**, *11*, 543. [CrossRef]
318. Shiratori, K.; Ohgami, K.; Ilieva, I.; Jin, X.-H.; Koyama, Y.; Miyashita, K.; Yoshida, K.; Kase, S.; Ohno, S. Effects of Fucoxanthin on Lipopolysaccharide-Induced Inflammation in Vitro and in Vivo. *Exp. Eye Res.* **2005**, *81*, 422–428. [CrossRef] [PubMed]
319. Yang, S.-P.; Wu, Z.-H.; Jian, J.-C.; Zhang, X.-Z. Effect of Marine Red Yeast *Rhodosporidium paludigenum* on Growth and Antioxidant Competence of *Litopenaeus vannamei*. *Aquaculture* **2010**, *309*, 62–65. [CrossRef]
320. Mussagy, C.U.; Khan, S.; Kot, A.M. Current Developments on the Application of Microbial Carotenoids as an Alternative to Synthetic Pigments. *Crit. Rev. Food Sci. Nutr.* **2022**, *62*, 6932–6946. [CrossRef]
321. Mapelli-Brahm, P.; Barba, F.J.; Remize, F.; Garcia, C.; Fessard, A.; Mousavi Khaneghah, A.; Sant'Ana, A.S.; Lorenzo, J.M.; Montesano, D.; Meléndez-Martínez, A.J. The Impact of Fermentation Processes on the Production, Retention and Bioavailability of Carotenoids: An Overview. *Trends Food Sci. Technol.* **2020**, *99*, 389–401. [CrossRef]
322. Amar, E.C.; Kiron, V.; Akutsu, T.; Satoh, S.; Watanabe, T. Resistance of Rainbow Trout *Oncorhynchus mykiss* to Infectious Hematopoietic Necrosis Virus (IHNV) Experimental Infection Following Ingestion of Natural and Synthetic Carotenoids. *Aquaculture* **2012**, *330–333*, 148–155. [CrossRef]
323. Dufossé, L.; Galaup, P.; Yaron, A.; Arad, S.M.; Blanc, P.; Chidambara Murthy, K.N.; Ravishankar, G.A. Microorganisms and Microalgae as Sources of Pigments for Food Use: A Scientific Oddity or an Industrial Reality? *Trends Food Sci. Technol.* **2005**, *16*, 389–406. [CrossRef]
324. Liu, X.; Xie, J.; Zhou, L.; Zhang, J.; Chen, Z.; Xiao, J.; Cao, Y.; Xiao, H. Recent Advances in Health Benefits and Bioavailability of Dietary Astaxanthin and Its Isomers. *Food Chem.* **2023**, *404*, 134605. [CrossRef]
325. Miyashita, K.; Beppu, F.; Hosokawa, M.; Liu, X.; Wang, S. Nutraceutical Characteristics of the Brown Seaweed Carotenoid Fucoxanthin. *Arch. Biochem. Biophys.* **2020**, *686*, 108364. [CrossRef] [PubMed]
326. Guedes, A.C.; Amaro, H.M.; Malcata, F.X. Microalgae as Sources of Carotenoids. *Mar. Drugs* **2011**, *9*, 625–644. [CrossRef]
327. Saleh, R.; Bearth, A.; Siegrist, M. "Chemophobia" Today: Consumers' Knowledge and Perceptions of Chemicals. *Risk Anal.* **2019**, *39*, 2668–2682. [CrossRef] [PubMed]

marine drugs

Review

Biological Properties and Health-Promoting Functions of Laminarin: A Comprehensive Review of Preclinical and Clinical Studies

Shanmugapriya Karuppusamy [1,*], Gaurav Rajauria [2,3], Stephen Fitzpatrick [4], Henry Lyons [4], Helena McMahon [3], James Curtin [5], Brijesh K. Tiwari [6] and Colm O'Donnell [1]

1 School of Biosystems and Food Engineering, University College Dublin, Belfield, D04 V1W8 Dublin, Ireland
2 Department of Biological and Pharmaceutical Sciences, Munster Technological University, Clash, V92 CX88 Tralee, Ireland
3 Circular Bioeconomy Research Group, Shannon Applied Biotechnology Centre, Munster Technological University, V92 CX88 Tralee, Ireland
4 Nutramara Ltd., Beechgrove House Strand Street, V92 FH0K Tralee, Ireland
5 School of Food Science and Environmental Health, College of Sciences and Health, Technological University Dublin, D01 K822 Dublin, Ireland
6 Teagasc Food Research Centre, Department of Food Chemistry and Technology, Ashtown, D15 KN3K Dublin, Ireland
* Correspondence: shanmugapriya.karuppusamy@ucd.ie; Tel.: +353-871-080-270

Citation: Karuppusamy, S.; Rajauria, G.; Fitzpatrick, S.; Lyons, H.; McMahon, H.; Curtin, J.; Tiwari, B.K.; O'Donnell, C. Biological Properties and Health-Promoting Functions of Laminarin: A Comprehensive Review of Preclinical and Clinical Studies. *Mar. Drugs* **2022**, *20*, 772. https://doi.org/10.3390/md20120772

Academic Editors: Elena Talero and Javier Ávila-Román

Received: 31 October 2022
Accepted: 6 December 2022
Published: 10 December 2022

Abstract: Marine algal species comprise of a large portion of polysaccharides which have shown multifunctional properties and health benefits for treating and preventing human diseases. Laminarin, or β-glucan, a storage polysaccharide from brown algae, has been reported to have potential pharmacological properties such as antioxidant, anti-tumor, anti-coagulant, anticancer, immunomodulatory, anti-obesity, anti-diabetic, anti-inflammatory, wound healing, and neuroprotective potential. It has been widely investigated as a functional material in biomedical applications as it is biodegradable, biocompatible, and is low toxic substances. The reported preclinical and clinical studies demonstrate the potential of laminarin as natural alternative agents in biomedical and industrial applications such as nutraceuticals, pharmaceuticals, functional food, drug development/delivery, and cosmeceuticals. This review summarizes the biological activities of laminarin, including mechanisms of action, impacts on human health, and reported health benefits. Additionally, this review also provides an overview of recent advances and identifies gaps and opportunities for further research in this field. It further emphasizes the molecular characteristics and biological activities of laminarin in both preclinical and clinical settings for the prevention of the diseases and as potential therapeutic interventions.

Keywords: laminarin; bioactive compounds; algal polysaccharide; biological activity; preclinical; clinical; human health; biomedical application

1. Introduction

Algal derived polysaccharides have gained much interest owing to their abundance as the main cell wall constituent as well as their unique physicochemical and biological properties [1]. Marine algal polysaccharides, including laminarin, fucoidan, ulvan, carrageenan, and alginate, have been widely explored for food, pharmaceutical and biomedical applications [2,3]. Polysaccharides are building blocks of about 30–50 monosaccharide units linked together by glycosidic bonds, together with many complex sugars to form cross-linked high–molecular weight biological macromolecules [4]. Initially, polysaccharides were used as thickening agents in industrial applications but recently have attracted much attention from the scientific community for their reported therapeutic potential [2,5]. The demand for novel, pure and highly biologically active polysaccharides is rapidly increasing, and their increased interest from researchers in exploiting natural algal resources. Algal biomass is a renewable bioresource and the recovered polysaccharides are safe, biodegradable, and biocompatible [2].

Laminarin is a relatively underexploited water-soluble polysaccharide present in brown algae that has been reported to exhibit potential therapeutic properties. Laminarin is increasingly being explored for the development of functional food and nutraceuticals [6–9], as well as a functional material in biomedical applications due to its biodegradable, biocompatible, and non-toxic nature [10,11] (Figure 1). This bioactive compound possesses the therapeutic potential to enhance and promote health and may help protect against diseases, including cardiovascular diseases, metabolic disorders, cancer, diabetes, obesity, anti-inflammatory activity, osteoarthritis, oral diseases, multiple sclerosis, as well as Alzheimer's and Parkinson's diseases, and vision-improving agents [12–17]. Several studies concluded that laminarin is also an excellent source of dietary fibre and a modulator of intestinal metabolism. Laminarin acts as a modulator through its effect on intestinal pH and short-chain fatty acids (SCFAs) production and is reported to enhance health, improve immunity, and treat and prevent diseases [6,9,12,13,16,18].

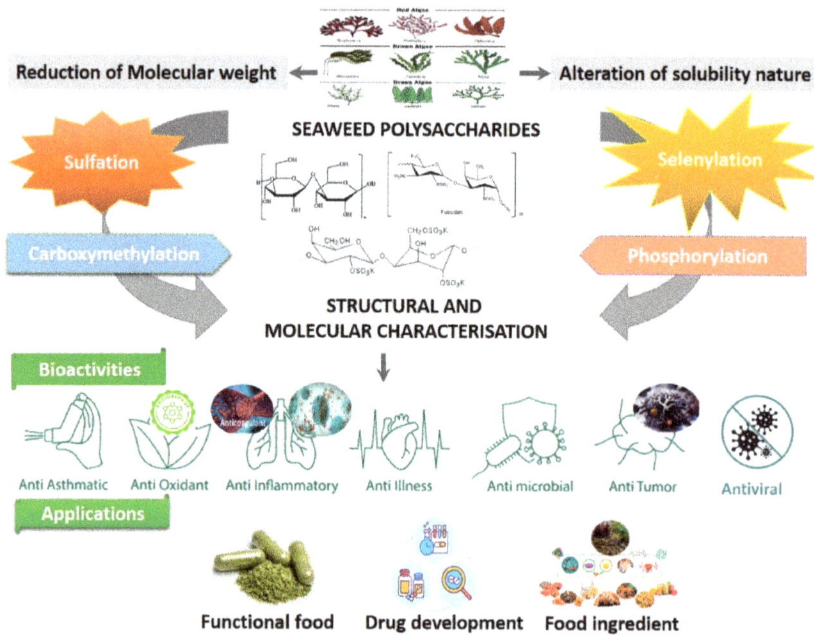

Figure 1. Seaweed polysaccharides have been investigated for commercial applications due to their reported properties and bioactivities.

Despite the reported benefits of laminarin, there are limited studies based on cell culture investigations or biomedical engineering applications, and no commercial biomedical products are available in the market [2]. A limited number of studies on preclinical and clinical trials of laminarin from different brown macroalgae are reported in the literature [3–5]. It is therefore timely to review the therapeutic potential, biomolecular functions, multifunctional bioactivities, and possible health applications of laminarin (Table 1). In this review, the molecular structure and mode of action of laminarin from brown macroalgae are reviewed using preclinical and clinical reported studies. This emphasizes the biological activities of laminarin reported in preclinical and clinical case studies for the prevention of diseases and potential therapeutic interventions. This review provides scientific knowledge to underpin future novel biomedical applications of laminarin.

Table 1. Bioactive algal polysaccharides and their reported biological activities with therapeutic and nutraceutical potential.

Bioactive Algal Polysaccharides and Their Sources with Yield	Bioactivity Potential	Therapeutic and Nutraceutical Applications	Ref.
Laminarin			
Laminaria digitata (51.8%)	Antitumor	Active ingredients	[3,10]
Saccharina latissima (19 ± 2.6%)	Antimicrobial	Photoprotection	[14]
	Wound healing	Antiaging products	[15,19]
	Immunomodulatory	Antioxidant	[7,17,20]
S. japonica	Antioxidant activity	Skin-benefiting activities	[11,21–23] [24,25]
Alaria angusta	Antibacterial and antifungal	Growth regulation	[26,27]
Undaria pinnatifida (3.2 ± 0.9%)	Modulatory effects on skin cells	Lipolytic activity	[28]
Dictyopteris delicatula	Hydro-gelling properties	Tissue engineering	[13,22]
Dictyota menstrualis	Anti-inflammatory	Cancer therapies	[29]
L. saccharina	Chemoprotective	Biofuel production	[30–32]
Ecklonia cava	Neuroprotective	Food industry	[33]
Ascophyllum nodosum	Antiviral	Antimicrobial	[34–36]
Sargassum spp. (13.47%)	Antiallergic	Health care and cosmetic uses	[37]
	Antipruritic	Boosts immunity	[38]
	Hepatoprotective	Prevents inflammation	[27,39,40]
	Anti-cholesterol	Biofertiliser	[41,42]
	Antidiabetic		[43,44]
Fucoidan			
S. japonica (1.26%),	Anti-tumor activity	Immunomodulatory effect	[10,35]
A. angusta, U. pinnatifida	Anti-cancer activity	Tumor destruction	[4,5]
(1.5 ± 0.3%), *Dictyopteris* spp.,	Anti-viral activity	Improved body weight and	[5,24]
L. saccharina	Immunomodulatory activity	fasting blood glucose	[5]
E. cava, A. nodosum,	Wound healing activity	Anti-proliferative	[7,19,36]
Cladosiphon okamuranus,	Anti-angiogenic activity	Antimetastatic	[13]
Fucus vesiculosus (18.22%),	Anti-allergy activity	Hypotriglyceridemic effects	[25,45,46]
L. japonica, F. evanescens (4.44%	Anticoagulant activity	Anti-inflammatory response	[24]
(4.7%, $CaCl_2$; 3.02% (5.11%,	Anti-diabetic activity	Neuroprotective effects	[41,42,47]
Ethanol))	Anti-hypolipidemic	Increase nitric oxide production.	[6]
	Anti-hyperglycemic activity	Activation of	[5]
	Antioxidant activity	PI3K/Akt/eNOS-dependent	[5,20]
	Cognitive protective activity	pathways.	[8,23]
Dictyota menstrualis, S. polycystum,	Anti-angiogenic activity		[7]
D. delicatula,	Antimicrobial activity		[22]
Turbinaria conoides, S. latissimi,	Anti-obesity activity		[45]
Spatoglossum asperum	Anti-inflammatory activity		[6,10]
Cystoseira sedoides	Anticancer	Anti-inflammatory	[5]
Coccophora langsdorfii		Anticancer	
Alginate			
L. digitata (51.8%),	Anti-obesity activity	Overweight and Obesity	[38]
A. nodosum,	Antiviral Activity	Inhibition of α-amylase,	[2,24]
L. hyperborea,	Anticancer Activities	pancreatic lipase and pepsin.	[4,10,45]
Macrocystis pyrifera,	Antidiabetic Activities	Applications such as feed	[18,47]
S. muticum (13.47%)	Antioxidant Activity	stabilizer, paper and welding rod	[13]
	Anti-inflammatory	coatings, and dye thickener in	[45,48]
	Anti-microbial	textile printing	[22,35]
	Anti-coagulant	Drug delivery	[42,49]
S. binderi (54%)	Anti-ageing	Pharmaceutical applications	[19,25]
U. pinnatifida (23.6 ± 1.2%)	Anti-obesity		

Table 1. *Cont.*

Bioactive Algal Polysaccharides and Their Sources with Yield	Bioactivity Potential	Therapeutic and Nutraceutical Applications	Ref.
Ulvan			
Ulva pertusa	Antioxidant activity	Influence plant immunity	[7,20]
Monostroma nitidum and *U. pertusa*	Immunomodulating activity	Triggering plant defense in several different plants	[13]
U. pertusa	Antihyperlipidemic activity	Treatment of gastric ulcers	[35]
Monostroma spp.	Antidiabetic activity		[47]
	Anticoagulant activity		[19]
M. nitidum	Anticancer activity	Preventive/Therapautic	[10]
	Neuroprotective activity		[12]
Chlorella Pyrenoidosa	Immunostimulation		[12]
U. intestinalis	Anti-inflammatory,		[24]
U. prolifera	Antiviral,		[42]
U. lactuca	Anticoagulant		
Ulva and *Enteromorpha* spp.	Gel-forming, skin protective and antioxidant properties	Skin aging products	[25,36]
Carrageenan		linebreak Enzymes hydrolyzing plant polysaccharides	[7]
Kappaphycus alvarezii, and	Anticoagulant activity		
Eucheuma dendiculatum (50–55%)	Antiviral activity	Food industry; mainly in dairy and meat applications	[19]
Chondrus crispus,	Cholesterol-lowering activity		[49]
Sarcothalia crispata	Anti-tumor activity	Thickening, gelling, and stabilizing agent.	[35]
E. cottonii (46.43%)	Immunomodulatory activity		[10]
Furcellaria lumbricalis and		Fertilizer	[13]
Coccotylus truncates (76.3%)		Production of nanoengineered injectable hydrogels in tissue regeneration therapy	[18,20]
Mastocarpus stellatus (28.65%)	Antioxidant activity		[25,48]
	Anti-allergic activity		
Agar			
Gracilaria lemaneiformis (29.7 ± 1.9%)	Gelling and stabilizing properties	Potential application in pharmaceutical industries	[7,48]
	Antioxidant		[20]
G. vermiculophylla (9.7–34.6%)	Antidiabetic activity	Gelling agent	[35,47]
		Food application, including bakery, confectionery, dairy products, canned meat and fish products	[13]
	Neuroprotective and anti-neurodegenerative properties		[18,42,49]
G. cornea		Preparation of bacteriological culture media	
		Support for the three-dimensional cultures of human and animal cells	[45,50]
	Scavenging properties, Antioxidant and gel-forming activity	Thickening agent	[25,36]
Gelidium amansii			

2. Structure and Molecular Characteristics of Laminarin

Structurally, laminarin is a water-soluble branched polysaccharide that consists of (1–3)- β-D-glucan with β (1–6)-linkages/branching of 20–25 glucose units, depending on the habitat, harvesting season and location (Figure 2) [48,51]. Brown macroalgae are reported to have around 350 mg/g laminarin content (on a dry basis), mostly present in the fronds part, which is influenced by algal species, harvesting season, geographic location, habitat, population age and method of extraction [45]. Laminarin possesses an average molecular weight of approximately 5 kDa depending on the degrees of polymerization. Additionally, depending on the type of sugar, the two forms of chains, M chains (end with terminal 1-O-substituted D-mannitol residues) or G chains (end with glucose residues), were observed at the reducing end. Rajauria et al. [52] identified the molecular weight of the purified laminarin in the range of 5.7–6.2 kDa. In particular, laminarin possesses a lower molecular weight than other polysaccharides reported in seaweed. The low molecular weight of laminarin has antioxidant activity due to carbonyl groups, which can improve lipid peroxidation.

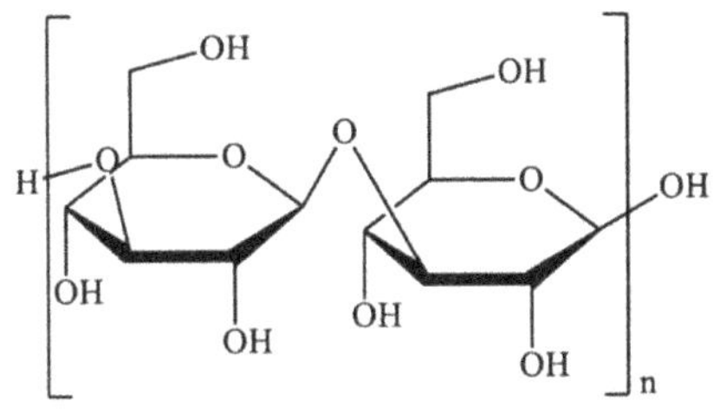

Figure 2. Structure of laminarin.

Laminarin has been extracted from different brown macroalgal species, including *Laminaria digitata, Saccharina latissima, Laminaria japonica, Ecklonia kurome,* and *Eisenia bicyclis,* and to a lesser extent in Ascophyllum, Fucus, and Undaria species obtained from Asian and European countries [3,21,53,54]. The biological activity of laminarin depends on the molecular size, extraction methods, type of sugar, type of linkage and molecular geometry. In addition, the structure and biological activities of laminarin are influenced by environmental factors. For instance, a decrease in nitrite and nitrate in water as a nitrogen source stimulates the synthesis of laminarin [26]. Laminarin solubilizes in aqueous media or organic solvents, and in cold or hot water. Therefore, different conventional extraction approaches such as grinding, precipitation in an acid or alkaline medium, ultrafiltration, and dialysis with different molecular weight cut-off membranes have been employed to extract laminarin [26,46,55]. Recently, emerging technologies, such as ultrasound-assisted, microwave-assisted, and hydrothermal-assisted extraction processes, have been employed for the fast and cleaner recovery of laminarin with enhanced bioactivity [45,52,53]. Some reports have suggested that specific chemical modifications or treatment with processing techniques could enhance the bioactivity of laminarin [56–59]. Various processes, e.g., irradiation, sulfation, reduction, and oxidation, have been investigated for the modification of laminarin structure to improve the physicochemical, biological, and mechanical properties. Sulfated laminarin was reported to possess antitumor activity in human colorectal adenocarcinoma cells [57], prostate cancer cells, human melanoma SK-MEL-28, and colon cancer DLD-1 cells [21], anticoagulant activity [58], degradation by gamma irradiation [59], inhibition of cell proliferation through the activation via both specific receptor-mediated and mitochondria-mediated apoptotic pathways, and antimetastatic potential [60,61].

3. Biological Effects of Laminarin

Many laminarin research studies [20,21,26,28] reported multifunctional biological properties for potential therapeutic applications. Chemical modification of the structure of algal polysaccharides can improve their solubility and biological properties, which facilitates their potential use in biomedical applications and clinical trials. While recent in vitro and in vivo studies highlighted the biological activities and functional and physicochemical properties of laminarin, further in-depth studies are still required to investigate mitochondrial pathways relevant to the inhibition of proliferation and induction of cell apoptosis [7,57]. While recent in vitro and in vivo studies highlighted the biological activities, and functional and physicochemical properties of laminarin, further in-depth studies are required to investigate mitochondrial pathways relevant to the inhibition of proliferation and induction of cell apoptosis [7,57]. The research studies on brown seaweed for various biological activities with respect to the diseases are listed in Table 2.

Table 2. Summary of preclinical studies with respect to biological activity and health benefits of laminarin extract or fraction (n = 17).

Source	Extraction, Purification, Yield and Characterisation	Study Design	Model, Administration Dose, Route and Period	Biological Activities	Effects/Outcome of Laminarin Extract or Fraction and Related Products	Ref.
Edible seaweed species including *Laminaria digitata*	50 mL of boiling absolute ethanol (>99.5%) was added to 1.25 g of dried seaweed powder, then centrifuged after which the supernatant was removed. Ethanol extracts and water extracts of *L. digitata* were freeze dried. On experimental days, dried seaweed extracts were reconstituted in appropriate buffer for experimentation.	The in vivo anti-hyperglycemic activity.	C57/BL6 mice 500 mg/kg Orally 0–105 min	Hyperglycemic	Extracts of *L. digitata* strongly inhibited DPP-4 activity. Medicinal foods or biotherapeutics to tackle type 2 diabetes mellitus Targeted GLP-1 secretion, DPP-4 activity or alpha-glucosidase activity.	[34]
Laminaria digitata	The formulated agar in a brown seaweed, *L. digitata* was prepared. 10 g of processed seaweed sample was transferred to the beaker and mixed up with 100 mL of high-pure Milli-Q water then autoclaved at 121 °C for 1 h for extraction of water-soluble polysaccharides. The yield of formulated agar in a brown seaweed, *L. digitata* was found to be 40%. Characterized by FTIR, and SEM analysis.	The anti-skin cancer effect of formulated agar (FA) from *L. digitata* on dimethyl benzanthracene induced skin cancer mice.	Female Swiss albino mice 60–120 mg/mL Orally Four weeks	Anticancer Antioxidant Anti-skin cancer agent	FA from brown seaweed, *L. digitata* is a potent anti-skin cancer agent and antioxidant. Enhanced the antioxidant system	[29]
Laminaria digitata	Laminarin extracted from *L. digitata* (purity of 95%) was obtained from Sigma-Aldrich (St. Louis, MO, USA) and dissolved in sterile water before use. The response to laminarin treatment (0.1, 0.25, 0.5, 1, and 2 mg/mL) were used.	The anticancer effects of laminarin, a beta-1, 3-glucan derived from brown algae, have been reported in ovarian cancer (OC), colon cancer, leukemia, and melanoma.	A zebrafish xenograft model zebrafish embryos 0.5, 1, and 2 mg/mL Orally 24 h–72 h	Anticancer	Laminarin derived from brown algae had anticancer effects in human OC cells. Prevented tumor formation within the embryo yolks A novel OC suppressor.	[14]

Table 2. *Cont.*

Source	Extraction, Purification, Yield and Characterisation	Study Design	Model, Administration Dose, Route and Period	Biological Activities	Effects/Outcome of Laminarin Extract or Fraction and Related Products	Ref.
Laminaria spp.	The laminarin-rich extract was obtained from Laminaria spp. using hydrothermal assisted extraction by applying optimized extraction conditions. The laminarin-rich extract was included in sufficient quantity to achieve a concentration of 200 ppm in the relevant treatment. The crude extract was partially purified in order to enhance the polysaccharide content by mixing the crude extract with pure ethanol followed by water and calcium chloride.	The effects of increasing dietary inclusion levels of laminarin in the first 14 d post-weaning on pig growth performance and weaning associated intestinal dysfunction.	Ninety-six healthy piglets 650 g/kg Lethal injection 28 days	Anti-obesity Gastrointestinal health	Laminarin-rich extract has potential to prevent pathogen proliferation. Enhanced the volatile fatty acid profile in the colon in a porcine model of colitis.	[62]
Laminaria spp.	The laminarin-rich extract was sourced from BioAtlantis Ltd (Tralee, Co. Kerry, Ireland). The extract was prepared by using water as an extraction solvent under optimum heating conditions. A single extraction was performed from Laminaria spp. to produce the extract. Appropriate quantity of laminarin as 100, 200 and 300 parts per million (ppm) was added.	The effects of dietary supplementation with laminarin on colonic health in pigs challenged with dextran sodium sulphate.	Forty-two healthy male pigs 200 ppm Orally 35 days	Anti-inflammatory	300 ppm of laminarin from a laminarin-rich extract as a dietary supplement, improved performance and prevented post-weaning intestinal dysfunction	[63]
Laminaria spp.	The laminarin enriched extract was sourced from BioAtlantis Ltd. The laminarin rich extract was obtained from Laminaria spp. using hydrothermal assisted extraction by applying optimised extraction conditions. The crude extract was partially purified in order to enhance polysaccharide content.	The effects of supplementing the diet of newly weaned pigs with 300 ppm of a laminarin rich extract, on animal performance, volatile fatty acids, and the intestinal microbiota using 16S rRNA gene sequencing	Fifty-four weaned pigs 300 ppm Orally 28 days	Anti-obesity Gut health	300 ppm of a laminarin rich macroalgal extract reduced post-weaning intestinal dysfunction in pigs. Promoted the proliferation of bacterial taxa Enhanced nutrient digestion by reducing the load of taxa	[64]

Table 2. *Cont.*

Source	Extraction, Purification, Yield and Characterisation	Study Design	Model, Administration Dose, Route and Period	Biological Activities	Effects/Outcome of Laminarin Extract or Fraction and Related Products	Ref.
Saccharina longicruris	Laminarin (TCI Shanghai, China) was dissolved in saline to prepared stock solution of 50 mg/mL. One g/kg of laminarin by intragastric administration and characterized by western blot assay.	The effect of laminarin on energy homeostasis, mice were orally administrated with laminarin to test food intake, fat deposition, and glucose homeostasis.	Homeostasis model C57/BL6 mice 50 mg/mL Orally Four weeks	Anti-obesity	Laminarin counteracts diet-induced obesity. Maintained glucose homeostasis. Promoted GLP-1 secretion via increasing intracellular calcium in enteroendocrine cells.	[31]
Laminaria digitata	Laminarin derived from *L. digitata* was purchased from Invivogen. The endotoxin levels in the purified laminarin were evaluated using a Limulus gametocyte lysate (LAL) assay kit (Lonza, Gampel, Switzerland).	The effects of laminarin on the maturation of dendritic cells and on the in vivo activation of anti-cancer immunity using homozygous transgenic mice (OT-I and II).	C57BL/6 mice, OT-I and OT-II TCR transgenic mice, and C57BL/6-Ly5.1 (CD45.1) congenic mice 25 mg/kg Intrasplenic B16-OVA injection 24 h	Immunomodulatory	The purified laminarin stimulated and inhibited tumor growth and metastasis Potential immune stimulatory molecule for use in cancer immunotherapy.	[30]
Laminaria spp.	The crude seaweed extract was derived from Laminaria spp. (BioAtlantis Ltd.). The laminarin content of the supplements and the feed samples was determined by spectrophotometry using a commercial assay kit (Megazyme Ireland Ltd., Bray, Ireland).	The potent anti-inflammatory activities of the algal polysaccharides laminarin and fucoidan in the gastrointestinal tract of pigs	Dextran sodium sulfate-challenged porcine model Thirty-five pigs 300 mg/kg Oral administration 56 days	Anti-inflammatory	Laminarin has potent anti-inflammatory activities in the gastrointestinal tract. Reduced Enterobacteriaceae in proximal colon digesta. Decreased in colonic IL-6 mRNA abundance for its inflammatory effect	[32]
Laminarin spp.	A high-fat diet (HFD) and 1% laminarin-supplemented water (HFL). Laminarin (Sigma) supplementation was terminated at the fourth week, followed by continuation HFD for an additional 2 weeks.	The anti-obesity effects of the potential prebiotic, laminarin, on mice, fed a high-fat diet.	Laminarin-supplemented high-fat diet (HFL) mice model Eighteen female BALB/c mice 45% (*w*/*v*) Orally 6 weeks	Anti-obesity effects Gut Health and microbiota	Laminarin (Sigma) reduced the adverse effects of a high-fat diet by shifting gut microbiota towards higher energy metabolism. Laminarin could be used to develop anti-obesity functional foods	[25]

Table 2. *Cont.*

Source	Extraction, Purification, Yield and Characterisation	Study Design	Model, Administration Dose, Route and Period	Biological Activities	Effects/Outcome of Laminarin Extract or Fraction and Related Products	Ref.
Laminaria digitata, L. hyperborea and *Saccharomyces cerevisiae*	Purified laminarin from *L. digitata* and *L. hyperborea* (990 g laminarin/kg) was sourced from Bioatlantis Limited and extracted. A basal diet supplemented with 250 parts per million (ppm) laminarin from *L. digitata*.	The effects of purified laminarin derived from *L. digitata, L. hyperborea* and *S. cerevisiae* on piglet performance in selected bacterial populations and intestinal volatile fatty acid (VFA) production	Thirty-two pigs 250 mg/kg Orally 28 days	Anti-inflammatory	Purified laminarin may be acting via a different mechanism from the insoluble β-glucan. Identified as natural biomolecules with immunomodulatory activity	[28]
Laminaria digitata	Purified laminarin (990 g/kg laminarin) was sourced from Bioatlantis Ltd., and extracted.A basal diet supplemented with 300, and 600 parts per million (ppm) laminarin from *L. digitata*.	The optimum inclusion level of laminarin derived from *L. digitata* on selected microbial populations, intestinal fermentation, cytokine and mucin gene expression in the porcine ileum and colon	Twenty-one Pigs 990 g/kg Orally 26 days	Gastrointestinal health	Dietary inclusion of 300 ppm purified laminarin appears to be the optimum dose. Reduced enterobacteriaceae populations Enhanced IL-6 and IL-8 cytokine expression in response to an ex-vivo LPS challenge	[65]
Laminaria digitata	Seaweed extracts derived from *L. digitata* were included at 2.8 g/kg. The extract contained both laminarin (112 g/kg), fucoidan (89 g/kg) and ash (799 g/kg) and was sourced from Bioatlantis Ltd.	The interactions between two different lactose (L) levels and seaweed extract containing laminarin and fucoidan derived from Laminaria spp. on growth performance, coefficient of total tract apparent digestibility (CTTAD) and faecal microbial populations in the weanling pig.	Two hundred and forty pigs 112 g/kg Orally 25 days	Intestinal digestibility	The inclusion of a high dietary concentration of laminarin increased the CTTAD of diet components and decreased the counts of *E. coli* in the faeces. Improved performance of pigs after weaning	[66]

Table 2. *Cont.*

Source	Extraction, Purification, Yield and Characterisation	Study Design	Model, Administration Dose, Route and Period	Biological Activities	Effects/Outcome of Laminarin Extract or Fraction and Related Products	Ref.
Laminaria hyperborea	The seaweed extract was extracted from Laminaria spp. The seaweed extract contained laminarin (112 g kg^{-1}), fucoidan (89 g kg^{-1}) and ash (799 g kg^{-1}) and was sourced from BioAtlantis Ltd.	The effect of dietary Laminaria-derived laminarin and fucoidan on nutrient digestibility, nitrogen utilisation, intestinal microflora and volatile fatty acid concentration in pigs	Thirty finishing boars/pigs 112 g/kg Orally 14 days	Anti-obesity effects Dietary and gut health	Reduced intestinal *Enterobacterium* spp. and increased in *Lactobacilli* spp. The dietary laminarin may provide a dietary means to improve gut health in pigs	[37]
Laminaria spp.	The seaweed extract was extracted from Laminaria spp. Seaweeds extracts of 1, 2 and 4 g/kg were used. The seaweed extract was sourced from BioAtlantis Ltd.	The interaction between different levels of lactose and seaweed extract derived from Laminaria spp. on growth performance and nutrient digestibility of weanling pigs.	384 piglets 1,2,4 g/kg Orally 21 days	Intestinal health	Pigs responded differently to the inclusion levels of seaweed extract at each level of lactose supplementation.	[67]
Laminaria spp.	Laminarin (purity of 90%). was provided by Goëmar (St Malo, France). The molecular weight was 4500 to 5500 g/mol and the purity was 90%. laminarin treated rats (LAM) received the same diet containing 5 g/100 g laminarin for 4 days followed by a dietary treatment of 10 g/100 g LAM for 21 days.	The hypothesis stated that LAM, a β (1–3) polysaccharide extracted from brown algae, can modulate the response to systemic inflammation.	Male Wistar rats 5 g/100 g for 4 days followed by 10 g/100 g Intraperitoneal 21 days	Hepatoprotective	The effects could be due to a direct effect of purified laminarin on immune cells, or to an indirect effect through their dietary fibre properties. Decreased serum monocytes number, TNF-α and nitrite.	[39]
Laminaria saccharina	Brown seaweeds, *L. saccharina* (Algoplus, Roscoff, France) were collected. Laminarin was extracted from ground algae (15–25 g) by addition of absolute ethanol, hot H_2SO_4, and hot HCl. The various methods of extraction of laminarin was implemented by partial characterisation.	The various methods of extraction of laminarin were compared by partial characterization and, on the other hand, to study the fate of this polysaccharide and its effects in the gastrointestinal tract in order to determine its potential as a dietary fibre in human nutrition.	Wistar rats 15–25 g (*w*/*v*) Orally 18 days.	Anti-obesity effects Gut health	Laminarin can be considered a dietary fibre No increased in the intestinal transit and stool output. Laminarin resisted hydrolysis in the human upper gastrointestinal tract.	[68]

3.1. Antioxidant

Rajauria et al. [52] reported that the molecular weight of laminarin influences antioxidant potential. In particular, crude extracts of laminarin have higher activity than purified and commercial products. Rajauria et al. [69] reported that laminarin was screened for its potential antioxidant capacity and found to have a significant radical scavenging capacity against free radicals and metal ions. These results showed that macroalgae are a rich source of natural antioxidants [9,19,53,70] in food and cosmetics [11,21].

Garcia-Vaquero et al. [53] investigated the antioxidant activity in laminarin from *Laminaria digitata* using DPPH and FRAP methods. The antioxidant and antimicrobial activities of crude laminarin extract were also examined by Kadam et al. [22] who confirmed a higher inhibition rate in scavenging of free radicals as antioxidant potential. Choi et al. [59] confirmed the therapeutic potential due to the antioxidant property of laminarin and outlined the interconnection with anti-inflammatory potential and a role in the activation of an immune response. Liu et al. [71] reported antioxidant activities against oxidative damage caused by reactive oxygen species (ROS) and free radicals. Preclinical studies were conducted to examine the antioxidant potential of laminarin using an animal model. Cheng et al. [72] showed using experimental rat studies that laminarin is a pulmonary oxidation and lipid peroxidation, protective agent. Another study by Jiang et al. [73] conducted in porcine early-stage embryos demonstrated significantly increased intracellular glutathione levels, cleavage, hatching, and blastocyst formation rates by maintaining the mitochondrial function, up-regulating differentiation and pluripotency-related genes. In addition, antimicrobial activity was also observed against different microorganisms [54,58]. Recent research has focused on antioxidant activity for the development of functional food products [19,70].

3.2. β-Glucan Related Receptors

Laminarin has been widely studied and utilized for various applications because of its molecular interaction with the glucan-specific pattern recognition receptor, Dectin-1 [74]. However, in recent years, laminarin has been used as a ligand for pattern recognition receptors and modulates innate immunity as immunoregulatory potential. Laminarin sulfate mimics the effects on smooth muscle cell proliferation and basic fibroblast growth factor-receptor binding and mitogenic activity [61]. Specifically, C-type lectin receptor (CLR) Dectin-1 acts as a receptor responsible for binding fungal β-glucans and eliciting innate immune responses [75]. Laminarin stimulates antitumor and antimicrobial activity by binding to receptors, such as complement receptor 3 and β-glucan receptor, as well as dectin-1 on macrophages and white blood cells, which provides new insights into the innate immune recognition of β-glucans [59,76]. In certain cases, variants of β-glucans polysaccharides downregulated autoimmune inflammation and can be mediated by cell surface receptors [77].

3.3. Immunomodulatory

The chemical characterization and quantification of laminarin to evaluate biochemical and ecological potential are considered unique methods, based on the physicochemical properties [78]. In this aspect, hydrogen bonds are resistant to hydrolysis in the small intestine and laminarin is considered as a dietary fibre possessing immunomodulating and anticoagulant properties [68,79]. Laminarin can interact with a specific receptor of the immune system for biological properties and immunomodulatory action [23]. Laminarin enhances the immune system with a high accumulation of B cells and helper T cells [23]. Recent research suggested that laminarin is an immune stimulatory molecule for cancer immunotherapy applications [12,21,68,78]. In this aspect, protein exhibited host–pathogen interactions and immunomodulatory potential, whereas lipids including long-chain fatty acids and short-chain fatty acids improve the healthy metabolism in cardiovascular and obesity-related diseases [6]. In addition, vitamins can also trigger the metabolic pathways in human health [2,12,29–32,47]. Due to the properties of proteins, lipids, and vitamins, laminarin possesses immunostimulatory activity. Kalasariya et al. [24] reported that laminarin promotes the adhesion of human skin fibroblast cells for wound healing and human osteoblast cells for bone formation in in vitro studies, demonstrating its immunostimulatory properties. In addition, laminarin was shown to inhibit tumor growth and metastasis due to its immunostimulatory potential [21,30].

The therapeutic potential of laminarin for the treatment of diseases was investigated in clinical trials conducted in pigs. The study by Rattigan et al. [62] reported that laminarin has the potential to prevent pathogen inflammation and proliferation and to enhance the fatty acids in the colon. Sweeney et al. [80] studied the immune response in the intestinal health of chicks by measuring growth performance, small intestinal morphology and function, and immune response. Similar results were also confirmed for gastrointestinal tract (GIT) health [67,81] in pig studies which showed a complex interaction between host genetics, environmental factors, and the gut microbiome [82] for inflammatory bowel diseases (IBD), including ulcerative colitis (UC) and Crohn's disease. In addition, Vigors et al. [64] showed that dietary supplementation of laminarin also improves nutrient digestion, volatile fatty acids, and the intestinal microbiota using 16S rRNA gene sequencing in post-weaned pigs. Ostrzenski et al. [83] demonstrated the safety and efficacy of laminarin for resectoscopic cervical trauma in a non-pregnant patient population. Zaharudin et al. [33] showed that laminarin possesses hyperglycaemic and glycaemic potential in healthy adults using three-way blinded cross over trials. In addition, clinical studies conducted on humans reported that laminarin helps in improving postprandial hyperglycaemic and appetite control in healthy and normal-weight adults.

The anti-inflammatory activity of laminarin was shown using the gene expression profiles of anti- and pro-inflammatory markers [28,80], lower secretion of inflammatory cells in liver tissue and inflammatory mediators due to the β-glucan on immune cells and dietary fibre properties [68]. In addition, anti-inflammatory activity can be easily increased by the release of significant inflammatory mediators in laminarin such as hydrogen peroxide, calcium, nitric oxide, monocyte chemotactic protein-1, vascular endothelial growth factor, leukaemia inhibitory factor and granulocyte-colony-stimulating factor with enhancing expression of signal transducer and activator of transcription 1 (STAT1), STAT3, c-Jun, c-Fos and cyclooxygenase-2 mRNA. Rattigan et al. [62] showed that laminarin possesses anti-inflammatory potential by preventing pathogenic proliferation. Thus, it can also enhance diarrhoeal scores, body weight loss, and clinical variables linked with dextran sodium sulfate in pigs. A similar result was supported by O'Shea et al. [32] with systematic inflammation of laminarin. According to recent studies, laminarin can be used as a therapeutic agent for anti-inflammatory and immunostimulatory activities.

3.4. Wound Healing

Laminarin is considered an anticancer, antioxidant and anti-skin cancer agent [24]. Kadam et al. [54] reported that laminarin has potential application in wound healing as an effective antimicrobial agent. Zargarzadeh et al. [3] showed that marine-derived polysaccharides are considered multifunctional supporting biomaterials that can stimulate the healing process owing to their physicochemical and biological potential. Kadam et al. [26] reported that laminarin also promotes cell adhesion of human skin fibroblast cells for wound healing and proliferation in osteoblast cells for bone formation. Laminarin from Alaria species has been reported to promote cell adhesion of human skin fibroblast cells and cell proliferation in human osteoblast cells [84]. These studies on laminarin support its potential use in clinical investigations in cancer treatment. Due to its non-toxic, hydrophilic, and biodegradable properties, laminarin has potential as a wound-healing agent in modern medicine.

Sellimi et al. [17] reported that laminarin improves wound contraction, accelerates re-epithelization, collagen deposition, reconstitution of the skin tissue, and increases fibroblast and vascular densities in rats for wound healing effects. In addition, the beneficial role of laminarin promotes the healing process and skin regeneration. However, laminarin also has antibacterial and antioxidant properties which protect against free radical-mediated oxidative damage effect. Laminarin showed anticoagulant activity after structural modifications, such as sulphation, reduction, or oxidation. Laminarin was reported to be an effective polysaccharide in the prevention and treatment of cerebrovascular diseases due to its anticoagulant activity [3,6,34]. Kadam et al. [26] confirmed that laminarin possesses anti-coagulant activity due to its diverse biological properties. In addition, laminarin sulphate has been demonstrated in preclinical and clinical applications for anticoagulant activity, wound healing, angiogenesis, atherosclerosis, and cerebrovascular diseases.

Zargarzadeh et al. [3] noted that laminarin has been used in tissue engineering with cell function, which leads to tissue regeneration. The reduction of cholesterol levels by the active role of laminarin thereby lowers systolic blood pressure and levels of total cholesterol, free cholesterol, triglyceride, and phospholipid in the liver [26,35]. The clinical studies on laminarin extracted from marine brown seaweeds with respect to potential therapeutic applications are listed in Table 3.

Table 3. Summary of clinical studies with respect to biological activity and health benefits of laminarin extract or fraction and related products ($n = 4$).

Source	Extraction, Purification, Yield and Characterisation	Study Design	Model, Administration Dose, Route and Period	Biological Activities	Effects/Outcome of Laminarin Extract or Fraction and Related Products	Ref.
Laminaria digitata and *Undaria pinnatifida*	The dried seaweed was soaked in 200 mL of water for 10 min, then rinsed and drained to remove excess water. Finally, they were cut into pieces and added with 0.5 g iodine enriched salt (6.5 mg iodine). The yield of dietary fibre (g/serving) about 1.8.	The effects of two brown edible seaweeds, *L. digitata* and *U. pinnatifida*, on postprandial glucose metabolism and appetite following a starch load in a human meal study as first trail.	A randomized, Three-way, blinded cross-over trial Twenty healthy participants 5 g (w/v) Orally 180 min	Hyperglycaemic	Concomitant ingestion of brown seaweeds may improve postprandial glycaemic and appetite control in healthy and normal weight adults, depending on the dose. Brown seaweeds inhibited the postprandial glucose and insulin response.	[33]
Laminaria digitata and *Eisenia bicyclis*	Laminarins from *L. digitata* and *Eisenia bicyclis* (Purity of 95%) were obtained from Sigma-Aldrich (St. Louis, MO, USA). Laminarin from *L. digitata* was reduced to laminaritol, to reduce background responses in the bicinchoninic acid (BCA) reducing sugar by enzyme kinetics assay. Purified by Crystallization, NMR, and electrophoresis and used for the human gut microbiota (HGM) study.	The symbiont Bacteroides uniform deploys a single, exemplar polysaccharide utilization locus (PUL) to access yeast β (1, 3)-glucan, brown seaweed β (1, 3)-glucan (laminarin), and cereal mixed-linkage β (1, 3)/β (1, 4)-glucan as first trial.	2441 adults	Anti-obesity effects Human gut microbiota	Fine-grained knowledge of PUL function Metabolic network analysis and proactive manipulation of the HGM. Purified Laminarin provided a validated set of molecular markers to identify β (1, 3)-glucan utilization potential among members of the HGM.	[85]

Table 3. *Cont.*

Source	Extraction, Purification, Yield and Characterisation	Study Design	Model, Administration Dose, Route and Period	Biological Activities	Effects/Outcome of Laminarin Extract or Fraction and Related Products	Ref.
Laminaria digitata	The purified laminarin (LAM) from the brown algae *L. digitata* were prepared. The basal diet supplemented with 250 ppm purified LAM; basal diet supplemented with a seaweed extracts.	The effects of supplementing the diet with seaweed extracts on growth performance, small intestinal morphology and function, immune response and *Campylobacter jejuni* colonisation following an experimental challenge in young chicks as first trial.	Hundred-and thirty-five-day-old male Ross chicks 250 ppm Orally 13 days	Intestinal health	Supplementation with laminarin improved growth rate, positively modified small intestinal architecture and Impacted the intestinal immune response.	[80]
Laminaria digitata	A group of 30 patients were dilated preoperatively with Laminaria tents, prepared from *L. digitata*.	The safety and efficacy of *L. digitata* was examined to dilate the cervix before resectoscopic surgery as randomized trial.	Thirty patients Randomized trial	Resectoscopic cervical trauma	More than 8-mm cervical dilatation	[83]

3.5. Obesity

The tremendous influence of the human gut microbiota on health and intestinal disease is reported in recent studies [15,17,86,87]. Therefore, scientists can target gut microbiota-mediated immune systems for the treatment of cancer, diabetes, obesity, and cardiovascular diseases [86,87]. Earlier findings suggested that dietary fibre has a prebiotic effect because it supplies carbon to the fermentation pathways in the colon, thus supporting digestive health in humans and animals [36]. Brown seaweeds are a rich source of dietary fibre with an estimated content of 25–70% DW [79]. In particular, laminarin is highly resistant to hydrolysis in the upper gastrointestinal tract (GIT) because it is easily stabilized with its complex structures by inter-chain hydrogen bonds, and hence is considered as a good dietary fibre. Laminarin has the ability to stimulate and enhance the immune system in humans and is considered a biological response modifier [88]. In particular, laminarin from Laminaria digitata was shown to alter the gut microbiota composition through metagenomic compositional analysis and short-chain fatty acid (SCFA) analysis [15,17].

Zou et al. [81] demonstrated anti-obesity effects in pigs. A GIT pig clinical model was selected due to its similarity both morphologically and physiologically to humans. Many anti-obesity studies were carried out using laminarin in pigs [31,62–64,89], mice [25] and rats [68] studies. These studies confirmed that laminarin could stabilize diet-induced obesity and improve intestinal health, prevent post-weaning intestinal dysfunction, and maintain glucose homeostasis. Strain et al. [82] stated that laminarin can be considered a dietary supplement and anti-obesity functional food. Laminarin reduces the adverse effects of a high-fat diet by shifting gut microbiota towards higher energy metabolism as confirmed by 16S rRNA gene (V4) amplicon sequencing. Similar improvements in gut health were observed by Smith et al. [65], O'Doherty et al. [66], and Lynch et al. [37].

Preclinical studies using in vitro and in vivo models investigated laminarin properties using metabolic network analysis and proactive manipulation of the human gut microbiota [82,90]. Various human clinical trials were conducted to study the anti-obesity effects of laminarin. Odunsi et al. [38] in a randomized, placebo-controlled study evaluated the anti-obesity effects of laminarin for weight loss and demonstrated its effects as a short-term weight loss treatment for obese and overweight patients. Déjean et al. [85] conducted clinical studies on adults and employed metabolic network analysis and proactive manipulation to demonstrate the potential anti-obesity effects of laminarin and the effects on human gut microbiota.

3.6. Diabetes

Studies by Calderwood et al. [34] confirmed that laminarin acts as medicinal food or bio-therapeutic for type 2 diabetes mellitus by modulating the signaling pathways. Laminarin is a bioactive compound which possesses anti-hyperglycemic activity, can inhibit the cholesterol levels in serum, lower systolic blood pressure levels and stimulate the immune system [50]. The antidiabetic potential of laminarin from marine brown algae should be explored in drug development and nutraceutical applications. Hyperglycaemia can be reduced by inhibiting the carbohydrate-hydrolysing enzymes, such as α-amylase and α-glucosidase, for targeting type 2 diabetes mellitus. In addition to the reported in vitro antidiabetic activity of laminarin, the in vivo antidiabetic potential was also evaluated using animal models to confirm it's hypoglycaemic effect through the inhibitory action on α-glucosidase and α-amylase enzymes [33].

Gunathilaka et al. [41] studied the antidiabetic potential of laminarin isolated from brown macroalgae and proved the inhibitory activity of rat lens aldose reductase enzyme in the presence of active porphyrin derivatives. Marine brown algae possess antidiabetic potential using the inhibitory activity of DPP-4 enzymes in a dose-dependent manner which is involved in glucose metabolism. Zaharudin et al. [33] reported that laminarin improves postprandial glycaemic and appetite control in healthy and normal-weight adults through clinical studies. However additional in-depth clinical studies are required to further investigate the biological activity of laminarin.

3.7. Cancer

Remya et al. [91] reported that laminarin showed anti-tumor activity against retinoblastoma Y79 cells and enhanced the activation of an immune response. Induction of apoptosis was also marked by the percentage of cells arrested in the G2/M phase using flow cytometry analysis and was further confirmed by a DNA fragmentation study. Similar studies were reported in other cancer cells, such as human melanoma SK-MEL-28 & SK-MEL-5 and colon cancer DLD-1 cells [92]; human colon cancer (HT-29), NK92-MI cells, human breast cancer (T-47D) and human skin melanoma (SK-MEL-28) [24,37,39,93,94]; human colorectal adenocarcinoma (HCT 116), and breast adenocarcinoma (MDA-MB-231) cells [38,95].

Ji et al. [57] reported that the sulfated modification of laminarin structure using the chlorosulfonic acid-pyridine method enhances antitumor activity on LoVo cells. Moreover, laminarin was shown to inhibit LoVo human colon cancer cell proliferation and induce LoVo cell apoptosis through a mitochondrial pathway. A similar finding was reported by Kadam et al. [26]. Menshova et al. [21] reported that the high molecular weight laminarin (19–27 kDa) from *Eisenia bicyclis* inhibited the colony formation of human melanoma SK-MEL-28 and colon cancer DLD-1 cells. In addition, the increase in 1, 6-linked glucose residues and the decrease of the molecular weight improved anticancer effect in laminarin.

George et al. [29] reported on the role of beta-1, 3-glucan derived from brown algae, which they showed had antitumor potential in hepatocellular carcinoma, colon cancer, leukemia, and melanoma. Their conclusions were also supported by Bae et al. [14]. Sulphated modification of laminarin structure enhances antitumor activity [57]. Laminarin also suppresses the formation of risk factors including SCFAs, indole compounds and ammonia for human colon cancer [57]. According to recent studies, laminaria supported the antitumor activity in brown macroalgae, which enhanced the activation of immune responses, cytotoxicity, apoptotic cell death, cell cycle arrest, colony formation, proliferation, and migration [21,24,37,38,93].

4. Therapeutic Effects of Laminarin as a Bioactive Agent

4.1. Antioxidant Activity

Laminarin is a dual regulator of apoptosis and cell proliferation, similar to β-Glucan [96]. Marine-derived antioxidant polysaccharides have three distinct mechanisms, including scavenging the ROS, regulating the antioxidant system or oxidative stress-mediated signaling pathways [97] (Figure 3a). The mechanism of β-glucan can activate different signaling pathways to regulate tumor cell proliferation, cell apoptosis, and cell cycle arrest, which can involve antioxidant, antitumor, and anti-inflammatory and immunostimulatory properties. Laminarin possesses potent antioxidant activity [52,59,69]. The mechanisms of antioxidant action in β-glucan with various pathways are shown in Figure 3b.

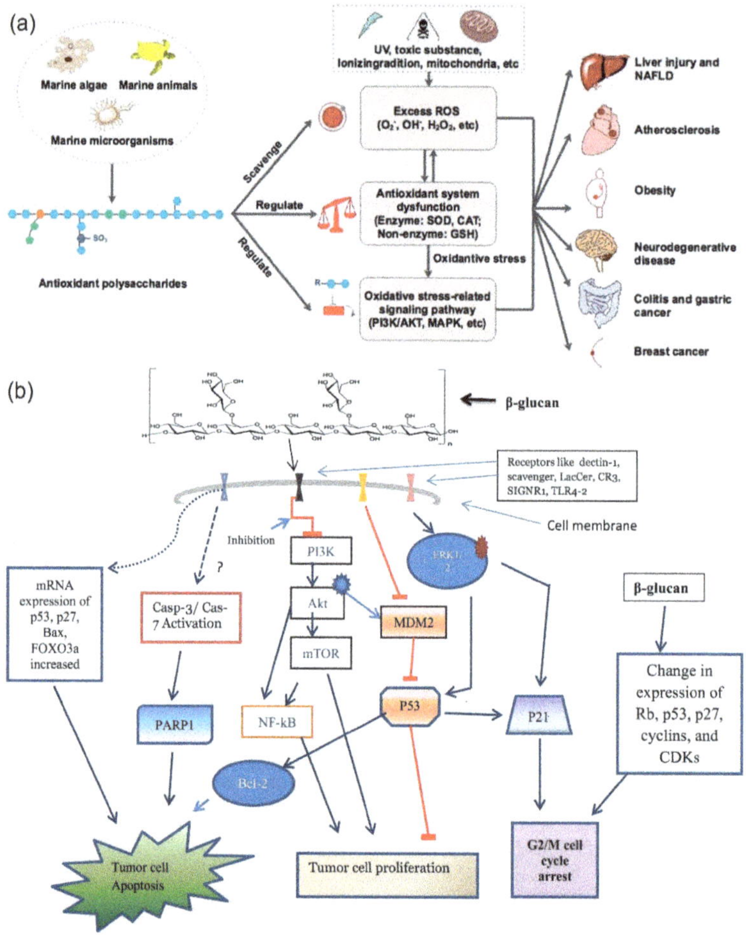

Figure 3. Antioxidant mechanisms. (**a**) Overview of marine-derived polysaccharides in alleviating oxidative stress-mediated diseases. Reprinted with permission from reference [97]. Copyright 2022, Zhong et al. (**b**) Antioxidant action of β-glucan. Reprinted with permission from reference [96]. Copyright 2022, Wani et al.

4.2. Immunomodulatory Activity

Laminarin has been reported to exert an immunostimulatory activity by the gene expression involved in inflammation and immune response, thereby it can stimulate the release of inflammatory mediators (Figure 4a). The specific receptors for β-glucan on dendritic cells (dectin-1), as well as interactions with other receptors, by innate immune cells (e.g., Toll-like receptors, complement receptor-3) are involved in the immune response to act as suitable therapeutic agents. Many studies have reported that β-glucan has similar biological activities to laminarin. Similar to β-glucan, laminarin has been reported to have activated innate with adaptive immunity, which induces humoral and cell-mediated immune responses and stimulates the production of proinflammatory molecules, such as complement components, IL-1α/β, TNF-α, IL-2, IFN-γ, and eicosanoids, as well as IL-10, IL-4, and proliferation of monocytes and macrophages (Figure 4b) [74,98].

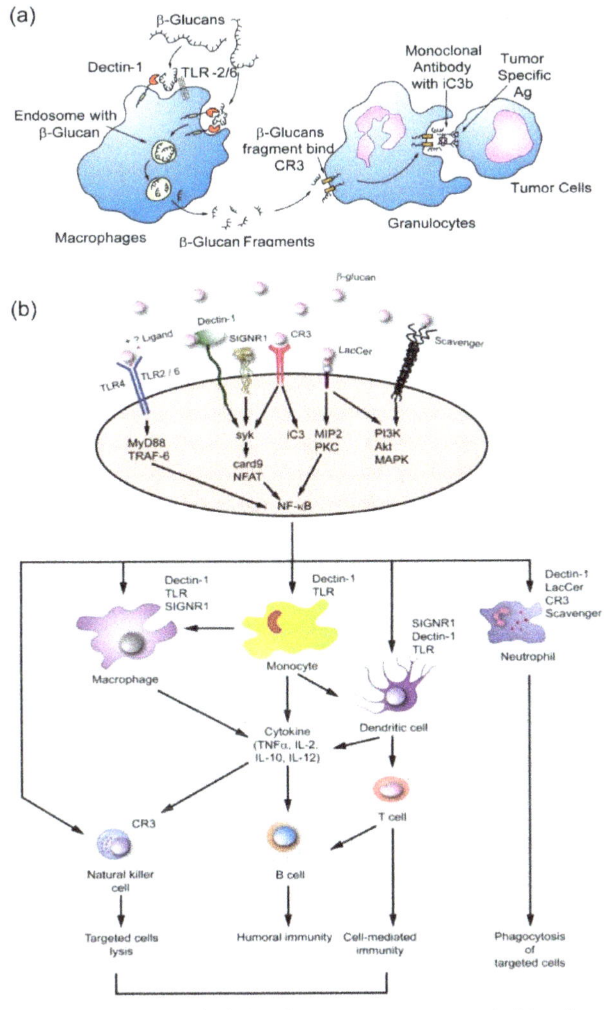

Figure 4. Immunomodulatory mechanism of β-glucan. (**a**) The uptake and subsequent actions of β-glucan on immune cells (**b**) Immune activation induced on a variety of membranes. Reprinted with permission from reference [74]. Copyright 2022, Chan et al.

The anti-inflammatory mechanism of laminarin is to stimulate the induction of innate immunity with certain ligands such as β-glucan. In particular, the antigens present in the tumor environment can stimulate an immune response in tumor cells and inhibit inflammation. Many preclinical and clinical studies have shown that β-glucan can enhance the antimicrobial activity of inflammatory macrophages, monocytes, and neutrophils, resulting in the maturation of target cells and an increased proinflammatory cytokine and chemokine release, stimulation of adaptive immune cells, including CD4+ T cells, CD8+ cytotoxic T lymphocytes (CTL) and B cells through the secretion of pro-inflammatory cytokines by T cells. In other cases, it also causes cell apoptosis with the release of ROS in the tumor microenvironment resulting in the destruction of tumor cells from oxidative stress [75]. Therefore, laminarin is considered as an immune-modulator agent and can be used as a synergic treatment in cancer and other inflammatory-related diseases.

Laminarin possesses anti-coagulant and inflammatory activities due to the modulation of innate immunity in specific metabolic pathways [99]. Furthermore, laminarin can reduce systolic blood pressure, cholesterol absorption in the gut, as well as cholesterol and total lipid levels in both serum and liver (Figure 5a). Microfold cells (M cells) are involved in transportation of antigen and particles to immune cells for modulating adaptive and innate immune responses. The activation of innate immunity by β-glucan is initiated by its binding the specific β-glucan receptor dectin-1 on macrophages. Thereby, it increases the inflammatory activity, enters the bone marrow, and stimulates the production of immune cells (Figure 5b).

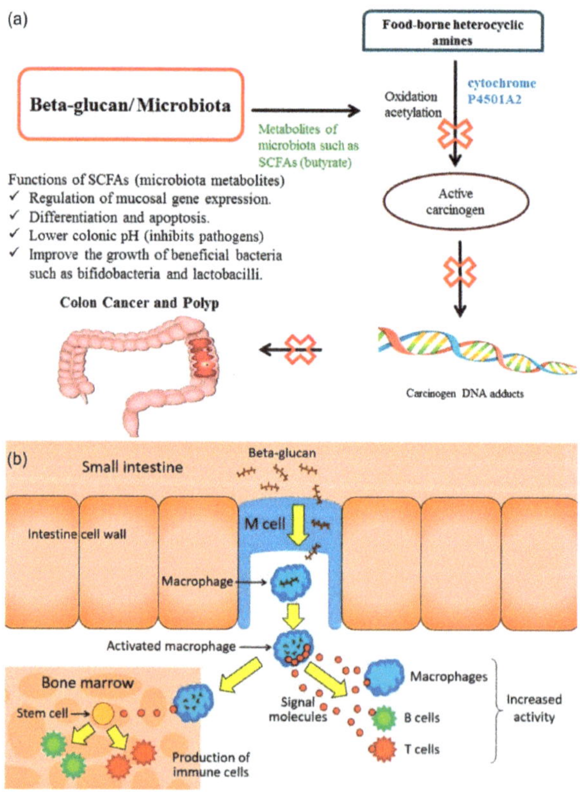

Figure 5. Mode of action in intestinal gut health (**a**) Action of absorption of β-glucan (**b**) influence of β-glucan in colon cancer via the gut microbiota. Reprinted with permission from reference [99]. Copyright 2022, Jayachandran et al.

4.3. Wound Healing

β-glucan was shown to have potential as a complementary therapy to manage various skin diseases and conditions due to its pluripotent activity [77]. The various cellular and extracellular matrix components and cells (keratinocytes, fibroblasts, endothelial cells, mast cells, nerve cells and leucocyte subtypes) participate differently in the three overlapping phases (inflammation, cell proliferation and tissue remodeling) in the healing process. Due to the antibacterial activity of laminarin, it also acts as a suitable wound healing agent with great stability. It can be mediated primarily by cell surface receptors, including immunocytes and cutaneous cells. The indirect signaling pathway through various cytokines of macrophages and direct signaling pathway on keratinocytes and fibroblasts cells are the two modes of immunostimulatory mechanisms enhancing wound healing. At the same time, growth factors from activated macrophages support cellular proliferation, angiogenesis, and reepithelialisation and an increase in wound tensile strength. Laminarin can accelerate the healing process in chronic and acute wounds by prolonging the inflammatory phase. The macrophages in granulation tissue were stimulated by β-glucan. Therefore, it acts as a source of growth factors and inflammatory cytokines (IL-6, IL-1, and TNFα), this pro-inflammatory event being mediated by the Dectin-1 receptor. In recent research, algal polysaccharides were investigated using green synthesis technology for their healing effects [100].

4.4. Anti-Obesity Activity

Laminarin significantly decreases high-fat diet-induced body weight gain and fat deposition. It also reduces blood glucose level and glucose tolerance. It enhances serum glucagon-like peptide-1 (GLP-1) content and the mRNA expression level of proglucagon and prohormone convertase 1 in the ileum. The mechanisms involved in anti-obesity activity have potential to be used to treat obesity and to maintain glucose homeostasis. β-Glucan is predominantly present in the cells in the cell walls of cereals, yeast, bacteria, and fungi, with significantly differing physicochemical properties [101]. However, laminarin is a type of β-glucan isolated from brown seaweed composed of D-glucose with β-(1, 3) linkages (Figure 6). However, laminarin promoted GLP-1 secretion and c-Fos protein expression dose-dependently. The expression of c-Fos protein is activated and regulated by mitogen-activated protein kinases (MAPKs) and the protein kinase C (PKC) signaling pathways [102]. Furthermore, glucose homeostasis and insulin sensitivity were improved. Laminarin promotes GLP-1 content via increasing intracellular calcium in enteroendocrine cells. Therefore, the long-term effects of laminarin counteract diet-induced obesity and improve glucose homeostasis by GLP-1 secretion [31].

4.5. Anti-Diabetic Activity

Hyperglycaemia can be reduced by inhibiting carbohydrate-hydrolysing enzymes such as α-amylase and α-glucosidase into glucose subunits. The inhibition of enzymes including aldose reductase, angiotensin-converting enzymes, dipeptidyl peptidase-4, and protein tyrosine phosphatase 1B can be exploited for the development of a therapeutic strategy [41,103]. In particular, dipeptidyl peptidase-4 increases GLP-1 levels to maintain hyperglycaemic conditions in patients with type 2 diabetes mellitus as compared with other pigments from microalgae and cyanobacteria as natural anti-hyperglycemic agents [31,104] (Figure 7a). Therefore, it acts as a therapeutic target for the development of antidiabetic drugs and different signaling pathways (Figure 7b). Recent studies reported that laminarin also causes inhibitory action involved in the antidiabetic mechanisms [31,43,44,105].

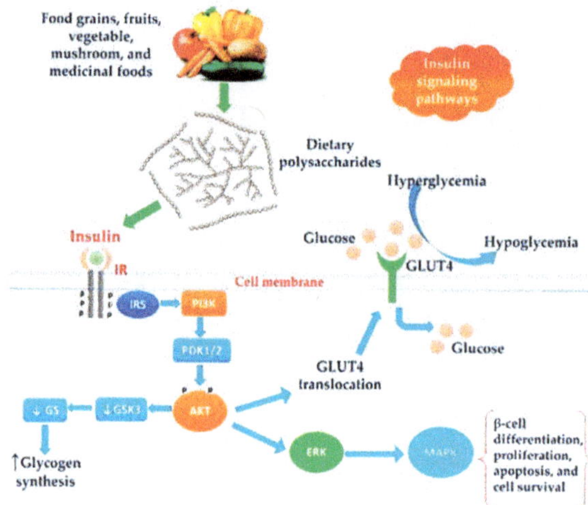

Figure 6. Schematic representation of the antiobesity action of dietary polysaccharides on the insulin signaling pathway. Reprinted with permission from reference [44]. Copyright 2022, Ganesan and Xu et al.

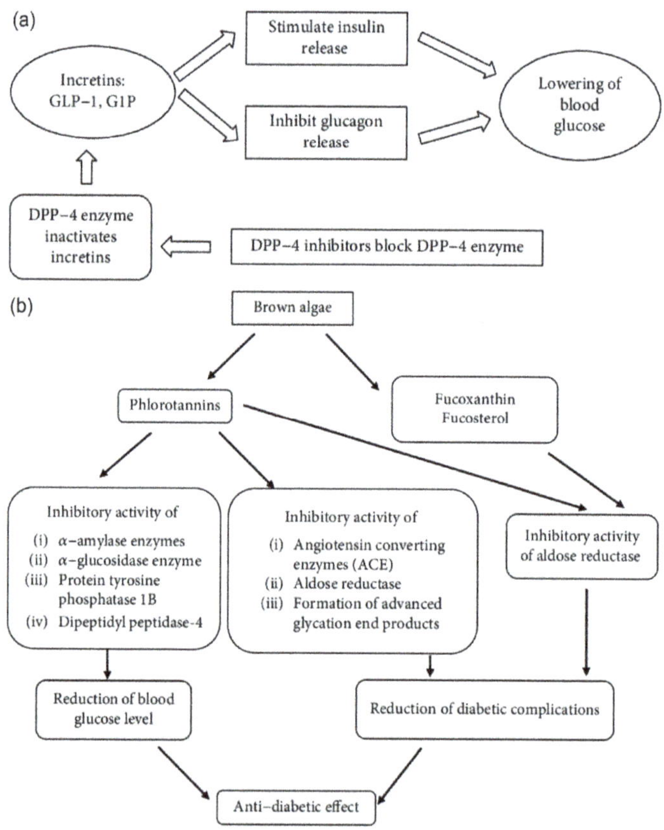

Figure 7. Antidiabetic mechanisms (**a**) Action of DPP-4 inhibitors on glucose homeostasis (**b**) Different antidiabetic mechanisms of active agents of brown algae. Reprinted with permission from reference [41]. Copyright 2022, Gunathilaka et al.

4.6. Antitumor Activity

Antitumor activity can inhibit cell apoptosis through a signaling pathway such as the mitochondrial pathway. Laminarin and laminarin sulfate possess direct cytotoxic activity against different types of tumor cells and have been shown to directly destroy cancer cells in studies using in vitro and in vivo models. Notably, laminarin also enhances T and B cells, macrophages, NK cells, and other immune cells, which can stimulate different complementary pathways and characterized by secretion of higher levels of anti-inflammatory cytokines for the regulation of the immune system. Algal polysaccharides help in the induction of apoptosis, cell proliferation, tumor angiogenesis, regulation of immune function and improve the effects of chemotherapy drugs on tumor cells in cancer therapy [101,106]. Laminarin was shown to be a potential therapeutic agent for human colon cancer by inhibiting cell proliferation due to the sulfated modification changing the molecular structure and spatial conformation of the polysaccharide, leading to changes in biological activity. The hydroxyl group of a sugar unit can be replaced by a sulfate group in laminarin. Thus, the conformation of the sugar chains changes, and the formation of a non-covalent bond further adds to the advantages of laminarin for antitumor activity. The mechanism of antitumor activity for algal polysaccharides is outlined in Figure 8.

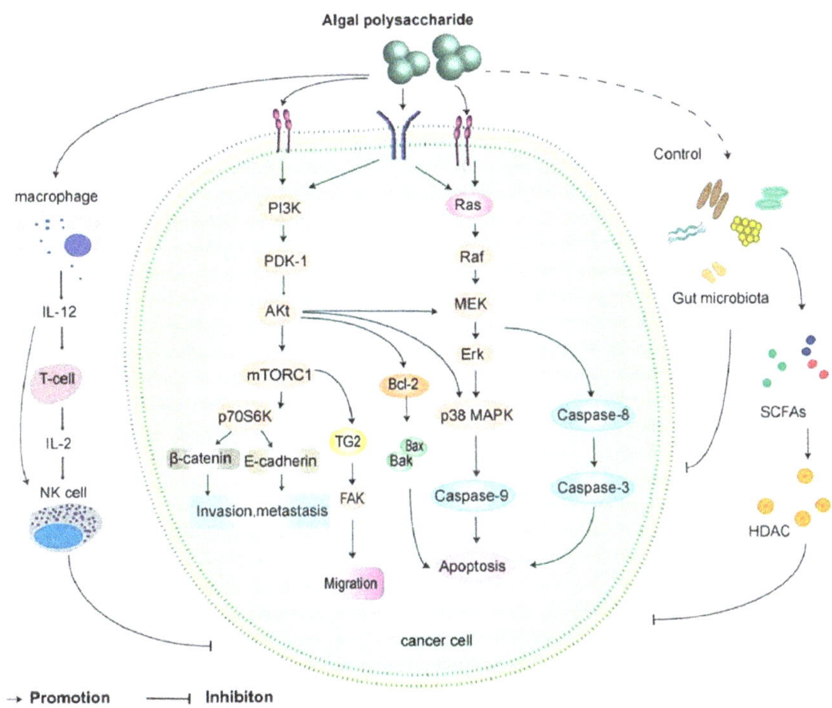

Figure 8. Anticancer mechanism of algal polysaccharide. 1: immunomodulation; 2: cytotoxicity; 3: cell cycle arrest; 4: NO-dependent pathway; 5: mitochondrial disruption. Reprinted from with permission from reference [101]. Copyright 2022, Ouyang et al.

5. Conclusions and Future Prospects

Research on the antioxidant, anti-tumor, anti-coagulant, and immunomodulatory activities of laminarin have shown beneficial effects in preclinical and clinical studies. Recent laminarin research also supported other potential activities including anti-obesity, anti-diabetic, anti-inflammatory, wound healing and hepatoprotective functions which have not been adequately characterized to date. Many research studies have demonstrated the biological activities of laminarin using structural modification and molecular weight determination. Further in-depth studies are required to facilitate the development of functional food and nutraceuticals incorporating laminarin.

Preclinical research on laminarin using in vitro and in vivo studies is very limited. Further clinical studies are also required to examine signaling pathways to facilitate the development of therapeutics, functional foods, and nutraceuticals. Further research is required to investigate the potential of laminarin in GI health and gut microbiome applications.

Laminarin is an underexploited polysaccharide in brown macroalgae of the marine environment with strong potential in therapeutic and nutraceutical applications. This review summarizes the preclinical and clinical laminarin studies and highlights opportunities and research priorities to facilitate the development of new therapeutics, functional foods, and nutraceuticals.

Author Contributions: S.K.: conceptualization, methodology, formal analysis, investigation, writing; G.R.: conceptualization, formal analysis, supervision; S.F.: conceptualization, resources; H.L.: conceptualization, resources; H.M.: conceptualization, resources; J.C.: conceptualization, formal analysis, resources, supervision. B.K.T.: conceptualization, formal analysis, resources, supervision, project administration. C.O.: conceptualization, formal analysis, validation, data curation, funding acquisition, supervision, project administration. All authors have read and agreed to the published version of the manuscript.

Funding: This research was funded by Enterprise Ireland and the European Union's Horizon 2020 Research and Innovation Programme under the Marie Skłodowska-Curie Career FIT PLUS grant agreement No. 68550 under project grant number MSCA20210184.

Institutional Review Board Statement: Not applicable.

Informed Consent Statement: Not applicable.

Data Availability Statement: Not applicable.

Acknowledgments: Shanmugapriya Karuppusamy is a Marie Skłodowska-Curie Career-FIT Plus Fellow at University College Dublin. She wishes to acknowledge the Open-Access Publication Policy of Marie Skłodowska-Curie grant Enterprise Ireland and the European Union's Horizon 2020 Research and Innovation Programme under the Marie Skłodowska-Curie Career FIT PLUS fellowship for dissemination of work to wider community. All authors acknowledge the funding received from enterprise Ireland for financial support for this work.

Conflicts of Interest: The authors declare no conflict of interest.

References

1. Alba, K.; Kontogiorgos, V. Seaweed polysaccharides (agar, alginate carrageenan). In *Encyclopedia of Food Chemistry*; Elsevier: Amsterdam, The Netherlands, 2018; pp. 240–250.
2. Nigam, S.; Singh, R.; Bhardwaj, S.K.; Sami, R.; Nikolova, M.P.; Chavali, M.; Sinha, S. Perspective on the Therapeutic Applications of Algal Polysaccharides. *J. Polym. Environ.* **2022**, *30*, 785–809. [CrossRef] [PubMed]
3. Zargarzadeh, M.; Amaral, A.J.; Custódio, C.A.; Mano, J.F. Biomedical applications of laminarin. *Carbohydr. Polym.* **2020**, *232*, 115774. [CrossRef] [PubMed]
4. Usman, A.; Khalid, S.; Usman, A.; Hussain, Z.; Wang, Y. Algal polysaccharides, novel application, and outlook. In *Algae Based Polymers, Blends, and Composites*; Elsevier: Amsterdam, The Netherlands, 2017; pp. 115–153.
5. Xu, S.-Y.; Huang, X.; Cheong, K.-L. Recent advances in marine algae polysaccharides: Isolation, structure, and activities. *Mar. Drugs* **2017**, *15*, 388. [CrossRef] [PubMed]
6. Gabbia, D.; De Martin, S. Brown seaweeds for the management of metabolic syndrome and associated diseases. *Molecules* **2020**, *25*, 4182. [CrossRef] [PubMed]
7. Jönsson, M.; Allahgholi, L.; Sardari, R.R.; Hreggviðsson, G.O.; Nordberg Karlsson, E. Extraction and modification of macroalgal polysaccharides for current and next-generation applications. *Molecules* **2020**, *25*, 930. [CrossRef]
8. Ganesan, A.R.; Tiwari, U.; Rajauria, G. Seaweed nutraceuticals and their therapeutic role in disease prevention. *Food Sci. Hum. Wellness* **2019**, *8*, 252–263. [CrossRef]
9. Wang, X.; Sun, G.; Feng, T.; Zhang, J.; Huang, X.; Wang, T.; Xie, Z.; Chu, X.; Yang, J.; Wang, H. Sodium oligomannate therapeutically remodels gut microbiota and suppresses gut bacterial amino acids-shaped neuroinflammation to inhibit Alzheimer's disease progression. *Cell Res.* **2019**, *29*, 787–803. [CrossRef]

10. Cotas, J.; Pacheco, D.; Gonçalves, A.M.; Silva, P.; Carvalho, L.G.; Pereira, L. Seaweeds' nutraceutical and biomedical potential in cancer therapy: A concise review. *J. Cancer Metastasis Treat.* **2021**, *7*, 13. [CrossRef]
11. Mourelle, M.L.; Gómez, C.P.; Legido, J.L. Role of Algal Derived Compounds in Pharmaceuticals and Cosmetics. In *Recent Advances in Micro and Macroalgal Processing: Food and Health Perspectives*; Wiley-Blackwell: Hoboken, NJ, USA, 2021; pp. 537–603.
12. Dhahri, M.; Alghrably, M.; Mohammed, H.A.; Badshah, S.L.; Noreen, N.; Mouffouk, F.; Rayyan, S.; Qureshi, K.A.; Mahmood, D.; Lachowicz, J.I. Natural Polysaccharides as Preventive and Therapeutic Horizon for Neurodegenerative Diseases. *Pharmaceutics* **2022**, *14*, 1. [CrossRef]
13. Otero, P.; Carpena, M.; Garcia-Oliveira, P.; Echave, J.; Soria-Lopez, A.; Garcia-Perez, P.; Fraga-Corral, M.; Cao, H.; Nie, S.; Xiao, J. Seaweed polysaccharides: Emerging extraction technologies, chemical modifications and bioactive properties. *Crit. Rev. Food Sci. Nutr.* **2021**. [CrossRef]
14. Bae, H.; Song, G.; Lee, J.-Y.; Hong, T.; Chang, M.-J.; Lim, W. Laminarin-derived from brown algae suppresses the growth of ovarian cancer cells via mitochondrial dysfunction and ER stress. *Mar. Drugs* **2020**, *18*, 152. [CrossRef]
15. Bishehsari, F.; Engen, P.A.; Preite, N.Z.; Tuncil, Y.E.; Naqib, A.; Shaikh, M.; Rossi, M.; Wilber, S.; Green, S.J.; Hamaker, B.R. Dietary fiber treatment corrects the composition of gut microbiota, promotes SCFA production, and suppresses colon carcinogenesis. *Genes* **2018**, *9*, 102. [CrossRef]
16. Haase, S.; Haghikia, A.; Gold, R.; Linker, R.A. Dietary fatty acids and susceptibility to multiple sclerosis. *Mult. Scler. J.* **2018**, *24*, 12–16. [CrossRef]
17. Sellimi, S.; Maalej, H.; Rekik, D.M.; Benslima, A.; Ksouda, G.; Hamdi, M.; Sahnoun, Z.; Li, S.; Nasri, M.; Hajji, M. Antioxidant, antibacterial and *in vivo* wound healing properties of laminaran purified from Cystoseira barbata seaweed. *Int. J. Biol. Macromol.* **2018**, *119*, 633–644. [CrossRef]
18. Rengasamy, K.R.; Mahomoodally, M.F.; Aumeeruddy, M.Z.; Zengin, G.; Xiao, J.; Kim, D.H. Bioactive compounds in seaweeds: An overview of their biological properties and safety. *Food Chem. Toxicol.* **2020**, *135*, 111013. [CrossRef]
19. Wang, H.-M.D.; Li, X.-C.; Lee, D.-J.; Chang, J.-S. Potential biomedical applications of marine algae. *Bioresour. Technol.* **2017**, *244*, 1407–1415. [CrossRef]
20. Kumar, Y.; Tarafdar, A.; Badgujar, P.C. Seaweed as a source of natural antioxidants: Therapeutic activity and food applications. *J. Food Qual.* **2021**, 1–17. [CrossRef]
21. Menshova, R.V.; Ermakova, S.P.; Anastyuk, S.D.; Isakov, V.V.; Dubrovskaya, Y.V.; Kusaykin, M.I.; Um, B.-H.; Zvyagintseva, T.N. Structure, enzymatic transformation and anticancer activity of branched high molecular weight laminaran from brown alga *Eisenia bicyclis*. *Carbohydr. Polym.* **2014**, *99*, 101–109. [CrossRef]
22. Kadam, S.U.; Álvarez, C.; Tiwari, B.K.; O'Donnell, C.P. Extraction of biomolecules from seaweeds. In *Seaweed Sustainability*; Elsevier: Amsterdam, The Netherlands, 2015; pp. 243–269.
23. Charoensiddhi, S.; Abraham, R.E.; Su, P.; Zhang, W. Seaweed and seaweed-derived metabolites as prebiotics. *Adv. Food Nutr. Res.* **2020**, *91*, 97–156.
24. Kalasariya, H.S.; Yadav, V.K.; Yadav, K.K.; Tirth, V.; Algahtani, A.; Islam, S.; Gupta, N.; Jeon, B.-H. Seaweed-based molecules and their potential biological activities: An eco-sustainable cosmetics. *Molecules* **2021**, *26*, 5313. [CrossRef]
25. Nguyen, S.G.; Kim, J.; Guevarra, R.B.; Lee, J.-H.; Kim, E.; Kim, S.-I.; Unno, T. Laminarin favorably modulates gut microbiota in mice fed a high-fat diet. *Food Funct.* **2016**, *7*, 4193–4201. [CrossRef] [PubMed]
26. Kadam, S.U.; Tiwari, B.K.; O'Donnell, C.P. Extraction, structure and biofunctional activities of laminarin from brown algae. *Int. J. Food Sci. Technol.* **2015**, *50*, 24–31. [CrossRef]
27. Kadam, S.U.; Álvarez, C.; Tiwari, B.K.; O'Donnell, C.P. Processing of seaweeds. In *Seaweed Sustainability*; Elsevier: Amsterdam, The Netherlands, 2015; pp. 61–78.
28. Sweeney, T.; Collins, C.; Reilly, P.; Pierce, K.; Ryan, M.; O'Doherty, J. Effect of purified β-glucans derived from *Laminaria digitata*, *Laminaria hyperborea* and *Saccharomyces cerevisiae* on piglet performance, selected bacterial populations, volatile fatty acids and pro-inflammatory cytokines in the gastrointestinal tract of pigs. *Br. J. Nutr.* **2012**, *108*, 1226–1234. [CrossRef] [PubMed]
29. George, J.; Thabitha, A.; Vignesh, N.; Manigandan, V.; Saravanan, R.; Daradkeh, G.; Qoronfleh, M.W. Antiskin Cancer and Antioxidant Activities of Formulated Agar from Brown Seaweed *Laminaria digitata* (Hudson) in Dimethyl Benzanthracene-Induced Swiss Albino Mice. *Int. J. Polym. Sci.* **2021**, 9930777. [CrossRef]
30. Song, K.; Xu, L.; Zhang, W.; Cai, Y.; Jang, B.; Oh, J.; Jin, J.-O. Laminarin promotes anti-cancer immunity by the maturation of dendritic cells. *Oncotarget* **2017**, *8*, 38554. [CrossRef]
31. Yang, L.; Wang, L.; Zhu, C.; Wu, J.; Yuan, Y.; Yu, L.; Xu, Y.; Xu, J.; Wang, T.; Liao, Z. Laminarin counteracts diet-induced obesity associated with glucagon-like peptide-1 secretion. *Oncotarget* **2017**, *8*, 99470. [CrossRef]
32. O'Shea, C.; O'Doherty, J.; Callanan, J.; Doyle, D.; Thornton, K.; Sweeney, T. The effect of algal polysaccharides laminarin and fucoidan on colonic pathology, cytokine gene expression and Enterobacteriaceae in a dextran sodium sulfate-challenged porcine model. *J. Nutr. Sci.* **2016**, *5*, e15. [CrossRef]
33. Zaharudin, N.; Tullin, M.; Pekmez, C.T.; Sloth, J.J.; Rasmussen, R.R.; Dragsted, L.O. Effects of brown seaweeds on postprandial glucose, insulin and appetite in humans—A randomized, 3-way, blinded, cross-over meal study. *Clin. Nutr.* **2021**, *40*, 830–838. [CrossRef]

34. Calderwood, D.; Rafferty, E.; Fitzgerald, C.; Stoilova, V.; Wylie, A.; Gilmore, B.F.; Castaneda, F.; Israel, A.; Maggs, C.A.; Green, B.D. Profiling the activity of edible European macroalgae towards pharmacological targets for type 2 diabetes mellitus. *Appl. Phycol.* **2021**, *2*, 10–21. [CrossRef]
35. Holdt, S.L.; Kraan, S. Bioactive compounds in seaweed: Functional food applications and legislation. *J. Appl. Phycol.* **2011**, *23*, 543–597. [CrossRef]
36. Carlson, J.L.; Erickson, J.M.; Lloyd, B.B.; Slavin, J.L. Health effects and sources of prebiotic dietary fiber. *Curr. Dev. Nutr.* **2018**, *2*, nzy005. [CrossRef]
37. Lynch, M.B.; Sweeney, T.; Callan, J.J.; O'Sullivan, J.T.; O'Doherty, J.V. The effect of dietary Laminaria-derived laminarin and fucoidan on nutrient digestibility, nitrogen utilisation, intestinal microflora and volatile fatty acid concentration in pigs. *J. Sci. Food Agric.* **2010**, *90*, 430–437. [CrossRef]
38. Odunsi, S.T.; Vázquez-Roque, M.I.; Camilleri, M.; Papathanasopoulos, A.; Clark, M.M.; Wodrich, L.; Lempke, M.; McKinzie, S.; Ryks, M.; Burton, D. Effect of alginate on satiation, appetite, gastric function, and selected gut satiety hormones in overweight and obesity. *Obesity* **2010**, *18*, 1579–1584. [CrossRef]
39. Neyrinck, A.M.; Mouson, A.; Delzenne, N.M. Dietary supplementation with laminarin, a fermentable marine β (1–3) glucan, protects against hepatotoxicity induced by LPS in rat by modulating immune response in the hepatic tissue. *Int. Immunopharmacol.* **2007**, *7*, 1497–1506. [CrossRef]
40. An, E.-K.; Hwang, J.; Kim, S.-J.; Park, H.-B.; Zhang, W.; Ryu, J.-H.; You, S.; Jin, J.-O. Comparison of the immune activation capacities of fucoidan and laminarin extracted from *Laminaria japonica*. *Int. J. Biol. Macromol.* **2022**, *208*, 230–242. [CrossRef]
41. Gunathilaka, T.L.; Samarakoon, K.; Ranasinghe, P.; Peiris, L.D.C. Antidiabetic potential of marine brown algae—A mini review. *J. Diabetes Res.* **2020**, 1230218. [CrossRef]
42. Hentati, F.; Tounsi, L.; Djomdi, D.; Pierre, G.; Delattre, C.; Ursu, A.V.; Fendri, I.; Abdelkafi, S.; Michaud, P. Bioactive Polysaccharides from Seaweeds. *Molecules* **2020**, *25*, 3152. [CrossRef]
43. Jayapala, N.; Perumal, M.K.; Baskaran, R.; Vallikannan, B. Pharmacological Importance of Bioactive Molecules of Seaweeds. In *Sustainable Global Resources of Seaweeds*; Springer: Berlin/Heidelberg, Germany, 2022; Volume 2, pp. 597–613.
44. Ganesan, K.; Xu, B. Anti-Diabetic Effects and Mechanisms of Dietary Polysaccharides. *Molecules* **2019**, *24*, 2556. [CrossRef]
45. Dobrinčić, A.; Balbino, S.; Zorić, Z.; Pedisić, S.; Bursać Kovačević, D.; Elez Garofulić, I.; Dragović-Uzelac, V. Advanced technologies for the extraction of marine brown algal polysaccharides. *Mar. Drugs* **2020**, *18*, 168. [CrossRef]
46. García-Vaquero, M.; Rajauria, G.; O'Doherty, J.V.; Sweeney, T. Polysaccharides from macroalgae: Recent advances, innovative technologies and challenges in extraction and purification. *Food Res. Int.* **2017**, *99*, 1011–1020. [CrossRef]
47. Cho, C.-H.; Lu, Y.-A.; Kim, M.-Y.; Jeon, Y.-J.; Lee, S.-H. Therapeutic Potential of Seaweed-Derived Bioactive Compounds for Cardiovascular Disease Treatment. *Appl. Sci.* **2022**, *12*, 1025. [CrossRef]
48. Rioux, L.-E.; Turgeon, S.L. Seaweed carbohydrates. In *Seaweed Sustainability*; Elsevier: Amsterdam, The Netherlands, 2015; pp. 141–192.
49. Chudasama, N.A.; Sequeira, R.A.; Moradiya, K.; Prasad, K. Seaweed polysaccharide-based products and materials: An assessment on their production from a sustainability point of view. *Molecules* **2021**, *26*, 2608. [CrossRef] [PubMed]
50. Patil, N.P.; Le, V.; Sligar, A.D.; Mei, L.; Chavarria, D.; Yang, E.Y.; Baker, A.B. Algal polysaccharides as therapeutic agents for atherosclerosis. *Front. Cardiovasc. Med.* **2018**, *5*, 153. [CrossRef] [PubMed]
51. Stone, B.A. Chemistry of β-glucans. In *Chemistry, Biochemistry, and Biology of 1-3 Beta Glucans and Related Polysaccharides*; Elsevier: Amsterdam, The Netherlands, 2009; pp. 5–46.
52. Rajauria, G.; Ravindran, R.; Garcia-Vaquero, M.; Rai, D.K.; Sweeney, T.; O'Doherty, J. Molecular characteristics and antioxidant activity of laminarin extracted from the seaweed species *Laminaria hyperborea*, using hydrothermal-assisted extraction and a multi-step purification procedure. *Food Hydrocoll.* **2021**, *112*, 106332. [CrossRef]
53. Garcia-Vaquero, M.; Rajauria, G.; Tiwari, B.; Sweeney, T.; O'Doherty, J. Extraction and yield optimisation of fucose, glucans and associated antioxidant activities from *Laminaria digitata* by applying response surface methodology to high intensity ultrasound-assisted extraction. *Mar. Drugs* **2018**, *16*, 257. [CrossRef]
54. Kadam, S.U.; O'Donnell, C.P.; Rai, D.K.; Hossain, M.B.; Burgess, C.M.; Walsh, D.; Tiwari, B.K. Laminarin from Irish brown seaweeds *Ascophyllum nodosum* and *Laminaria hyperborea*: Ultrasound assisted extraction, characterization and bioactivity. *Mar. Drugs* **2015**, *13*, 4270–4280. [CrossRef]
55. Imbs, T.I.; Ermakova, S.P.; Malyarenko, O.S.; Isakov, V.V.; Zvyagintseva, T.N. Structural elucidation of polysaccharide fractions from the brown alga *Coccophora langsdorfii* and *in vitro* investigation of their anticancer activity. *Carbohydr. Polym.* **2016**, *135*, 162–168. [CrossRef]
56. Sterner, M.; Gröndahl, F. Extraction of laminarin from Saccharina latissima seaweed using cross-flow filtration. *J. Appl. Phycol.* **2021**, *33*, 1825–1844. [CrossRef]
57. Ji, C.-F.; Ji, Y.-B.; Meng, D.-Y. Sulfated modification and anti-tumor activity of laminarin. *Exp. Ther. Med.* **2013**, *6*, 1259–1264. [CrossRef]
58. Shanmugam, M.; Mody, K. Heparinoid-active sulphated polysaccharides from marine algae as potential blood anticoagulant agents. *Curr. Sci.* **2000**, *79*, 1672–1683.
59. Choi, J.-i.; Kim, H.-J.; Lee, J.-W. Structural feature and antioxidant activity of low molecular weight laminarin degraded by gamma irradiation. *Food Chem.* **2011**, *129*, 520–523. [CrossRef]

60. Déléris, P.; Nazih, H.; Bard, J.-M. Seaweeds in human health. In *Seaweed in Health and Disease Prevention*; Academic Press: Cambridge, MA, USA, 2016; pp. 319–367.
61. Miao, H.Q.; Elkin, M.; Aingorn, E.; Ishai-Michaeli, R.; Stein, C.A.; Vlodavsky, I. Inhibition of heparanase activity and tumor metastasis by laminarin sulfate and synthetic phosphorothioate oligodeoxynucleotides. *Int. J. Cancer* **1999**, *83*, 424–431. [CrossRef]
62. Rattigan, R.; O'Doherty, J.V.; Vigors, S.; Ryan, M.T.; Sebastiano, R.S.; Callanan, J.J.; Thornton, K.; Rajauria, G.; Margassery, L.M.; Dobson, A.D. The effects of the marine-derived polysaccharides laminarin and chitosan on aspects of colonic health in pigs challenged with dextran sodium sulphate. *Mar. Drugs* **2020**, *18*, 262. [CrossRef]
63. Rattigan, R.; Sweeney, T.; Maher, S.; Thornton, K.; Rajauria, G.; O'Doherty, J. Laminarin-rich extract improves growth performance, small intestinal morphology, gene expression of nutrient transporters and the large intestinal microbial composition of piglets during the critical post-weaning period. *Br. J. Nutr.* **2020**, *123*, 255–263. [CrossRef]
64. Vigors, S.; O'Doherty, J.V.; Rattigan, R.; McDonnell, M.J.; Rajauria, G.; Sweeney, T. Effect of a laminarin rich macroalgal extract on the caecal and colonic microbiota in the post-weaned pig. *Mar. Drugs* **2020**, *18*, 157. [CrossRef]
65. Smith, A.; O'doherty, J.; Reilly, P.; Ryan, M.; Bahar, B.; Sweeney, T. The effects of laminarin derived from *Laminaria digitata* on measurements of gut health: Selected bacterial populations, intestinal fermentation, mucin gene expression and cytokine gene expression in the pig. *Br. J. Nutr.* **2011**, *105*, 669–677. [CrossRef]
66. O'Doherty, J.; Dillon, S.; Figat, S.; Callan, J.; Sweeney, T. The effects of lactose inclusion and seaweed extract derived from *Laminaria* spp. on performance, digestibility of diet components and microbial populations in newly weaned pigs. *Anim. Feed. Sci. Technol.* **2010**, *157*, 173–180. [CrossRef]
67. Gahan, D.; Lynch, M.; Callan, J.; O'sullivan, J.; O'Doherty, J. Performance of weanling piglets offered low-, medium-or high-lactose diets supplemented with a seaweed extract from *Laminaria* spp. *Animal* **2009**, *3*, 24–31. [CrossRef]
68. Devillé, C.; Damas, J.; Forget, P.; Dandrifosse, G.; Peulen, O. Laminarin in the dietary fibre concept. *J. Sci. Food Agric.* **2004**, *84*, 1030–1038. [CrossRef]
69. Rajauria, G. In-vitro antioxidant properties of lipophilic antioxidant compounds from 3 brown seaweed. *Antioxidants* **2019**, *8*, 596. [CrossRef]
70. Rajauria, G.; Foley, B.; Abu-Ghannam, N. Characterization of dietary fucoxanthin from *Himanthalia elongata* brown seaweed. *Food Res. Int.* **2017**, *99*, 995–1001. [CrossRef] [PubMed]
71. Liu, X.; Yuan, W.; Sharma-Shivappa, R.; van Zanten, J. Antioxidant activity of phlorotannins from brown algae. *Int. J. Agric. Biol. Eng.* **2017**, *10*, 184–191. [CrossRef]
72. Cheng, D.; Liang, B.; Li, M.; Jin, M. Influence of laminarin polysaccahrides on oxidative damage. *Int. J. Biol. Macromol.* **2011**, *48*, 63–66. [CrossRef] [PubMed]
73. Jiang, H.; Liang, S.; Yao, X.-R.; Jin, Y.-X.; Shen, X.-H.; Yuan, B.; Zhang, J.-B.; Kim, N.-H. Laminarin improves developmental competence of porcine early-stage embryos by inhibiting oxidative stress. *Theriogenology* **2018**, *115*, 38–44. [CrossRef] [PubMed]
74. Chan, G.C.-F.; Chan, W.K.; Sze, D.M.-Y. The effects of β-glucan on human immune and cancer cells. *J. Hematol. Oncol.* **2009**, *2*, 1–11. [CrossRef]
75. Cognigni, V.; Ranallo, N.; Tronconi, F.; Morgese, F.; Berardi, R. Potential benefit of β-glucans as adjuvant therapy in immuno-oncology: A review. *Explor. Target. Antitumor Ther.* **2021**, *2*, 122–138. [CrossRef]
76. Usoltseva, R.V.; Belik, A.A.; Kusaykin, M.I.; Malyarenko, O.S.; Zvyagintseva, T.N.; Ermakova, S.P. Laminarans and 1,3-β-D-glucanases. *Int. J. Biol. Macromol.* **2020**, *163*, 1010–1025. [CrossRef]
77. Majtan, J.; Jesenak, M. β-Glucans: Multi-functional modulator of wound healing. *Molecules* **2018**, *23*, 806. [CrossRef]
78. Graiff, A.; Ruth, W.; Kragl, U.; Karsten, U. Chemical characterization and quantification of the brown algal storage compound laminarin—A new methodological approach. *J. Appl. Phycol.* **2016**, *28*, 533–543. [CrossRef]
79. Praveen, M.A.; Parvathy, K.K.; Balasubramanian, P.; Jayabalan, R. An overview of extraction and purification techniques of seaweed dietary fibers for immunomodulation on gut microbiota. *Trends Food Sci. Technol.* **2019**, *92*, 46–64. [CrossRef]
80. Sweeney, T.; Meredith, H.; Vigors, S.; McDonnell, M.J.; Ryan, M.; Thornton, K.; O'Doherty, J.V. Extracts of laminarin and laminarin/fucoidan from the marine macroalgal species *Laminaria digitata* improved growth rate and intestinal structure in young chicks but does not influence *Campylobacter jejuni* colonisation. *Anim. Feed. Sci. Technol.* **2017**, *232*, 71–79. [CrossRef]
81. Ryan, M.T.; Collins, C.B.; O'Doherty, J.; Sweeney, T. Effects of dietary β-glucans supplementation on cytokine expression in porcine liver. *J. Anim. Sci.* **2012**, *90*, 40–42. [CrossRef]
82. Strain, C.R.; Collins, K.C.; Naughton, V.; McSorley, E.M.; Stanton, C.; Smyth, T.J.; Soler-Vila, A.; Rea, M.C.; Ross, P.R.; Cherry, P. Effects of a polysaccharide-rich extract derived from Irish-sourced Laminaria digitata on the composition and metabolic activity of the human gut microbiota using an *in vitro* colonic model. *Eur. J. Nutr.* **2020**, *59*, 309–325. [CrossRef]
83. Ostrzenski, A. Resectoscopic cervical trauma minimized by inserting *Laminaria digitata* preoperatively. *Int. J. Fertil. Menopausal Stud.* **1994**, *39*, 111–113.
84. Rinaudo, M. Main properties and current applications of some polysaccharides as biomaterials. *Polym. Int.* **2008**, *57*, 397–430. [CrossRef]
85. Déjean, G.; Tamura, K.; Cabrera, A.; Jain, N.; Pudlo, N.A.; Pereira, G.; Viborg, A.H.; Van Petegem, F.; Martens, E.C.; Brumer, H. Synergy between cell surface glycosidases and glycan-binding proteins dictates the utilization of specific beta (1, 3)-glucans by human gut Bacteroides. *mBio* **2020**, *11*, e00095-20. [CrossRef]

86. Wasko, N.J.; Nichols, F.; Clark, R.B. Multiple sclerosis, the microbiome, TLR2, and the hygiene hypothesis. *Autoimmun. Rev.* **2020**, *19*, 102430. [CrossRef]
87. Cryan, J.F.; O'Riordan, K.J.; Sandhu, K.; Peterson, V.; Dinan, T.G. The gut microbiome in neurological disorders. *Lancet Neurol.* **2020**, *19*, 179–194. [CrossRef]
88. Deville, C.; Gharbi, M.; Dandrifosse, G.; Peulen, O. Study on the effects of laminarin, a polysaccharide from seaweed, on gut characteristics. *J. Sci. Food Agric.* **2007**, *87*, 1717–1725. [CrossRef]
89. Takei, M.N.; Kuda, T.; Taniguchi, M.; Nakamura, S.; Hajime, T.; Kimura, B. Detection and isolation of low molecular weight alginate-and laminaran-susceptible gut indigenous bacteria from ICR mice. *Carbohydr. Polym.* **2020**, *238*, 116205. [CrossRef]
90. Cui, Y.; Zhu, L.; Li, Y.; Jiang, S.; Sun, Q.; Xie, E.; Chen, H.; Zhao, Z.; Qiao, W.; Xu, J. Structure of a laminarin-type β-(1→ 3)-glucan from brown algae *Sargassum henslowianum* and its potential on regulating gut microbiota. *Carbohydr. Polym.* **2021**, *255*, 117389. [CrossRef] [PubMed]
91. Remya, R.; Rajasree, S.R.; Suman, T.; Aranganathan, L.; Gayathri, S.; Gobalakrishnan, M.; Karthih, M. Laminarin based AgNPs using brown seaweed *Turbinaria ornata* and its induction of apoptosis in human retinoblastoma Y79 cancer cell lines. *Mater. Res. Express* **2018**, *5*, 035403. [CrossRef]
92. Ermakova, S.; Men'shova, R.; Vishchuk, O.; Kim, S.-M.; Um, B.-H.; Isakov, V.; Zvyagintseva, T. Water-soluble polysaccharides from the brown alga *Eisenia bicyclis*: Structural characteristics and antitumor activity. *Algal Res.* **2013**, *2*, 51–58. [CrossRef]
93. Zhu, X.; Zhu, R.; Jian, Z.; Yu, H. Laminarin enhances the activity of natural killer cells in immunosuppressed mice. *Cent. -Eur. J. Immunol.* **2019**, *44*, 357. [CrossRef] [PubMed]
94. Usoltseva, R.V.; Anastyuk, S.D.; Shevchenko, N.M.; Zvyagintseva, T.N.; Ermakova, S.P. The comparison of structure and anticancer activity *in vitro* of polysaccharides from brown algae *Alaria marginata* and *A. angusta*. *Carbohydr. Polym.* **2016**, *153*, 258–265. [CrossRef]
95. Malyarenko, O.S.; Usoltseva, R.V.; Shevchenko, N.M.; Isakov, V.V.; Zvyagintseva, T.N.; Ermakova, S.P. *In vitro* anticancer activity of the laminarans from Far Eastern brown seaweeds and their sulfated derivatives. *J. Appl. Phycol.* **2017**, *29*, 543–553. [CrossRef]
96. Wani, S.M.; Gani, A.; Mir, S.A.; Masoodi, F.A.; Khanday, F.A. β-Glucan: A dual regulator of apoptosis and cell proliferation. *Int. J. Biol. Macromol.* **2021**, *182*, 1229–1237. [CrossRef]
97. Zhong, Q.; Wei, B.; Wang, S.; Ke, S.; Chen, J.; Zhang, H.; Wang, H. The antioxidant activity of polysaccharides derived from marine organisms: An overview. *Mar. Drugs* **2019**, *17*, 674. [CrossRef]
98. Vannucci, L.; Krizan, J.; Sima, P.; Stakheev, D.; Caja, F.; Rajsiglova, L.; Horak, V.; Saieh, M. Immunostimulatory properties and antitumor activities of glucans. *Int. J. Oncol.* **2013**, *43*, 357–364. [CrossRef]
99. Jayachandran, M.; Chen, J.; Chung, S.S.M.; Xu, B. A critical review on the impacts of β-glucans on gut microbiota and human health. *J. Nutr. Biochem.* **2018**, *61*, 101–110. [CrossRef]
100. Younis, N.S.; Mohamed, M.E.; El Semary, N.A. Green synthesis of silver nanoparticles by the *Cyanobacteria synechocystis* sp.: Characterization, antimicrobial and diabetic wound-healing actions. *Mar. Drugs* **2022**, *20*, 56. [CrossRef]
101. Ouyang, Y.; Qiu, Y.; Liu, Y.; Zhu, R.; Chen, Y.; El-Seedi, H.R.; Chen, X.; Zhao, C. Cancer-fighting potentials of algal polysaccharides as nutraceuticals. *Food Res. Int.* **2021**, *147*, 110522. [CrossRef]
102. Wan, Y.; Xu, X.; Gilbert, R.G.; Sullivan, M.A. A Review on the Structure and Anti-Diabetic (Type 2) Functions of β-Glucans. *Foods* **2021**, *11*, 57. [CrossRef]
103. Unnikrishnan, P.; Suthindhiran, K.; Jayasri, M. Antidiabetic potential of marine algae by inhibiting key metabolic enzymes. *Front. Life Sci.* **2015**, *8*, 148–159. [CrossRef]
104. Ghosh, T.; Bhayani, K.; Paliwal, C.; Maurya, R.; Chokshi, K.; Pancha, I.; Mishra, S. Cyanobacterial pigments as natural anti-hyperglycemic agents: An *in vitro* study. *Front. Mar. Sci.* **2016**, *3*, 146. [CrossRef]
105. Shalaby, N.M.; Abd-Alla, H.I.; Aly, H.F.; Albalawy, M.A.; Shaker, K.H.; Bouajila, J. Preliminary in vitro and in vivo evaluation of antidiabetic activity of *Ducrosia anethifolia* Boiss. and its linear furanocoumarins. *BioMed Res. Int.* **2014**, 831841.
106. Paliwal, C.; Nesamma, A.A.; Jutur, P.P. Industrial scope with high-value biomolecules from microalgae. In *Sustainable Downstream Processing of Microalgae for Industrial Application*; CRC Press: Boca Raton, FL, USA, 2019; pp. 83–98.

MDPI
St. Alban-Anlage 66
4052 Basel
Switzerland
www.mdpi.com

Marine Drugs Editorial Office
E-mail: marinedrugs@mdpi.com
www.mdpi.com/journal/marinedrugs

www.ingramcontent.com/pod-product-compliance
Lightning Source LLC
Chambersburg PA
CBHW040829250926
51850CB00007BA/1297
* 9 7 8 3 7 2 5 8 0 3 3 3 0 *